Applied Optimization Methods for Wireless Networks

Written in a unique style, this book is a valuable resource for faculty, graduate students, and researchers in the communications and networking area whose work interfaces with optimization. It teaches you how various optimization methods can be applied to solve complex problems in wireless networks. Each chapter reviews a specific optimization method and then demonstrates how to apply the theory in practice through a detailed case study taken from state-of-the-art research.

You will learn various tips and step-by-step instructions for developing optimization models, reformulations, and transformations, particularly in the context of cross-layer optimization problems in wireless networks involving flow routing (network layer), scheduling (link layer), and power control (physical layer). Throughout, a combination of techniques from both Operations Research and Computer Science disciplines provides a holistic treatment of optimization methods and their applications.

Each chapter includes homework exercises, with Powerpoint slides available online for students and instructors, and a solutions manual for instructors only. To access the materials, please visit www.cambridge.org/hou.

Y. Thomas Hou is a Professor in the Bradley Department of Electrical and Computer Engineering, Virginia Tech, Blacksburg, Virginia, USA. He is an IEEE Fellow for contributions to modeling and optimization of wireless networks.

Yi Shi is a Senior Research Scientist at Intelligent Automation Inc., Rockville, Maryland, USA and an Adjunct Assistant Professor in the Bradley Department of Electrical and Computer Engineering, Virginia Tech, Blacksburg, Virginia, USA.

Hanif D. Sherali is a University Distinguished Professor Emeritus in the Grado Department of Industrial and Systems Engineering, Virginia Tech, Blacksburg, Virginia, USA. He is an elected member of the U.S. National Academy of Engineering.

Applied Optimization Methods for Wireless Networks

Y. Thomas Hou
Yi Shi
Hanif D. Sherali

Virginia Tech, Blacksburg, Virginia, USA

CAMBRIDGE
UNIVERSITY PRESS

Shaftesbury Road, Cambridge CB2 8EA, United Kingdom

One Liberty Plaza, 20th Floor, New York, NY 10006, USA

477 Williamstown Road, Port Melbourne, VIC 3207, Australia

314–321, 3rd Floor, Plot 3, Splendor Forum, Jasola District Centre, New Delhi – 110025, India

103 Penang Road, #05–06/07, Visioncrest Commercial, Singapore 238467

Cambridge University Press is part of Cambridge University Press & Assessment,
a department of the University of Cambridge.

We share the University's mission to contribute to society through the pursuit of
education, learning and research at the highest international levels of excellence.

www.cambridge.org
Information on this title: www.cambridge.org/9781107018808

First published 2014

A catalogue record for this publication is available from the British Library

Library of Congress Cataloging-in-Publication data
Hou, Y. Thomas.
Applied optimization methods for wireless networks / Y. Thomas Hou, Virginia Polytechnic and
State University, Yi Shi, Intelligent Automation Inc., Hanif D. Sherali, Virginia Polytechnic and
State University.
    pages cm
Includes bibliographical references and index.
ISBN 978-1-107-01880-8 (hardback)
1. Wireless communication systems.   2. Mathematical optimization.
I. Shi, Yi. (Electrical engineer)   II. Sherali, Hanif D., 1952–   III. Title.
TK5103.2.H68 2014
004.601´1–dc23   2013044705

ISBN   978-1-107-01880-8   Hardback

To our parents
and
Our wives Tingting, Meiyu, and Semeen

Contents

Preface

Reasons for writing the book   In recent years, there has been a growing trend in applying optimization approaches to study wireless networks. Such an approach is usually necessary when the underlying goal is to pursue fundamental performance limits or theoretical results. This book is written to serve this need and is mainly targeted to graduate students who are conducting theoretical research in wireless networks using optimization-based approaches. This book will also serve as a very useful reference for researchers who wish to explore various optimization techniques as part of their research methodologies.

To prepare a graduate student in either electrical and computer engineering (ECE) or computer science (CS) to conduct fundamental research in wireless networks, an ideal roadmap would include a series of graduate courses in operations research (OR) and CS, in addition to traditional communications and networking courses in ECE. These OR and CS courses would include (among others) linear programming, nonlinear programming, integer programming from OR, and complexity theory and algorithm design and analysis from CS. Today, these courses are typically offered as core courses within the respective disciplines. Instructors in OR and CS departments typically have little knowledge of wireless networks and are unable to make a connection between the mathematical tools and techniques in these courses and problem-solving skills in wireless networks. ECE/CS students often find it difficult to see how these courses would benefit their research in wireless networks. Due to this gap between teaching scopes and learning expectations, we find that the learning experience of our ECE/CS students in these courses is passive (or "blind") at best, as they do not have a clear picture of how these courses will benefit their research.

One approach to bridge this gap is to offer a course that reviews a collection of mathematical tools from OR and CS (with a focus on optimization techniques) and shows how they can be used to address some challenging problems in wireless networks. This book is written for this purpose.

Each chapter in this book starts with a brief pointer to the underlying optimization technique (with references to tutorials or textbooks so that students can do an in-depth study in a formal course or on their own). The chapter then immediately delves into a detailed case study in wireless networks to which the technique will be applied. The focus in each chapter is to show how the underlying technique can be used to solve a challenging problem in wireless networks. To achieve this goal, we offer details on how to formulate a research problem into a formal optimization model, reformulate or transform it in order to improve mathematical tractability, and apply the underlying optimization technique (with necessary customizations that are specific to the underlying problem) to derive an optimal or near-optimal solution.

We have taught this course a number of times to ECE and CS graduate students at Virginia Tech, using chapters from this book. The response from the students has been overwhelmingly positive. In particular, we find that:

- For a graduate student (regardless of whether they have taken related OR or CS courses), this course opens a new landscape or perspective on what optimization techniques are available and how they can be applied to solve hard problems in wireless networks;
- For those graduate students who are currently taking or will take the aforementioned OR and CS courses, this course will help them better appreciate the mathematical techniques in such OR and CS courses. The student will also have a better purpose and a stronger motivation when she takes these core courses in her future study.

We recognize that a single-volume book cannot possibly cover all techniques exhaustively. Neither is it our intention to cover everything in one book. Nevertheless, we have organized this book into four parts, where every chapter focuses on a single technique. We hope this organization will serve our purpose of offering a first course on this important subject of *Applied Optimization Methods in Wireless Networks*. Our experience shows that after taking this course, students become substantially more mature mathematically. Most of them are able to consciously develop their learning paths into many areas in OR and CS not covered in this book in order to further expand their own mathematical capabilities. This is an important ingredient in their life-long learning and discovery.

Finally, the idea of having a book that offers a systematic coverage of optimization techniques and their applications in wireless networks is a very natural one. Unfortunately (and quite surprisingly), after a rather thorough survey of the market (when we presented our initial proposal to our publisher), we found that there were hardly any such books available. The closest book that we can find that by Dimitri Bertsekas: *Network Optimization: Continuous and Discrete Models* (Athena Scientific, 1998). But that book still falls short in showing students suitable case studies that are relevant to modern wireless networks.

On the other hand, most other books addressing network optimization follow a problem-oriented approach (vs. our method-oriented approach). They do not offer a systematic treatment of the underlying optimization techniques like we do in this book. To make this point clear, we quote the following text from the preface of the book *Combinatorial Optimization in Communication Networks*, edited by Maggie Xiaoyan Cheng, Yingshu Li, and Ding-Zhu Du (Springer, 2006), to explain why the problem-oriented approach was adopted by most authors:

Two approaches were considered: optimization method oriented (starting from combinatorial optimization methods and finding appropriate network problems as examples) and network problem oriented (focusing on specific network problems and seeking appropriate combinatorial optimization methods to solve them). We finally decided to use the problem-oriented approach, mainly because of the availability of papers: most papers in the recent literature appear to address very specific network problems, and combinatorial optimization comes as a convenient problem solver.

Such a problem-oriented approach offers a convenient way of composing a book quickly (i.e., by assembling some research papers in the literature into an edited volume). But books based on such a problem-oriented approach, although useful as a reference book, do not teach graduate students optimization techniques in a systematic manner. This critical dearth in the existing literature was our main motivation for writing this book and bringing it to the community.

Acknowledgments

This book is the fruit of close collaboration among the three authors for more than ten years. We would like first to thank the former and current members of our research group. In particular, many thanks to: Jia (Kevin) Liu, whose work led to Chapter 3, Sushant Sharma, whose work led to Chapter 4, Liguang Xie, whose work led to Chapter 7, Dr. Shiwen Mao, whose work led to Chapter 11. We want to thank Huacheng Zeng, Liguang Xie, Xu Yuan, and Canming Jiang for their help in proofreading some of the chapters. They also contributed to the preparation and revision of the solution manual and Powerpoint Slides. Without their help, this book would not have reached its current shape. Some other former and current members of our group, whose names were not mentioned above but who contributed to this book in many other ways, include Sastry Kompella, Cunhao Gao, Tong Liu, Xiaojun Wang, Xiaolin Cheng, Dr. Rongbo Zhu, Dr. Lili Zhang, and Dr. Wangdong Qi.

We also want to thank the students in our ECE/CS 6570 class (Advanced Foundations of Networking) over the years, who offered valuable feedback to different versions of this book and helped us gauge the best match of such materials for a graduate course in networking. In particular, those students who took ECE/CS 6570 in Fall 2012 directly contributed to proofreading the final book manuscript and their feedback is greatly appreciated.

We want to thank Dr. Phil Meyler, Publishing Development Director of Cambridge University Press, who showed a genuine interest in the initial conception of this book and encouraged us to move forward for a book proposal. He has also been extremely patient with us when we requested a one year extension for the final delivery of our manuscript. We thank him for his trust, patience, and understanding, which allowed us to work on our schedule to bring this book to reality. Looking back, we feel really lucky that we chose the best publisher for this book. During the manuscript preparation stage, we worked with three different Assistant Editors of Cambridge University Press – Elizabeth Horne, Kirsten Bot, and Sarah Marsh. We thank all three of them, who worked diligently with us at each step along the way to make this book a polished product.

Tom Hou would like to thank Scott Midkiff, who recruited him to join the Electrical and Computer Engineering Department at Virginia Tech in 2002. Over the years, Scott has been a great colleague, a close friend, a resourceful mentor, and, most recently, a supportive department head. The environment that Scott and the department were able to offer has been instrumental to Tom's success in research and scholarship.

Finally, we would like to thank the National Science Foundation (NSF) and the Office of Naval Research (ONR), whose funding support of our research over the years led to this book.

Copyright Permissions

A highlight of this book is to include, in each chapter, a comprehensive presentation of a case study that demonstrates how the particular optimization method can be applied in solving a wireless networking problem. These case studies are based on a number of papers written by the authors. The following list acknowledges these publications and their respective journals and publishers on a chapter-by-chapter basis. Portions of these papers have been adapted with permission from the publishers as required. All rights are reserved as stipulated by the various copyright agreements.

Chapter 2
Y.T. Hou, Y. Shi, and H.D. Sherali, "Rate allocation and network lifetime problems for wireless sensor networks," *IEEE/ACM Transactions on Networking*, vol. 16, no. 2, pp. 321–334, April 2008. Copyright © 2008 by, and with kind permission from, IEEE.

Chapter 3
J. Liu, Y.T. Hou, Y. Shi, and H.D. Sherali, "Cross-layer optimization for MIMO-based wireless ad hoc networks: routing, power allocation, and bandwidth allocation," *IEEE Journal on Selected Areas in Communications*, vol. 26, no. 6, pp. 913–926, August 2008. Copyright © 2008 by, and with kind permission from, IEEE.

Chapter 4
S. Sharma, Y. Shi, Y.T. Hou, and S. Kompella, "An optimal algorithm for relay node assignment in cooperative ad hoc networks," *IEEE/ACM Transactions on Networking*, vol. 19, issue 3, pp. 879–892, June 2011. Copyright © 2011 by, and with kind permission from, IEEE.

Chapter 5
Y. Shi, Y.T. Hou, and H. Zhou, "Per-node based optimal power control for multi-hop cognitive radio networks," *IEEE Transactions on Wireless*

Communications, vol. 8, no. 10, pp. 5290–5299, October 2009. Copyright © 2009 by, and with kind permission from, IEEE.

Chapter 6
Y. Shi, Y.T. Hou, S. Kompella, and H.D. Sherali, "Maximizing capacity in multi-hop cognitive radio networks under the SINR model," *IEEE Transactions on Mobile Computing*, vol. 10, no. 7, pp. 954–967, July 2011. Copyright © 2011 by, and with kind permission from, IEEE.

Chapter 7
L. Xie, Y. Shi, Y.T. Hou, and H.D. Sherali, "Making sensor networks immortal: an energy-renewal approach with wireless power transfer," *IEEE/ACM Transactions on Networking*, vol. 20, issue 6, pp. 1748–1761, December 2012. Copyright © 2012 by, and with kind permission from, IEEE.

Chapter 8
Y. Shi and Y.T. Hou, "Optimal base station placement in wireless sensor networks," *ACM Transactions on Sensor Networks*, vol. 5, issue 4, article 32, November 2009. Copyright © 2009 by, and with kind permission from, ACM.

Chapter 9
Y. Shi and Y.T. Hou, "Some fundamental results on base station movement problem for wireless sensor networks," *IEEE/ACM Transactions on Networking*, vol. 20, issue 4, pp. 1054–1067, August 2012. Copyright © 2012 by, and with kind permission from, IEEE.

Chapter 10
Y.T. Hou, Y. Shi, and H.D. Sherali, "Spectrum sharing for multi-hop networking with cognitive radios," *IEEE Journal on Selected Areas in Communications*, vol. 26, no. 1, pp. 146–155, January 2008. Copyright © 2008 by, and with kind permission from, IEEE.

Chapter 11
S. Mao, Y.T. Hou, X. Cheng, H.D. Sherali, S.F. Midkiff, and Y.-Q. Zhang, "On routing for multiple description video over wireless ad hoc networks," *IEEE Transactions on Multimedia*, vol. 8, no. 5, pp. 1063–1074, October 2006. Copyright © 2006 by, and with kind permission from, IEEE.

1

Introduction

The present moment is the only moment available to us, and it is the door to all moments.

Thich Nhat Hanh

1.1 Book overview

The goal of this book is to offer a course to graduate students by demonstrating an important set of mathematical tools (with a focus on optimization techniques) and to show how they can be used to address some challenging problems in wireless networks.

Book organization   This book consists of four parts.

- Part I, consisting of three chapters, is devoted to optimization and designing algorithms that can offer optimal solutions.
- Part II, consisting of five chapters, is devoted to techniques that can offer provably near-optimal solutions.
- Part III, consisting of two chapters, is devoted to some highly effective heuristics.
- Part IV, consisting of only one chapter, is devoted to some miscellaneous topics in the broader context of wireless network optimizations. This part will be expanded in a future edition.

Structure of each chapter   Each chapter starts with a brief overview of a particular optimization technique and subsequently is followed by a comprehensive coverage of a case study in wireless networking. The goal of giving a pointer to the underlying theory at the beginning of each chapter is to offer a direction to students on what they should explore further in formal course work or textbooks in these areas. These pointers are not meant to be comprehensive tutorials, each of which could constitute a book on its own. The first section of each chapter is not meant to be a replacement of formal courses in operations research (OR) or computer science (CS) and certainly does not offer a short cut for students in their formal education in these subject areas.

Instead, the focus of each chapter is on how to apply the technique under discussion to solve a challenging problem in wireless networking. Each chapter is self-contained and shows all the details involved in problem formulation, reformulation, and customization of optimization techniques in order to devise a final solution. Our guiding principles in the choice of a case study in each chapter are the following:

- To reflect state-of-the-art wireless network research rather than discuss some classic but outdated problems that are no longer of current research interest. In this spirit, the problems that we chose are mainly in the context of multi-hop wireless networks, with the underlying wireless technologies being cognitive radio (CR), multiple-input multiple-output (MIMO), and cooperative communications (CC), among others.
- To offer a reasonable level of difficulty or challenge in each case study rather than a simple problem that the students would hardly encounter in their research. This is in contrast to simple and diverse examples that are typically presented in standard optimization or algorithmic textbooks. In this book, we want to show readers how the method introduced in each chapter can be applied to solve a hard wireless networking problem, which they are likely to encounter in research. For this purpose, each chapter is dominated by a case study in terms of length coverage. We believe this approach will help readers better appreciate the underlying method and to gain a better understanding on its application in practice.
- To offer only essential background on the underlying wireless communication technology that is needed in formulating the problem rather than a comprehensive overview of the technology. This is because the main goal of this book is to learn various optimization techniques and apply each one to solve a wireless networking problem as a case study rather than offering a comprehensive tutorial on various wireless communication technologies. Therefore, we decided to minimize the coverage on wireless technologies and offer references that the readers can study further on their own.
- To help readers develop strong problem formulation and reformulation skills. This can only be taught with examples with sufficient sophistication and complexity. We believe such formulation/reformulation skills are important for research and thus want to teach the readers such skills in detail in each case study.

Key characteristics of the book

- Presents a collection of useful optimization techniques (one technique in each chapter), with an emphasis on how each technique is put into action to solve challenging wireless networking problems.

- Combines techniques from both operations research (OR) and computer science (CS) disciplines, with a strong focus on solving optimization problems in wireless networks.
- Shows various tricks and step-by-step details on how to develop optimization models and reformulations, particularly in the context of cross-layer optimization problems involving flow routing (network layer), scheduling (link layer), and power control (physical layer).
- Discusses case studies that focus on multi-hop wireless networks (e.g., ad hoc and sensor networks) and incorporates a number of advanced physical layer technologies such as MIMO, cognitive radio (CR), and cooperative communications (CC).
- Contains problem sets at the end of each chapter. PowerPoint slides for each chapter are available to both the students and instructors. A solutions manual is available to the instructors.

1.2 Book outline

This book has four parts. Part I of this book, consisting of Chapters 2 to 4, is devoted to optimization and designing algorithms that can offer optimal solutions.

- Chapter 2 reviews linear programming (LP) and shows how it can be employed to solve certain problems in wireless networks. Although the LP methodology itself is rather basic and straightforward, special care is still needed to ensure that it is used correctly, as we demonstrate in the case study in this chapter. The case study is rather interesting as it shows that even in LP-based problem formulations, deep insights can be gained once we dig deep into it. In particular, the case study considers lexicographic max-min (LMM) rate allocation and node lifetime problems in a wireless sensor network (WSN). We introduce the *parametric analysis* (PA) technique, which is very useful in its own right. The concept of LMM is also important, and can be employed as a fairness criterion for other problems in wireless networking research. Through the case study in this chapter, the readers will gain a rather deep understanding of LP and its applications in wireless networks.
- Chapter 3 reviews convex programming, which is a popular and powerful tool for studying nonlinear optimization problems. Once a problem is shown to be a convex program, then there are standard solution techniques and we may even directly apply a solver to obtain an optimal solution. For many cross-layer convex optimization problems, the research community is more interested in exploring a solution in its dual domain. There are two reasons for this approach. First, many cross-layer problems, once properly formulated in the dual domain, can be decomposed into sub-problems, each of which may be decoupled from the other layers. Such a

layering-based decomposition offers better insights and interpretations for the underlying problems. Second, once a problem is decomposed in its dual domain, the solution may be implemented in a distributed fashion, which is a highly desirable feature for networking researchers. In the case study of this chapter, we study a cross-layer optimization problem for a multi-hop MIMO network, which involves variables at the transport, network, link, and physical layers. We show that the problem can be formulated as a convex program. By studying the problem in its dual domain, we show that the dual problem can be decomposed into two subproblems: one subproblem solely involving variables at the transport and network layers and the other problem involving variables at the link and physical layers. We describe how the dual problem can be solved by a cutting-plane method and how the solution to the primal problem can be recovered from the solution to the dual problem.

- Chapter 4 illustrates how an optimization problem can be solved by clever algorithmic techniques from CS. For certain problems, general optimization methods from OR may not always be the best approach. In fact, a formulation following OR's optimization approach may lead to a solution with nonpolynomial-time complexity. But a solution with nonpolynomial-time complexity does not mean that the problem is not in P. In fact, we may well develop a different algorithm to solve the problem with polynomial-time complexity. This is what we illustrate in this chapter, where we develop a polynomial-time algorithm. In the case study, we consider a relay node assignment problem in cooperative communications (CC). Our objective is to assign a set of available relay nodes to different source–destination pairs so as to maximize the minimum data rate among all the pairs. Following the OR optimization approach, we show that the problem can be formulated as a mixed-integer linear programming (MILP) problem, which is NP-hard in general. But this does not mean that the problem is NP-hard. Instead, by following a CS algorithm design approach, we develop a polynomial-time exact algorithm for this problem. A novel idea in this algorithm is a linear marking mechanism, which is able to achieve polynomial-time complexity at each iteration. We give a formal proof of the optimality of the algorithm.

Part II of this book, consisting of Chapters 5 to 9, is devoted to techniques that can offer provably near-optimal solutions.

- Chapter 5 presents the branch-and-bound framework and shows how it can be applied to solve discrete and combinatorial optimization problems. Such problems are typically considered most difficult in nonconvex optimization and the branch-and-bound framework offers a general-purpose and effective approach. The effectiveness of branch-and-bound resides in the careful design of each component of its framework, such as computation of a lower bound, local search of an upper bound, and selection of partitioning variables (in the case of a minimization problem). It should be noted that the worst-case complexity of a branch-and-bound-based method remains exponential, although a judicious design of each component could achieve reasonable computational times in practice. In the case study, we consider a per-node power control problem for a multi-hop CRN. This problem has a large design space that involves a tight coupling relationship among power control, scheduling, and flow routing, which is typical for a cross-layer optimization problem. We develop a mathematical model and a problem formulation, which is a mixed-integer nonlinear programming (MINLP) problem. We show how to apply the branch-and-bound framework to design a solution procedure.

- Chapter 6 presents the Reformulation-Linearization Technique (RLT) for deriving tight linear relaxations for any monomial. Simply put, RLT can be applied to any polynomial term of the form $\prod_{i=1}^{n}(x_i)^{c_i}$ in variables x_i, where the c_i-exponents are constant integers. Given such generality, RLT is a powerful tool in deriving tight linear relaxations. In the case study, we consider a throughput maximization problem in a multi-hop CRN under the signal-to-interference-and-noise-ratio (SINR) model. We develop a mathematical formulation for joint optimization of power control, scheduling, and flow routing. We present a solution procedure based on the branch-and-bound framework and apply RLT to derive tight linear relaxations for a product of variables. In this case study, we also learn how to identify the core optimization space for the underlying problem and how to exploit different physical interpretations of the core variables in developing a solution.

- Chapter 7 presents a linear approximation algorithm, which is a powerful method to tackle certain nonlinear optimization problems. We show how such an approach could be employed to solve a nonlinear programming (NLP) problem in a wireless sensor network (WSN). In addition to the linear approximation technique, the problem in the case study is interesting on its own, and shows how the so-called wireless energy transfer technology can be employed to address network lifetime problems in a WSN.

- Chapter 8 shows how to design a polynomial-time approximation algorithm to provide an $(1 - \varepsilon)$-optimal solution to a nonconvex optimization problem. The case study focuses on a classic base station placement problem in a WSN. The design of the $(1 - \varepsilon)$-optimal approximation algorithm is based on several clever techniques such as discretization of cost parameters (and distances), partitioning of the search space into a finite number of subareas, and representation of subareas with fictitious points (with tight bounds on costs). These three techniques can be exploited to develop approximation algorithms for other problems. We prove that the approximation algorithm is $(1 - \varepsilon)$-optimal.

- Chapter 9 is a sequel to Chapter 8. Again, our interest is on the design of a $(1 - \varepsilon)$-optimal approximation algorithm for a mobile base station problem. But the problem is much harder than that in the last chapter. By allowing the base station to be mobile, both the location of the base station and the multi-hop flow routing in the network are time-dependent. To address this problem, we show that as far as the network lifetime objective is concerned, we can *transform* the time-dependent problem to a location (space)-dependent problem. In particular, we show that flow routing only depends on the base station location, regardless of *when* the base station visits this location. Further, the specific time instances for the base station to visit a location are not important, as long as the total sojourn time for the base station to be present at this location is the same. This result allows us to focus on solving a location-dependent problem. Based on the above result, we further show that to obtain a $(1 - \varepsilon)$-optimal solution to the location-dependent problem, we only need to consider a finite set of points within the smallest enclosing disk for the mobile base station's location. Here, we follow the same approach as that in Chapter 8, i.e., discretization of energy cost through a geometric sequence, division of a disk into a finite number of subareas, and representation of each subarea with a fictitious cost point (FCP). Then we can find the optimal sojourn time for the base station to stay at each FCP (as well as the corresponding flow routing solution) so that the overall network lifetime (i.e., sum of the sojourn times) is maximized via a single LP problem. We prove that the proposed solution can guarantee that the achieved network lifetime is at least $(1 - \varepsilon)$ of the maximum (unknown) network lifetime. This chapter offers some excellent examples on how to transform a problem from time domain to space domain and how to prove results through construction.

Part III, consisting of Chapters 10 and 11, is devoted to some highly effective heuristics.

- Chapter 10 presents an effective approach to address a class of mixed-integer optimization problems. The technique, called sequential fixing (SF), is designed to iteratively determine (fix) binary integer variables. It is a heuristic procedure and has polynomial-time complexity. Its performance is typically measured by comparing its solution value to some performance bound, e.g., a lower bound for a minimization problem, or an upper bound for a maximization problem. Based on our own experience, we find that this SF technique is very efficient and can offer highly competitive solutions. As a case study, we study an optimization problem in a multi-hop CRN. Since the problem formulation is an MINLP model, we develop a lower bound to estimate the optimal objective value. Subsequently, we present an SF algorithm for this optimization problem. Numerical examples show that the solutions produced by this SF algorithm can offer objective values that are very close to the computed lower bounds, thus confirming their near-optimality.
- Chapter 11 presents metaheuristic methods, which are an important class of heuristic methods and have been applied to solve some very complex problems in wireless networks. In this chapter, we give a review of some well-

known metaheuristic methods (e.g., basic local search, simulated annealing (SA), tabu search (TS), and genetic algorithms (GA)). In the case study, we focus on developing a GA-based method to solve a multi-path routing problem for MD video. We find that a GA-based solution is eminently suitable to address this particular problem, which involves complex objective functions and exponential solution space. By exploiting the survival-of-the-fittest principle, a GA-based solution is able to evolve to a population of better solutions after each iteration and eventually offers a near-optimal solution.

Part IV, currently consisting of only Chapter 12, is devoted to some miscellaneous topics in the broader context of wireless network optimizations. This part will be further expanded to include other topics in a future book edition.

- Chapter 12 presents an asymptotic capacity analysis for wireless ad hoc networks. Such an analysis addresses an achievable per-node throughput when the number of nodes goes to infinity. We focus on so-called random networks, where each node is randomly deployed and each node has a randomly chosen destination node. In this asymptotic capacity analysis, the results are derived in the form of $\Omega(\cdot)$, $O(\cdot)$, and $\Theta(\cdot)$ and the underlying analysis is very different from what we do in the other chapters, which focus on optimization problems for *finite-sized* networks. We show that the asymptotic capacity analysis heavily depends on the underlying interference model. In this chapter, we consider three interference models (i.e., the protocol model, the physical model, and the generalized physical model) and show how to develop asymptotic capacity bounds for each model.

1.3 How to use this book

This book is written as a textbook and is mainly aimed at graduate students (particularly students in electrical and computer engineering) pursuing advanced research and study in wireless networks. The book could be adopted for a second or third graduate course in networking. The prerequisites for this course are a graduate course in networking.

At Virginia Tech, we have been this book in an Electrical and Computer Engineering (ECE) and Computer Science (CS) cross-listed course titled "ECE/CS 6570: Advanced Foundations of Networking." The prerequisite for this course is a graduate course in networking. We cover Chapters 2, 3, 4, 5, 8, 10, and 11 in a one semester course and the response from the students has been overwhelmingly positive! For each chapter, we have prepared PowerPoint slides, which are available to both the students and instructors. We have also prepared a solutions manual for all end-of-chapter problems, which is only available to the instructors. To access these materials, please visit www.cambridge.org/hou.

In addition to its primary role as a graduate textbook, researchers in academia who are active in conducting research in wireless networks will find this book a very useful reference to expand their toolboxes in problem solving. Further, researchers and engineers in industry and government laboratories who perform active research in wireless networks will also find this book to be a useful reference.

PART

I

Methods for Optimal Solutions

Linear programming and applications

The real voyage of discovery consists not in seeking new landscapes but in having new eyes.

Marcel Proust

2.1 Review of key results in linear programming

Linear programming (LP) is a problem consisting of a linear objective function and a set of linear constraints (equations or inequalities) with real variables. Such a problem aims to maximize or minimize a specific objective function by systematically choosing the values of the real variables within an allowed set (i.e., solution space). In an LP problem, both the objective function to be optimized and all the constraints restricting the variables are linear.

A general form of an LP problem is as follows:

$$\text{Maximize} \quad c_1 x_1 + c_2 x_2 + \ldots + c_n x_n$$
$$\text{subject to} \quad a_{11} x_1 + a_{12} x_2 + \ldots + a_{1n} x_n \leq b_1$$
$$a_{21} x_1 + a_{22} x_2 + \ldots + a_{2n} x_n \geq b_2$$
$$a_{31} x_1 + a_{32} x_2 + \ldots + a_{3n} x_n = b_3$$
$$\vdots$$
$$a_{m1} x_1 + a_{m2} x_2 + \ldots + a_{mn} x_n = b_m$$
$$x_j \geq 0 \quad (1 \leq j \leq n).$$

The function $c_1 x_1 + c_2 x_2 + \ldots + c_n x_n$ is the objective function to be maximized, where $c_1, c_2, \ldots, c_n$ are constant coefficients, and $x_1, x_2, \ldots, x_n$ are the so-called *decision variables* to be determined. The equality (or inequality) $\sum_{j=1}^{n} a_{ij} x_j = (\text{or} \leq \text{or} \geq) b_i$ is the ith constraint, where a_{ij} is a constant coefficient. The constant column vector with the ith component b_i, $1 \le i \le m$, is the right-hand-side (RHS) vector. The restrictions $x_j \ge 0, 1 \le j \le n$ are called nonnegativity constraints. Sometimes, they may be replaced by general lower and upper bounds on the decision variables. Observe that a maximization problem can be easily converted into a minimization problem and conversely, note that

$$\text{Maximum } c_1 x_1 + c_2 x_2 + \ldots + c_n x_n$$
$$= -\text{Minimum } - c_1 x_1 - c_2 x_2 - \ldots - c_n x_n.$$

In a general form of an LP problem, both equalities and inequalities may appear. Performing a simple manipulation on the inequalities (by adding or subtracting nonnegative slack variables, see [10] for more details), any LP problem can be transformed to an equivalent *standard form* as follows:

$$\text{Maximize} \quad \mathbf{c}^T \mathbf{x}$$
$$\text{subject to } \mathbf{Ax} = \mathbf{b} \qquad \qquad (2.1)$$
$$\mathbf{x} \ge \mathbf{0},$$

where $\mathbf{x}$ is the vector of nonnegative continuous variables, $\mathbf{c}$ is the vector of coefficients in the objective function, $\mathbf{A}$ is the matrix of coefficients in the constraints, and $\mathbf{b}$ is the RHS vector for the constraints. This simple transformation is convenient for developing a general algorithmic procedure to solve and analyze LPs [10].

An LP problem can be solved optimally in a time complexity of $O(n_A{}^3)$ (e.g., using *interior point methods* [10]), where n_A is the number of variables in the standard form representation. Therefore, once we are able to formulate a problem as an LP, we can solve this problem in polynomial time by employing open-source solvers (e.g., GLPK [52]) or commercial solvers (e.g., CPLEX [31] and Lindo [95]).

For an LP problem in the standard form, there is another associated LP problem called the *dual problem*, which is given as follows:

$$\text{Minimize} \quad \mathbf{w}^T \mathbf{b}$$
$$\text{subject to } \mathbf{A}^T \mathbf{w} \ge \mathbf{c},$$

where the vector of dual variables $\mathbf{w}$ is unrestricted in sign [10]. The dual problem has some important properties and economic interpretation, and can also be used to obtain the solution to the original LP problem (e.g., with the dual simplex method or the primal-dual algorithm [10]). Based on the duality relationship in LP theory, both the original LP and its dual LP can be solved simultaneously by standard LP techniques in polynomial time.

We want to emphasize that all the variables in an LP problem should be *real* (continuous) variables. If an optimization problem has a linear objective function, linear constraints, some real variables and some integer/binary variables, then this optimization problem is a *mixed-integer linear programming* (MILP), which is not an LP problem. A special case of MILPs is *integer linear program* (ILP), where all the decision variables are integer/binary restricted. Unlike LP, there does not exist a polynomial-time algorithm to solve an MILP or an ILP optimally.

In the rest of this chapter, we offer a case study on how LP can be applied to address some interesting but difficult problems from wireless sensor networks (WSNs).

2.2 Case study: Lexicographic max-min rate allocation and node lifetime problems

Consider a WSN consisting of battery-powered nodes for data collection. Although there have been significant improvements in processor design and computing, advances in battery technology still lag behind, making energy resource considerations the fundamental challenge in WSNs. Consequently, there have been active research efforts on performance limits of WSNs. These performance limits include, among others, *network capacity* (e.g., [78]) and *network lifetime* (e.g., [24]). Network capacity typically refers to the maximum amount of bit volume that can be successfully delivered to the base station ("sink node") by all the nodes in the network, while network lifetime refers to the maximum time limit for which all the nodes in the network remain alive until one or more nodes drain up their energy.

In this chapter, we consider an overarching problem that encompasses both performance metrics. In particular, we study the network capacity problem under a given network lifetime requirement. Specifically, for a WSN where each node is provisioned with an initial energy, if *all* nodes are required to live up to a certain lifetime, what is the maximum amount of bit volume that can be generated by the entire network? At first glance, it appears desirable to maximize the sum of rates for all the nodes in the network, subject to the condition that each node can meet the network lifetime requirement. Mathematically, this problem can be formulated as a linear programming (LP) problem (see Section 2.3.2) within which the objective function is defined as the sum of rates over all the nodes in the network and the constraints require that (i) flow balance is preserved at each node, and (ii) the energy constraint at each node is met for the given network lifetime requirement. However, the solution to this problem shows (see Section 2.7) that there exists a severe bias in rate allocation among the nodes, despite that the sum of bit rates is maximized. In particular, those nodes that consume the least amount of power for a unit of data on their data path toward the base station are allocated with much more bit rates than other nodes in the network. Consequently, the data collection behavior for the entire network only favors certain nodes that have this property, while other nodes will be unfavorably penalized with much smaller bit rates.

The fairness issue associated with the network capacity maximization objective calls for careful consideration in rate allocation among the nodes. In this chapter, we study the rate allocation problem in an energy-constrained sensor network for a given network lifetime requirement. Our objective is to achieve a certain measure of optimality in the rate allocation that takes into account both fairness and bit rate maximization. We employ the so-called *Lexicographic Max-Min* (LMM) criterion [99], which maximizes the bit rates for *all* the nodes under the given energy constraint and network lifetime requirement. At the first level, the smallest rate among all the nodes is maximized. Then, we continue to maximize the second smallest rate level and so forth. The LMM rate allocation criterion is appealing since it addresses both fairness and efficiency (i.e., bit rate maximization) in an energy-constrained network.

A naive approach to the LMM rate allocation problem would be to apply a max-min-like iterative procedure. Under this approach, successive LPs are employed to calculate the maximum rate at each level based on the available energy for the remaining nodes, until all nodes use up their energy. We call this approach "serial LP with energy reservation" (SLP-ER). We will show that, although SLP appears intuitive, it is likely to offer an *incorrect* solution. To understand how this could happen, we must understand a fundamental difference between the LMM rate allocation problem described here and the classical max-min rate allocation in [14]. Under the LMM rate allocation problem, the rate allocation is implicitly *coupled* with a flow routing problem, while under the classical max-min rate allocation problem, there is no routing involved since the routes for all flows are given. As it turns out, for the LMM rate allocation problem, any iterative rate allocation approach that requires energy reservation at each iteration is incorrect. This is because, unlike max-min, which addresses only the rate allocation problem with fixed routes and yields a unique solution at each iteration, for the LMM rate allocation problem, there usually exist *nonunique* flow routing solutions corresponding to the same rate allocation at each level. Each of these flow routing solutions will yield *different* available energy levels on the remaining nodes for future iterations and so forth. This will lead to a different rate allocation vector, which may not coincide with the optimal LMM rate allocation vector.

In this chapter, we show a correct approach to solve the LMM rate allocation problem. Our approach employs the so-called *parametric analysis* (PA) technique [10] in LP to determine the minimum set of nodes at each rate level that must deplete their energy. We call this approach *serial LP with PA* (SLP-PA). We also extend the PA technique for the LMM rate allocation problem to address the so-called maximum node lifetime curve problem in [20], which we call the LMM node lifetime problem. More importantly, we show that there exists a simple and elegant *mirror* relationship between the LMM rate allocation problem and the LMM node lifetime problem. As a result, it is sufficient to solve only one of these two problems.

The remainder of this chapter is organized as follows. Section 2.3 introduces the network and energy model, and formulates the LMM rate allocation problem. Section 2.4 describes the SLP-PA algorithm for the LMM rate allocation problem. Section 2.5 introduces the LMM node lifetime problem and shows how to apply the SLP-PA algorithm to solve it. Section 2.6 reveals an interesting mirror relationship between the LMM rate allocation problem and the LMM node lifetime problem. Section 2.7 presents numerical results. Section 2.8 summarizes this chapter.

2.3 System modeling and problem formulation

We consider a two-tier architecture for WSNs. Figs. 2.1(a) and (b) show the physical and hierarchical network topologies for such a network, respectively. There are three types of nodes in the network – namely, *micro-sensor nodes* (MSNs), *aggregation and forwarding nodes* (AFNs) – and a *base station* (BS). The MSNs can be application-specific sensor nodes (e.g., temperature, chemical, video) and they constitute the lower-tier data collection network. They are deployed in groups (or clusters) at strategic locations for surveillance and monitoring applications. The MSNs are small and low cost. The objective of an MSN is simple: once triggered by an event, it starts to collect sensing data and sends it to the local AFN (in one hop).

For each cluster of MSNs, there is one AFN, which is different from an MSN in terms of physical properties and functions. The primary functions of an AFN are: (1) *data aggregation* (or "fusion") for data flows from the local cluster of MSNs, and (2) *forwarding* (or relaying) the aggregated information to the next hop AFN (ultimately to the base station). Although an AFN is expected to be provisioned with much more energy than an MSN, it also consumes energy at a substantially higher rate (due to data transmission over longer distances). Consequently, an AFN has a limited lifetime. Upon depletion of energy at an AFN, we expect that the *coverage* for the particular area under surveillance is lost, despite the fact that some of the MSNs within the cluster may still have remaining energy.

The third component in the two-tier architecture is the base station. The base station is, essentially, the *sink* node for data streams from all the AFNs in the network. We assume that there is sufficient energy resource available at the base station and thus there is no energy constraint at the base station. In summary, from networking perspective, the lower-tier MSNs are for data acquisition, while the upper-tier AFNs are for transporting data to the base station.

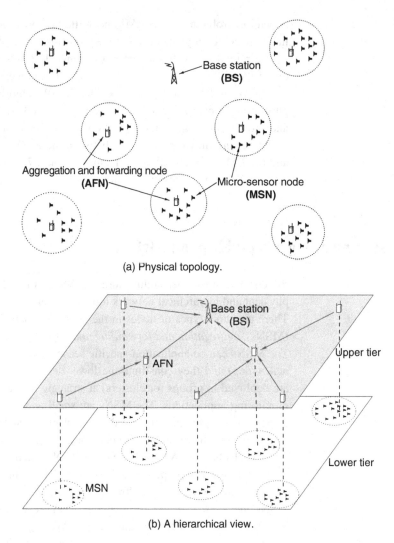

Figure 2.1 Reference architecture for two-tier WSNs.

(a) Physical topology.

(b) A hierarchical view.

2.3.1 Energy model

Table 2.1 lists the notation used in this chapter. We focus on energy consumption at AFNs. For AFN i, we assume that the aggregated bit rate collected locally (after data fusion) is r_i, $i = 1, 2, \ldots, N$. These collected local bit streams must be relayed to the base station. Our objective is to maximize the r_i-values according to the LMM criterion (see Definition 2.1) under a given network lifetime requirement.

For an AFN, we assume that energy consumption due to transmission and reception is the dominant source of energy consumption [4]. The power dissipation at a radio transmitter can be modeled as

$$u_{ij}^t = C_{ij} \cdot f_{ij}, \tag{2.2}$$

Table 2.1 Notation.

General notation for the LMM rate and LMM lifetime problems	
B	The base station
C_{ij} (or C_{iB})	The power consumption coefficient for transmitting data from AFN i to AFN j (or the base station B)
e_i	The initial energy at AFN i
f_{ij} (or f_{iB})	Data rate from AFN i to AFN j (or the base station B)
n	The number of distinct elements in the sorted LMM-optimal rate/lifetime vector
N	Total number of AFNs in the network
S_i	The minimum set of nodes that reach their energy constraint limits at the ith level
$\hat{S}_i$	The set of all possible AFNs that may reach their energy constraint limits at the ith level, $S_i \subseteq \hat{S}_i$
V_{ij} (or V_{iB})	Total data volume from AFN i to AFN j (or the base station B)
α	Path-loss index
β_1, β_2	Constant terms in transmission power model
ρ	The power consumption coefficient for receiving data
Symbols used for the LMM rate problem	
g_i	The ith element in the sorted LMM-optimal rate vector, where $g_1 \leq g_2 \leq \ldots \leq g_N$
r_i	The local bit rate collected at AFN i
T	The network lifetime requirement
δ_i	$= \lambda_i - \lambda_{i-1}$, the difference between λ_i and λ_{i-1}
λ_i	The ith rate level in the sorted LMM-optimal rate vector, i.e., $\lambda_1(=g_1) < \lambda_2 < \ldots < \lambda_n(=g_N)$
Symbols used for the LMM lifetime problem	
r_i	The rate requirement at AFN i
t_i	The node lifetime at AFN i
ζ_i	$= \mu_i - \mu_{i-1}$, the difference between μ_i and μ_{i-1}
μ_i	The ith drop point in the sorted LMM-optimal lifetime vector, i.e., $\mu_1(=\tau_1) < \mu_2 < \ldots < \mu_n(=\tau_N)$
τ_i	The ith element in the sorted LMM-optimal lifetime vector, where $\tau_1 \leq \tau_2 \leq \ldots \leq \tau_N$

where u^t_{ij} is the power dissipated at AFN i when it is transmitting to node j, f_{ij} is the rate from AFN i to node j, C_{ij} is the power consumption cost of radio link $i \rightarrow j$ and is given by

$$C_{ij} = \beta_1 + \beta_2 \cdot d^\alpha_{ij}, \tag{2.3}$$

where β_1 and β_2 are two constant terms, d_{ij} is the distance between these two nodes, and α is the path-loss index. Typical values for these parameters are $\beta_1 = 50$ nJ/b, $\beta_2 = 0.0013$ pJ/b/m^4 [67], and $\alpha = 4$ [129]. Since the power level of an AFN's transmitter can be used to control the distance coverage of an AFN (e.g., [127; 131; 166]), different network flow routing topologies can be formed by adjusting the power level of each AFN's transmitter.

The power dissipation at a receiver can be modeled as

$$u_i^r = \rho \cdot \sum_{k \neq i} f_{ki}, \tag{2.4}$$

where $\sum_{k \neq i} f_{ki}$ (in b/s) is the rate of the received data stream at AFN i. A typical value for the parameter ρ is 50 nJ/b [67].

To ensure data transmissions at different AFNs do not interfere with each other, we assume that a scheduling mechanism has been employed. The details of such a scheduling mechanism is beyond the scope of this chapter.

2.3.2 The LMM rate allocation problem

Before we formulate the LMM rate allocation problem, we revisit the maximum capacity problem (with "bias" in rate allocation) that was discussed in Section 2.2. For a network with N AFNs, suppose that the rate of AFN i is r_i, and that the initial energy at this node is e_i ($i = 1, 2, \ldots, N$). For a given network lifetime requirement T (i.e., each AFN must remain alive for at least the time duration T), the maximum sum of rates can be found by the following LP problem:

MaxCap Maximize $\qquad \sum_{i=1}^{N} r_i$

$$\text{subject to } f_{iB} + \sum_{j \neq i} f_{ij} - \sum_{k \neq i} f_{ki} = r_i \ (1 \leq i \leq N) \tag{2.5}$$

$$\sum_{k \neq i} \rho f_{ki} T + \sum_{j \neq i} C_{ij} f_{ij} T + C_{iB} f_{iB} T \leq e_i \ (1 \leq i \leq N) \tag{2.6}$$

$$f_{ij}, f_{iB} \geq 0 \qquad (1 \leq i, j \leq N, j \neq i).$$

The set of constraints in (2.5) are the flow balance equations. They state that the total bit rate transmitted by AFN i is equal to the total bit rate received by AFN i from other AFNs, plus the bit rate generated locally at this node (r_i). Note that we allow flow splitting at a node so as to achieve more flexibility in flow routing and load balancing in the network. The set of constraints in (2.6) are the energy constraints. They state that for a given network lifetime requirement T, the energy consumed for communications (i.e., transmitting and receiving) cannot exceed the initial available energy.

Note that f_{ki}, f_{ij}, f_{iB}, and r_i are variables and T is a constant (the given network lifetime requirement). Since MaxCap is an LP problem, it can be solved in polynomial time [10]. Unfortunately, as we will see in the numerical results (Section 2.7), the solution to this MaxCap problem favors those AFNs whose data paths consume the least amount of power toward the base station. Consequently, although the sum of rates is maximized over T, the specific bit rate allocation among the AFNs (i.e., r_is) favors those AFNs that have this property, while the other AFNs are unfavorably allocated with much smaller (even close to 0) data rates. As a result of this unfairness, the effectiveness of the sensor network in performing data collection is questionable.

To address this fairness issue, we can employ the so-called *lexicographic max-min* (LMM) rate allocation strategy [99], which bears some similarity to the max-min rate allocation in data networks [14].[1] Under LMM rate allocation, we start with the objective of maximizing the bit rate of *all* nodes until one or more nodes reach their limits for the given network lifetime requirement. Given that the first level of the smallest rate allocation is maximized, we continue to maximize the second level of rate for the remaining nodes that still have available energy, and so forth. More formally, denote $\mathbf{g} = [g_1, g_2, \ldots, g_N]$ as the sorted version (i.e., $g_1 \leq g_2 \leq \ldots \leq g_N$) of the rate vector $\mathbf{r} = [r_1, r_2, \ldots, r_N]$, with r_i corresponding to the rate of node i. We then have the following definition for LMM-optimal rate allocation:

Definition 2.1

LMM-optimal rate allocation   For a given network lifetime requirement T, a sorted rate vector $\mathbf{g} = [g_1, g_2, \ldots, g_N]$ yields an LMM-optimal rate allocation if and only if for any other sorted rate allocation vector $\hat{g} = [\hat{g}_1, \hat{g}_2, \ldots, \hat{g}_N]$ with $\hat{g}_1 \leq \hat{g}_2 \leq \ldots \leq \hat{g}_N$ there exists k, $1 \leq k \leq N$, such that $g_i = \hat{g}_i$ for $1 \leq i \leq k - 1$ and $g_k > \hat{g}_k$.

2.3.3 Two incorrect approaches

Before we present a correct solution to the optimal LMM rate allocation problem, we discuss two incorrect solutions and explain why they cannot provide an LMM-optimal solution.

Serial LP with energy reservation   Based on the LMM-optimal definition, we can calculate the first level optimal rate $\lambda_1 = g_1$ easily through the following LP problem:

$$\text{Maximize} \qquad\qquad \lambda_1$$

$$\text{subject to} \quad f_{iB} + \sum_{j \neq i} f_{ij} - \sum_{k \neq i} f_{ki} - \lambda_1 = 0 \quad (1 \leq i \leq N)$$

$$\sum_{k \neq i} \rho T f_{ki} + \sum_{j \neq i} C_{ij} T f_{ij} + C_{iB} T f_{iB} \leq e_i \ (1 \leq i \leq N)$$

$$f_{ij}, f_{iB} \geq 0 \qquad\qquad (1 \leq i, j \leq N, j \neq i).$$

Once we obtain a solution with the maximum λ_1, we can also calculate the energy consumption at each node under this flow routing solution. Then we can check whether or not a node has any remaining energy. If there are some nodes that still have remaining energy, then we can construct another LP problem to further increase their data rates (with a maximum rate increment of λ_2).

---

[1]   However, there is a significant difference between max-min and LMM, which we will discuss shortly.

This process terminates until all nodes use up their energy. Since energy is reserved at each node after each iteration, we call this naive approach *Serial LP with Energy Reservation* (SLP-ER).

SLP-ER cannot provide an LMM-optimal solution because there is a fundamental difference in the nature of the LMM rate allocation problem described here and the classical max-min rate allocation problem in [14]. The LMM rate allocation problem implicitly couples a flow routing problem (i.e., a determination of the f_{ij} and f_{iB} for the entire network), while the classical max-min rate allocation explicitly assumes that the routes for all the flows are given *a priori* and remain fixed. For the LMM rate allocation problem, starting from the first iteration, there usually exist *nonunique* flow routing solutions corresponding to the same maximum rate level. Consequently, each of these flow routing solutions, once chosen, will yield different remaining energy levels on the nodes for future iterations and so forth. This will lead to a nonunique rate allocation vector, which may not coincide with the LMM-optimal rate vector (see numerical examples in Section 2.7).

Serial LP with rate reservation   Instead of reserving energy based on a flow routing solution at each rate level, another approach is to only determine the set of nodes that use up their energy at this level. The final flow routing solution can be deferred to the last iteration. Since rate is reserved during each iteration, we call this approach *Serial LP with Rate Reservation* (SLP-RR).

SLP-RR cannot achieve an LMM-optimal rate vector either. This may be harder to understand. But it is still due to the existence of nonunique flow routing solutions at each rate level, whereby a node that uses up its energy in one flow routing solution may still have remaining energy in another flow routing solution. Therefore, if we reserve rates just based on one flow routing solution, the set of nodes thought to have used up their energy at this rate level may include some extra nodes that would otherwise be allocated with a larger data rate in another flow routing solution. Consequently, a solution obtained by SLP-RR may not be LMM-optimal.

2.4 A serial LP algorithm based on parametric analysis

In this section, we present a correct polynomial-time algorithm to solve the LMM rate allocation problem. We first introduce some of the notation used. Suppose that the rate vector $\mathbf{g} = [g_1, g_2, \ldots, g_N]$ is LMM-optimal, with $g_1 \leq g_2 \leq \ldots \leq g_N$. Note that the values of these N rates may not be distinct. To focus on those distinct rate levels, we remove any repetitive elements in this vector and rewrite it as $[\lambda_1, \lambda_2, \ldots, \lambda_n]$ such that $\lambda_1 < \lambda_2 < \ldots < \lambda_n$, where $\lambda_1 = g_1, \lambda_n = g_N$, and $n \leq N$. Now, for each $\lambda_i, i = 1, 2, \ldots, n$, denote S_i

as the corresponding set of nodes that use up their energy at this rate. Clearly, we have $\sum_{i=1}^{n} |S_i| = |S| = N$, where S denotes the set of all N nodes.

The essence to solving the LMM rate allocation problem is to find the correct values for λ_1, λ_2, ..., λ_n and the corresponding sets S_1, S_2, ..., S_n, respectively. This can be done iteratively. We first determine the rate level λ_1 and the corresponding set S_1, then determine the rate level λ_2 and the corresponding set S_2, and so on. In Section 2.4.1, we will show how to determine each rate level and in Section 2.4.2, we will show how to determine the corresponding node set.

2.4.1 Determining rate levels

Denote $\lambda_0 = 0$ and $S_0 = \emptyset$. For $l = 1, 2, \ldots, n$, suppose that we have already determined $\lambda_0, \lambda_1, \ldots, \lambda_{l-1}$ and the corresponding sets $S_0, S_1, \ldots, S_{l-1}$. The rate level λ_l can be found by the following optimization problem:

$$\text{Maximize} \qquad\qquad \delta_l$$

$$\text{subject to } f_{iB} + \sum_{j \neq i} f_{ij} - \sum_{k \neq i} f_{ki} - \delta_l = \lambda_{l-1} \ (i \notin \bigcup_{h=0}^{l-1} S_h) \qquad (2.7)$$

$$f_{iB} + \sum_{j \neq i} f_{ij} - \sum_{k \neq i} f_{ki} = \lambda_h \quad (i \in S_h, 1 \leq h < l) \quad (2.8)$$

$$(\sum_{k \neq i} \rho f_{ki} + \sum_{j \neq i} C_{ij} f_{ij} + C_{iB} f_{iB})T \leq e_i \quad (i \notin \bigcup_{h=0}^{l-1} S_h) \qquad (2.9)$$

$$(\sum_{k \neq i} \rho f_{ki} + \sum_{j \neq i} C_{ij} f_{ij} + C_{iB} f_{iB})T = e_i \quad (i \in S_h, 1 \leq h < l) \ (2.10)$$

$$f_{ij}, f_{iB} \geq 0 \qquad\qquad (1 \leq i, j \leq N, j \neq i).$$

Note that for $l = 1$, constraints (2.8) and (2.10) do not exist. For $2 \leq l \leq n$, constraints (2.8) and (2.10) are for those nodes that have already reached their LMM rate allocation during the previous $(l - 1)$ iterations. In particular, the set of constraints in (2.8) say that the sum of in-coming and local data rates is equal to the out-going data rates for each node with its LMM-optimal rate λ_h, $1 \leq h < l$. The set of constraints in (2.10) say that for those nodes that have already reached their LMM-optimal rates, the total energy consumed for communications has reached their initial energy provisioning. On the other hand, the constraints in (2.7) and (2.9) are for the remaining nodes that have not yet reached their LMM-optimal rates. Specifically, the set of constraints in (2.7) state that, for those nodes that have not yet reached their energy constraint levels, the sum of in-coming and local data rates is equal to the out-going data rates. Note that the objective function is to maximize the additional rate δ_l for those nodes. Furthermore, for those nodes, the set of constrains in (2.9) state that the total energy consumed for communications should be upper bounded by the initial energy provisioning.

To facilitate our later discussion on mirror results in Section 2.6, we further re-formulate the foregoing LP problem. In particular, we multiply both sides of (2.7) and (2.8) by T (which is a constant) and denote $V_{iB} = f_{iB}T$, $V_{ij} = f_{ij}T$, and $V_{ki} = f_{ki}T$. Intuitively, V_{ij} and V_{iB} represent the bit volume that is transferred from node i to j and from node i to B, respectively, during the lifetime T. We obtain the following LP formulation, which can be solved using standard LP techniques:

LMM-rate Maximize $\qquad\qquad \delta_l$

subject to $V_{iB} + \sum_{j \neq i} V_{ij} - \sum_{k \neq i} V_{ki} - \delta_l T = \lambda_{l-1} T \ (i \notin \bigcup_{h=0}^{l-1} S_h)$ (2.11)

$$V_{iB} + \sum_{j \neq i} V_{ij} - \sum_{k \neq i} V_{ki} = \lambda_h T \quad (i \in S_h, 1 \leq h < l)$$

$$\sum_{k \neq i} \rho V_{ki} + \sum_{j \neq i} C_{ij} V_{ij} + C_{iB} V_{iB} \leq e_i \ (i \notin \bigcup_{h=0}^{l-1} S_h)$$

$$\sum_{k \neq i} \rho V_{ki} + \sum_{j \neq i} C_{ij} V_{ij} + C_{iB} V_{iB} = e_i \ (i \in S_h, 1 \leq h < l)$$

$$V_{ij}, V_{iB} \geq 0 \qquad\qquad (1 \leq i, j \leq N, j \neq i).$$

Although a solution to the LMM rate problem gives the optimal solution for δ_l at iteration l, it remains to determine the *minimum* set of nodes corresponding to this δ_l, which is the key difficulty in the LMM rate allocation problem. Fortunately, this problem can be solved by the so-called *parametric analysis* (PA) technique in LP [10].

2.4.2 Determining minimum node set for a rate level

For the minimum node set for a rate level, we first need to know whether or not this set is unique. This is affirmed in the following lemma. Its proof can be found in [71].

Lemma 2.1
The minimum node set for each rate level under the LMM-optimal rate allocation is unique.

With this lemma in place, we now discuss how to determine the minimum node set S_l corresponding to the rate level λ_l. Denote $\hat{S}_l$ ($\neq \emptyset$) as the set of nodes for which the constraints (2.9) are *binding* at the lth iteration in LMM rate, i.e., $\hat{S}_l$ include all the nodes that achieve *equality* in (2.9) at iteration l. Although it is certain that at least one of the nodes in $\hat{S}_l$ belong to S_l (the minimum node set for rate λ_l), some nodes in $\hat{S}_l$ may still achieve greater rates under other flow routing solutions. In other words, if $|\hat{S}_l| = 1$, then we must have $S_l = \hat{S}_l$ otherwise, we must determine the *minimum* node set S_l ($\subseteq \hat{S}_l$) that achieves the LMM-optimal rate allocation.

The PA technique [10] is most suitable to address this problem. The main idea of PA is to investigate how an infinitesimal perturbation on some parameters of the LMM rate problem can affect the objective function. Specifically, node i belongs to the minimum node set S_l if and only if a small increase in node i's rate leads to a *decrease* in the objective value. In our problem, we consider a small increase on the right-hand-side (RHS) of (2.11), i.e., changing b_i in the standard form (2.1) to $b_i + \epsilon_i$, where $\epsilon_i > 0$. The physical meaning is that node i's data rate increased from $\delta_l + \lambda_{l-1}$ to $\delta_l + \lambda_{l-1} + \frac{\epsilon_i}{T}$. We solve the updated optimization problem with this new requirement on node i's data rate. Node i belongs to the minimum node set S_l if and only if the new δ_l is smaller, i.e., $\frac{\partial^+ \delta_l}{\partial \epsilon_i}(0) < 0$. The details of applying PA to determine the minimum node set S_l can be found in [71].

2.4.3 Optimal flow routing for LMM rate allocation

After we solve the LMM rate allocation problem iteratively using the procedure described in Sections 2.4.1 and 2.4.2, the corresponding optimal flow routing can be obtained by dividing the total bit volume on each link (V_{ij} or V_{iB}) by T, i.e.,

$$ f_{ij} = \frac{V_{ij}}{T} \quad \text{and} \quad f_{iB} = \frac{V_{iB}}{T}. \tag{2.12} $$

Although the LMM-optimal rate allocation is unique, it is important to note that the corresponding flow routing solution is *not* unique. This is because upon the completion of the LMM rate allocation problem (i.e., upon finding $[\lambda_1, \lambda_2, \dots, \lambda_n]$), there usually exist nonunique bit volume solutions (V_{ij} and V_{iB} values) corresponding to the same LMM-optimal rate allocation. This result is summarized in the following lemma:

Lemma 2.2

The optimal flow routing solution corresponding to the LMM rate allocation may not be unique.

We use the following example to illustrate the nonuniqueness of the optimal flow routing solution for an LMM rate allocation.

Example 2.1

Consider an eight-node network with a topology shown in Fig. 2.2. The base station B is located at the origin $(0, 0)$. There are two groups of nodes,

G_1 and G_2, in the network, with each group consisting of four nodes. Group G_1 consists of AFN$_1$ at (100, 0), AFN$_3$ at (0, 100), AFN$_5$ at (-100, 0), and AFN$_7$ at (0, -100), respectively (all in meters); Group G_2 consists of AFN$_2$ at (100, 100), AFN$_4$ at (-100, 100), AFN$_6$ at (-100, -100), and AFN$_8$ at (100, -100), respectively. Assume that all nodes have the same initial energy e. For a network lifetime requirement of T, we can calculate (via SLP-PA) that the final LMM-optimal rate allocation for all eight nodes are identical (perfect fairness), i.e., $r_1 = r_2 = \ldots = r_8$. We denote $r_i = g$ for $1 \leq i \leq 8$.

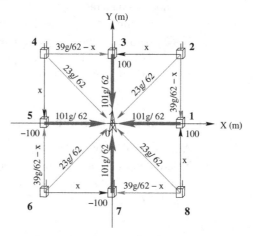

Figure 2.2 A simple example showing that the optimal flow routing to the LMM rate allocation is not unique. The range of x is $0 \leq x \leq \frac{39g}{62}$.

Upon the completion of the SLP-PA algorithm, we also obtain an optimal flow routing solution corresponding to this LMM-optimal rate g. This optimal flow routing solution has the following flows: $f_{21} = f_{43} = f_{65} = f_{87} = \frac{39}{62}g$, $f_{2B} = f_{4B} = f_{6B} = f_{8B} = \frac{23}{62}g$, and $f_{1B} = f_{3B} = f_{5B} = f_{7B} = \frac{101}{62}g$. We now show that the optimal flow routing solution is nonunique. Since the network has a symmetrical property, it can be easily verified that for any x, $0 \leq x \leq \frac{39}{62}g$, the LMM-optimal rate allocation can be achieved if the flow routing solution satisfies the following two conditions: (i) each node in G_2 (i.e., AFNs 2, 4, 6, and 8) sends a flow of x and a flow of $\frac{39}{62}g - x$ to its two neighboring G_1 nodes as shown in Fig. 2.2, and a remaining flow of $\frac{23}{62}g$ directly to the base station; and (ii) each node in G_1 (i.e., AFNs 1, 3, 5, and 7) sends a total amount of $\frac{101}{62}g$ to the base station, which includes x and $\frac{39}{62}g - x$ from its neighboring nodes, plus g from itself. Clearly, there are infinitely many flow routing solutions that meet these two conditions, each of which can be shown to yield the LMM-optimal rate allocation g with the given network lifetime requirement T.

2.4.4 Complexity analysis

We now analyze the complexity of the SLP-PA algorithm. First, we consider the complexity of finding each node's rate and the total bit volume transmitted along each link. At each stage, we solve an LP problem with a complexity of $O(n_A{}^3)$ [10], where n_A is the number of variables in the formulated LP (in the standard form). Since the number of variables is $O(N^2)$, the complexity of solving the LP problem is $O(N^6)$. After solving an LP problem at each stage, we need to determine whether or not a node that just reached its energy binding constraint belongs to the minimum node set for this stage. In [71], we showed that the complexity of this is $O(N^7)$. Hence, the complexity at each stage is $O(N^6) + O(N^7) = O(N^7)$. Since there are at most N stages, the overall complexity is $O(N^8)$.

The complexity in finding an optimal flow routing is bounded by the number of radio links in the network, which is $O(N^2)$. Hence, the overall complexity is $O(N^8) + O(N^2) = O(N^8)$. Note that the analysis here gives a worst-case time complexity. In practice, the run-time of SLP-PA is much faster.

2.4.5 Extension to variable bit rate

So far, we have considered the case that each AFN generates data at a constant rate. In practice, an AFN node may not always transmit data and may work in an on–off mode to conserve energy. In this case, it is necessary to construct an optimal flow routing solution for a variable bit rate source (where the on–off mode is a special case). In [70], we have developed techniques to construct an optimal flow routing solution for a variable bit rate, as long as the average rate is known. Such an average rate corresponds to the constant rate in this chapter. As a result, the case of the on–off mode (with a known average rate) can also be handled using the techniques described in [70].

2.5 SLP-PA for the LMM node lifetime problem

In this section, we show that the SLP-PA algorithm can be used to solve the so-called *maximum node lifetime curve problem* in [20], which we define as the *LMM node lifetime problem*. We also show that the SLP-PA algorithm is much more efficient than the one proposed in [20].

2.5.1 The LMM node lifetime problem

The LMM node lifetime problem can be described as follows. For a network with N AFNs, with a given local bit rate r_i (fixed) and initial energy

e_i for AFN i, $i = 1, 2, \ldots, N$, how can we maximize the network lifetime for *all* AFNs in the network? In other words, the LMM node lifetime problem not only considers how to maximize the lifetime until the first AFN runs out of energy, but also the lifetimes for the rest of the AFNs in the network.

More formally, denote the lifetime for each AFN i as t_i, $i = 1, 2, \ldots, N$. Note that the r_is are fixed here, while the t_is are the optimization variables, which are different from those in the LMM rate allocation problem in the previous section. Denote $[\tau_1, \tau_2, \ldots, \tau_N]$ as the *sorted* sequence of the t_i-values in nondecreasing order. Then, the LMM-optimal node lifetime can be defined as follows:

Definition 2.2

LMM-optimal node lifetime A sorted node lifetime vector $[\tau_1, \tau_2, \ldots, \tau_N]$ with $\tau_1 \leq \tau_2 \leq \ldots \leq \tau_N$ is LMM-optimal if and only if for any other sorted node lifetime vector $[\hat{\tau}_1, \hat{\tau}_2, \ldots, \hat{\tau}_N]$, with $\hat{\tau}_1 \leq \hat{\tau}_2 \leq \ldots \leq \hat{\tau}_N$, there exists a k, $1 \leq k \leq N$ such that $\tau_i = \hat{\tau}_i$ for $1 \leq i \leq k - 1$ and $\tau_k > \hat{\tau}_k$.

2.5.2 Solution

It should be clear that, under the LMM-optimal node lifetime objective, we must *maximize* the time until a set of nodes use up their energy (which is also called a *drop point* in [20]), while *minimizing* the number of nodes that drain up their energy at each drop point. We now show that the SLP-PA algorithm developed for the LMM rate allocation problem can be applied to solve the LMM node lifetime problem.

Suppose that $[\tau_1, \tau_2, \ldots, \tau_N]$ with $\tau_1 \leq \tau_2 \leq \ldots \leq \tau_N$ is LMM-optimal. To keep track of *distinct* node lifetimes (or drop points) in this vector, we remove all repetitive elements in the vector and rewrite it as $[\mu_1, \mu_2, \ldots, \mu_n]$ such that $\mu_1 < \mu_2 < \ldots < \mu_n$, where $\mu_1 = \tau_1$, $\mu_n = \tau_N$, and $n \leq N$. Corresponding to these drop points, denote $S_1, S_2, \ldots, S_n$ as the sets of nodes that drain up their energy at drop points $\mu_1, \mu_2, \ldots, \mu_n$, respectively. Then $|S_1| + |S_2| + \ldots + |S_n| = |S| = N$, where S denotes the set of all N AFNs in the network. The problem is to find the LMM-optimal values of $\mu_1, \mu_2, \ldots, \mu_n$ and the corresponding sets $S_1, S_2, \ldots, S_n$.

Similar to the LMM rate allocation problem, the LMM node lifetime problem can be formulated as an iterative optimization problem as follows. Denote $\mu_0 = 0$, $S_0 = \emptyset$, and $\zeta_l = \mu_l - \mu_{l-1}$. Starting from $l = 1$, we solve the following LP problem iteratively:

LMM lifetime Maximize ζ_l

subject to $\quad V_{iB} + \sum_{j \neq i} V_{ij} - \sum_{k \neq i} V_{ki} - \zeta_l r_i = \mu_{l-1} r_i \quad (i \notin \bigcup_{h=0}^{l-1} S_h) \quad$ (2.13)

$$V_{iB} + \sum_{j \neq i} V_{ij} - \sum_{k \neq i} V_{ki} = \mu_h r_i \qquad (i \in S_h, 1 \leq h < l)$$

$$\sum_{k \neq i} \rho V_{ki} + \sum_{j \neq i} C_{ij} V_{ij} + C_{iB} V_{iB} \leq e_i \qquad (i \notin \bigcup_{h=0}^{l-1} S_h)$$

$$\sum_{k \neq i} \rho V_{ki} + \sum_{j \neq i} C_{ij} V_{ij} + C_{iB} V_{iB} = e_i \qquad (i \in S_h, 1 \leq h < l)$$

$$V_{ij}, V_{iB}, \zeta_l \geq 0 \qquad (1 \leq i, j \leq N, j \neq i).$$

Comparing the above LMM lifetime problem to the LMM rate problem in Section 2.4.1, we find that they are mathematically identical. The only difference is that under the LMM lifetime problem, the local bit rates r_is are constants and the node lifetimes τ_is are variables, while under the LMM rate problem, the r_is are variables and the node lifetimes are all identical (T), $i = 1, 2, \ldots, N$. Since the mathematical formulation for the two problems are identical, the SLP-PA algorithm can be applied to solve the LMM node lifetime problem.

The only issue that we need to be concerned about is the optimal flow routing solution corresponding to the LMM-optimal lifetime vector. The optimal flow routing solution here is not as simple as that for the LMM rate allocation problem, which merely involves a simple division (see (2.12)). We refer readers to the appendix at the end of this chapter for an $O(N^4)$ algorithm to obtain an optimal flow routing solution for the LMM-optimal lifetime vector. Similar to Lemma 2.2, the optimal flow routing solution corresponding to the LMM node lifetime problem may not be unique.

2.6 A mirror result

In this section, we present an elegant result showing that there is a mirror relationship between the LMM rate allocation problem and the LMM node lifetime problem. As a result, it is only necessary to solve only one of the two problems and the results for the other problem can be obtained via simple algebraic calculations.

To start with, we denote $\mathcal{P}_R$ as the LMM rate allocation problem where we have N AFNs in the network and all nodes have a common lifetime requirement T (a given constant). Denote r_i as the LMM-optimal rate allocation for node i under $\mathcal{P}_R$, $i = 1, 2, \ldots, N$. Similarly, we denote $\mathcal{P}_L$ as the LMM node lifetime problem where all nodes have the same local bit rate R (constant). Denote t_i as the LMM node lifetime for node i under $\mathcal{P}_L$, $i = 1, 2, \ldots, N$. Then the following theorem shows how the solution to one problem can be used to obtain the solution to the other.

Table 2.2 Mirror relationship between the LMM rate allocation problem $\mathcal{P}_R$ and the LMM node lifetime problem $\mathcal{P}_L$.

$\mathcal{P}_R$	$\mathcal{P}_L$
r_i (optimization variable)	$r_i = R$ (constant)
$t_i = T$ (constant)	t_i (optimization variable)
Total bit volume at AFN i: $r_i \cdot T = t_i \cdot R$	

Theorem 2.1

Mirror relation   For a given node lifetime requirement T for all nodes under Problem $\mathcal{P}_R$ and a given local bit rate R for all nodes under Problem $\mathcal{P}_L$, we have the following relationship between the solutions to the LMM rate allocation problem $\mathcal{P}_R$ and the LMM node lifetime problem $\mathcal{P}_L$.

(i) Suppose that we have solved Problem $\mathcal{P}_R$ and obtained the LMM-optimal rate allocation r_i for each node i ($i = 1, 2, \ldots, N$). Then under $\mathcal{P}_L$, the LMM node lifetime t_i for node i is given by

$$t_i = \frac{r_i T}{R}. \tag{2.14}$$

(ii) Suppose that we have solved Problem $\mathcal{P}_L$ and obtained the LMM-optimal node lifetime t_i for each node i ($i = 1, 2, \ldots, N$). Then under $\mathcal{P}_R$, the LMM rate allocation r_i for node i is given by

$$r_i = \frac{t_i R}{T}. \tag{2.15}$$

Table 2.2 shows the mirror relationship between solutions to Problems $\mathcal{P}_R$ and $\mathcal{P}_L$.

Proof. We prove (i) and (ii) in Theorem 2.1 separately.

(i) We organize our proof into two parts. First, we show that the t_is are feasible node lifetimes in terms of flow balance and energy constraints at each node i ($i = 1, 2, \ldots, N$). Then we show that this solution indeed achieves the LMM-optimal node lifetime.

Feasibility   Since we have obtained the solution to Problem $\mathcal{P}_R$, we have one feasible flow routing solution for sending bit rates r_i, $i = 1, 2, \ldots, N$, to the base station. Under Problem $\mathcal{P}_R$, the bit volumes (V_{ij}s and V_{iB}s) must meet the following equalities under the LMM-optimal rate allocation:

$$V_{iB} + \sum_{1 \le j \le N, j \ne i} V_{ij} - \sum_{1 \le k \le N, k \ne i} V_{ki} = r_i T,$$

$$\sum_{1 \le k \le N, k \ne i} \rho V_{ki} + \sum_{1 \le j \le N, j \ne i} C_{ij} V_{ij} + C_{iB} V_{iB} = e_i.$$

Now, replacing $r_i T$ by $t_i R$, we see that the *same* bit volume solution under $\mathcal{P}_R$ yields a feasible bit volume solution to the node lifetime problem under $\mathcal{P}_L$. Consequently, we can use Algorithm 2.1 to obtain the flow routing solution to Problem $\mathcal{P}_L$ under the bit volume solution to Problem $\mathcal{P}_L$ and this verifies that $t_i, i = 1, 2, \ldots, N$, is a feasible solution to Problem $\mathcal{P}_L$.

Optimality   The proof is based on contradiction. But first, let's give some notations. To prove that the t_i-values obtained via (2.14) are indeed LMM-optimal for Problem $\mathcal{P}_L$, we sort $r_i, i = 1, 2, \ldots, N$, under Problem $\mathcal{P}_R$ in nondecreasing order and denote it as $[g_1, g_2, \ldots, g_N]$. We also introduce a node index $I = [i_1, i_2, \ldots, i_N]$ for $[g_1, g_2, \ldots, g_N]$. For example, $i_3 = 7$ means that g_3 actually corresponds to the rate of AFN 7, i.e., $g_3 = r_7$.

Since t_i is proportional to r_i through the relationship ($t_i = \frac{T}{R} \cdot r_i$), listing $t_i, i = 1, 2, \ldots, N$, according to $I = [i_1, i_2, \ldots, i_N]$ will yield a *sorted* (in *nondecreasing* order) lifetime list, denoted as $[\tau_1, \tau_2, \ldots, \tau_N]$. We now prove that if $[\tau_1, \tau_2, \ldots, \tau_N]$ is not LMM-optimal for Problem $\mathcal{P}_L$, then we will have that $[g_1, g_2, \ldots, g_N]$ is not LMM-optimal, which is a contradiction.

Suppose that $[\tau_1, \tau_2, \ldots, \tau_N]$ is not LMM-optimal for Problem $\mathcal{P}_L$. Assume that the LMM-optimal lifetime vector to Problem $\mathcal{P}_L$ is $[\hat{\tau}_1, \hat{\tau}_2, \ldots, \hat{\tau}_N]$ (sorted in nondecreasing order) with the corresponding node index being $\hat{I} = [\hat{i}_1, \hat{i}_2, \ldots, \hat{i}_N]$. Then, by Definition 2.2, there exists a k such that $\hat{\tau}_j = \tau_j$ for $1 \le j \le k - 1$ and $\hat{\tau}_k > \tau_k$.

We now claim that if $\hat{t}_i, i = 1, 2, \ldots, N$, is a feasible solution to Problem $\mathcal{P}_L$, then $\hat{r}_i$ obtained via $\hat{r}_i = \frac{\hat{t}_i R}{T}, i = 1, 2, \ldots, N$, is also a feasible solution to Problem $\mathcal{P}_R$. The proof to this claim follows identically as above. Using this result, we can obtain a corresponding feasible solution $[\hat{g}_1, \hat{g}_2, \ldots, \hat{g}_N]$ with $\hat{g}_i = \frac{\hat{t}_i R}{T}$ and the node index $\hat{I}$ for Problem $\mathcal{P}_R$. Hence, we have $\hat{g}_j = \frac{\hat{t}_j R}{T} = \frac{\tau_j R}{T} = g_j$ for $1 \le j \le k - 1$ but $\hat{g}_k = \frac{\hat{\tau}_k R}{T} > \frac{\tau_k R}{T} = g_k$. That is, $[g_1, g_2, \ldots, g_N]$ is not LMM-optimal and this leads to a contradiction.

(ii) The proof for this part is similar to that for (i) and is left as a homework exercise. $\square$

This mirror relationship offers important insights on system performance issues, in addition to providing solutions to the LMM rate allocation and the LMM node lifetime problems. For example, in Section 2.2, we pointed out the potential bias (fairness) issue associated with the network capacity maximization objective (i.e., sum of rates from all nodes). It is interesting to see that a similar fairness issue exists in the node lifetime problem. In particular, the objective of maximizing the *sum* of node lifetimes among all nodes also leads to a bias (or fairness) problem because this objective would only favor those nodes that consume energy at a small rate. As a result, certain nodes will have much larger lifetimes, while some other nodes will be penalized with much smaller lifetimes, although the sum of node lifetimes is maximized.

Figure 2.3 Network topologies for the 10-AFN and 20-AFN networks.

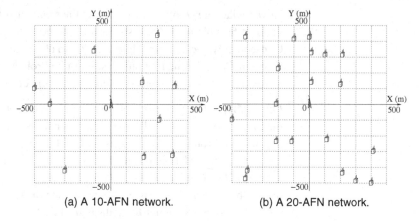

(a) A 10-AFN network.   (b) A 20-AFN network.

Table 2.3 Node coordinates for a 10-AFN network.

AFN i	(x_i, y_i) (in meters)	AFN i	(x_i, y_i) (in meters)
1	(400, −320)	6	(−500, 100)
2	(300, 440)	7	(−400, 0)
3	(−300, −420)	8	(420, 120)
4	(320, −100)	9	(200, 140)
5	(−120, 340)	10	(220, −340)

2.7 Numerical results

In this section, we use numerical results to illustrate our SLP-PA algorithm to the LMM rate allocation problem and compare it with other approaches. We also use numerical results to illustrate the mirror relationship between the LMM rate allocation problem and the LMM node lifetime problem.

We consider two network topologies, one with 10 AFNs and the other with 20 AFNs. Under both topologies, the base station B is located at the origin, while the locations for the 10 or 20 AFNs are randomly generated over a 1000 m × 1000 m square area. See Figs. 2.3(a) and (b) and Tables 2.3 and 2.4 respectively for the two network topologies. We set $\beta_1 = 50$ nJ/b, $\beta_2 = 0.0013$ pJ/b/m^4 [67], and $\alpha = 4$ [129] in this study.

2.7.1 SLP-PA algorithm to the LMM rate allocation problem

We will compare SLP-PA with the naive approach SLP-ER (see Section 2.3.2) that uses serial LP to "blindly" solve the LMM rate allocation problem and

Table 2.4 Node coordinates for a 20-AFN network.

AFN i	(x_i, y_i) (in meters)	AFN i	(x_i, y_i) (in meters)
1	(200, 130)	11	(110, −230)
2	(−400, −430)	12	(−210, 0)
3	(−100, 420)	13	(210, 320)
4	(0, 430)	14	(300, −480)
5	(−410, 440)	15	(−420, −470)
6	(−200, 230)	16	(−120, −240)
7	(400, −490)	17	(220, −440)
8	(410, −300)	18	(−220, −240)
9	(100, 310)	19	(−500, −110)
10	(10, 140)	20	(20, 330)

Table 2.5 Rate allocation under three approaches for the 10-AFN network.

i (sorted node index)	SLP-PA		SLP-ER		MaxCap	
	g_i (Kb/s)	AFN	g_i (Kb/s)	AFN	g_i (Kb/s)	AFN
1	0.1023	3	0.1023	1	0.0553	2
2	0.1023	6	0.1023	2	0.0627	3
3	0.1023	7	0.1023	3	0.0646	1
4	0.1536	5	0.1023	6	0.0658	6
5	0.2941	1	0.1023	7	0.1222	8
6	0.2941	2	0.1536	5	0.1653	10
7	0.2941	4	0.1536	8	0.1736	7
8	0.2941	8	0.1536	10	0.2628	5
9	0.2941	9	0.6563	4	0.3513	4
10	0.2941	10	0.6563	9	1.2398	9

performs energy reservation during each iteration. As discussed in Section 2.3.2, the SLP-ER approach will not give the correct final solution to the LMM rate allocation problem.

We will also compare our SLP-PA algorithm to the *Maximum-Capacity* (MaxCap) approach (see Section 2.3.2). As discussed in the beginning of Section 2.3.2, the rate allocation under the MaxCap approach can be extremely biased and in favor of only those AFNs that consume the least power along their data paths toward the base station.

10-AFN network We assume that the initial energy at each AFN is 50 KJ and that the network lifetime requirement is 100 days under the LMM rate allocation problem. The power consumption is for transmission and reception defined in (2.2) and (2.4), respectively.

Table 2.5 shows the rate allocation for the AFNs under each approach, which is also plotted in Fig. 2.4(a). The "sorted node index" corresponds to the sorted rates among the AFNs in nondecreasing order. Clearly, among the

Figure 2.4 Rate allocation under the SLP-PA, SLP-ER, and MaxCap approaches for a 10-AFN network and a 20-AFN network.

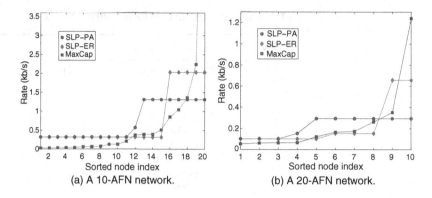

(a) A 10-AFN network.   (b) A 20-AFN network.

three rate allocation approaches, only the rate allocation under SLP-PA meets the LMM-optimal rate allocation definition (see Definition 2.1). Specifically, comparing SLP-PA with SLP-ER, we have $g_1^{\text{SLP-PA}} = g_1^{\text{SLP-ER}}$, $g_2^{\text{SLP-PA}} = g_2^{\text{SLP-ER}}$, $g_3^{\text{SLP-PA}} = g_3^{\text{SLP-ER}}$, and $g_4^{\text{SLP-PA}} > g_4^{\text{SLP-ER}}$; comparing SLP-PA with MaxCap, we have $g_1^{\text{SLP-PA}} > g_1^{\text{MaxCap}}$.

We also observe, as expected, a severe bias in the rate allocation under the MaxCap approach. In particular, g_{10}^{MaxCap} alone accounts for over 48% of the sum of total rates among all the AFNs. For the three approaches, we have $g_1^{\text{SLP-PA}} = g_1^{\text{SLP-ER}} > g_1^{\text{MaxCap}}$ and $g_{10}^{\text{SLP-PA}} < g_{10}^{\text{SLP-ER}} < g_{10}^{\text{MaxCap}}$. In other words, the rate allocation vector under the SLP-PA algorithm has the smallest rate difference between the smallest rate (g_1) and the largest rate (g_{10}), i.e., $g_{10} - g_1$, among the three approaches. In addition, although $g_1^{\text{SLP-PA}} = g_1^{\text{SLP-ER}}$ for the first level rate allocation, the minimum node set for $g_1^{\text{SLP-PA}}$ is smaller than the minimum node set for $g_1^{\text{SLP-ER}}$, i.e., $|S_1^{\text{SLP-PA}}| = 3 < |S_1^{\text{SLP-ER}}| = 5$. This confirms that the naive SLP-ER approach does not necessarily produce the correct solution to the LMM rate allocation problem.

20-AFN network   For the 20-AFN network (Table 2.4), we assume that the initial energy at each AFN is 50 KJ and that the network lifetime requirement under the LMM rate allocation problem is 100 days. Table 2.6 shows the sorted rate allocation under the three approaches, which are also displayed in Fig. 2.4(b). It can be easily verified that all the observations for the 10-AFN network also hold here.

2.7.2 Mirror results

We now use numerical results to verify the mirror relationship between the LMM rate allocation problem ($\mathcal{P}_R$) and the LMM node lifetime problem ($\mathcal{P}_L$)

Table 2.6 Rate allocation under three approaches for the 20-AFN network.

i (Sorted node index)	SLP-PA		SLP-ER		MaxCap	
	g_i (Kb/s)	AFN	g_i (Kb/s)	AFN	g_i (Kb/s)	AFN
1	0.3182	2	0.3182	2	0.0278	7
2	0.3182	7	0.3182	3	0.0282	15
3	0.3182	8	0.3182	4	0.0340	5
4	0.3182	11	0.3182	5	0.0374	2
5	0.3182	12	0.3182	6	0.0433	14
6	0.3182	14	0.3182	7	0.0648	19
7	0.3182	15	0.3182	8	0.0668	8
8	0.3182	16	0.3182	11	0.0760	17
9	0.3182	17	0.3182	12	0.1280	3
10	0.3182	18	0.3182	14	0.1301	4
11	0.3182	19	0.3182	15	0.2070	13
12	0.5694	5	0.3182	16	0.3714	20
13	1.3099	1	0.3182	17	0.3941	9
14	1.3099	3	0.3182	18	0.3948	18
15	1.3099	4	0.3182	19	0.5135	6
16	1.3099	6	2.0344	1	0.8524	16
17	1.3099	9	2.0344	9	1.0441	11
18	1.3099	10	2.0344	10	1.3588	1
19	1.3099	13	2.0344	13	2.2446	12
20	1.3099	20	2.0344	20	10.4362	10

(see Section 2.6). Again, we use the 10-AFN and 20-AFN network configurations in Figs. 2.3(a) and (b), respectively. For both networks, we assume that the initial energy at each AFN is 50 KJ and that the network lifetime requirement under the LMM rate allocation problem is $T = 100$ days. Under $\mathcal{P}_L$, we assume that the local bit rate for all AFNs is $R = 0.2$ Kb/s.

To verify the mirror relationship (Theorem 2.1), we perform the following calculations. First, we solve the LMM rate allocation problem ($\mathcal{P}_R$) and the LMM node lifetime problem ($\mathcal{P}_L$) *independently* with the above initial conditions using the SLP-PA algorithm. Consequently, we obtain the LMM-optimal rate allocation (r_i for each AFN i) under $\mathcal{P}_R$ and the LMM-optimal node lifetime (t_i for each AFN i) under $\mathcal{P}_L$. Then we compute $T \cdot r_i$ and $R \cdot t_i$ separately for each AFN i and examine if they are equal to each other.

The results for the LMM-optimal rate allocation (r_i, $i = 1, 2, \ldots, 10$) and the LMM-optimal node lifetime (t_i, $i = 1, 2, \ldots, 10$) for the 10-AFN network are shown in Table 2.7. We find that $T \cdot r_i$ and $R \cdot t_i$ are exactly equal for all AFNs, precisely as we would expect under Theorem 2.1. Similarly, the results for the 20-AFN network are shown in Table 2.8.

Table 2.7 Mirror relationship $T \cdot r_i = R \cdot t_i$ between the LMM rate allocation problem ($\mathcal{P}_R$) and the LMM node lifetime problem ($\mathcal{P}_L$) for the 10-AFN network.

AFN	$\mathcal{P}_R$ ($T = 100$ days)		$\mathcal{P}_L$ ($R = 0.2$ Kb/s)	
i	r_i	$T \cdot r_i$	t_i	$R \cdot t_i$
1	0.2941	29.41	147.07	29.41
2	0.2941	29.41	147.07	29.41
3	0.1023	10.23	51.17	10.23
4	0.2941	29.41	147.07	29.41
5	0.1536	15.36	76.79	15.36
6	0.1023	10.23	51.17	10.23
7	0.1023	10.23	51.17	10.23
8	0.2941	29.41	147.07	29.41
9	0.2941	29.41	147.07	29.41
10	0.2941	29.41	147.07	29.41

Table 2.8 Mirror relationship $T \cdot r_i = R \cdot t_i$ between the LMM rate allocation problem ($\mathcal{P}_R$) and the LMM node lifetime problem ($\mathcal{P}_L$) for the 20-AFN network.

AFN	$\mathcal{P}_R$ ($T = 100$ days)		$\mathcal{P}_L$ ($R = 0.2$ Kb/s)	
i	r_i	$T \cdot r_i$	t_i	$R \cdot t_i$
1	1.3099	130.99	654.94	130.99
2	0.3182	31.82	159.10	31.82
3	1.3099	130.99	654.94	130.99
4	1.3099	130.99	654.94	130.99
5	0.5694	56.94	284.71	56.94
6	1.3099	130.99	654.94	130.99
7	0.3182	31.82	159.10	31.82
8	0.3182	31.82	159.10	31.82
9	1.3099	130.99	654.94	130.99
10	1.3099	130.99	654.94	130.99
11	0.3182	31.82	159.10	31.82
12	0.3182	31.82	159.10	31.82
13	1.3099	130.99	654.94	130.99
14	0.3182	31.82	159.10	31.82
15	0.3182	31.82	159.10	31.82
16	0.3182	31.82	159.10	31.82
17	0.3182	31.82	159.10	31.82
18	0.3182	31.82	159.10	31.82
19	0.3182	31.82	159.10	31.82
20	1.3099	130.99	654.94	130.99

2.8 Chapter summary

In this chapter, we reviewed LP and showed how it can be employed to solve certain problems in wireless networks. Although the LP methodology itself is rather basic and straightforward, special care is still needed to ensure that it is used correctly, as we demonstrated this in the case study of this chapter. We introduced the *parametric analysis* (PA) technique, which is very useful in its own right. We refer readers to [10] for other advanced techniques in LP. We also introduced the concept of *lexicographical max-min* (LMM) rate allocation, which can be employed as a fairness criterion for other problems in wireless networking research. Our case study is rather interesting as it showed that even in LP-based problem formulations, deep insights could be gained once we dig deep into it. We hope the readers can exploit the LP method to its fullest extent in their research endeavor, as we have demonstrated in this case study.

Appendix: Optimal flow routing solution for LMM-optimal node lifetime

It is straightforward to develop an example similar to the one given in Section 2.4.3 that demonstrates the nonuniqueness of the flow routing schedule. Given that the optimal flow routing solution is nonunique, there are potentially many flow routing solutions that can achieve the LMM-optimal lifetime vector. In this appendix, we show a simple polynomial-time algorithm that provides an LMM-optimal flow routing solution.

The main task in this algorithm is to define flows from the bit volumes (V_{ij} and V_{iB} values), which are obtained upon the completion of the last iteration in the LMM rate problem with our SLP-PA algorithm. Note that the bit volumes obtained here represent the total amount of bit volume being transported between the nodes during $[0, \mu_n]$, where $\mu_n = \tau_N$ is the time when the last group of nodes drains up their energy. The main result here is that if we let the total amount of out-going flow at a node be distributed *proportionally to the bit volumes* on each out-going link for all the remaining alive nodes at each stage, then we can achieve the drop points $\mu_1, \mu_2, \ldots, \mu_n$ as well as the corresponding minimum node sets $S_1, S_2, \ldots, S_n$. The algorithm is formally described in Algorithm 2.1 and proof of its correctness can be found in [69].

As shown in this algorithm, for each time interval $(\mu_{l-1}, \mu_l]$, $l = 1, 2, \ldots, n$, we initialize U_l as the set of remaining alive nodes at this stage, which is represented by $U_l = S - \bigcup_{h=0}^{l-1} S_h$. For these nodes, we compute flow routings by starting with the "boundary" nodes and then move to the "interior" nodes. More precisely, we calculate the flow routing for a node i if and only if we have calculated the flow routing for each node m that has traffic going into node i. The out-going flow from node i is calculated by distributing the aggregated flow *proportionally* according to the overall bit volume along its out-going radio links. As an example, suppose that during $(\mu_4, \mu_5]$, node 2 receives an aggregated flow with rate 2 Kb/s and generates 0.4 Kb/s locally. Assume that $V_{24} = 100$ Kb, $V_{25} = 200$ Kb, and $V_{2B} = 300$ Kb over $[0, \mu_n]$.

Algorithm 2.1 An optimal flow routing solution

Upon the completion of the SLP-PA algorithm for the LMM-optimal lifetime vector, we have the drop points (in strictly increasing order) μ_1, μ_2, ..., μ_n, the corresponding minimum node sets S_1, S_2, ..., S_n, and the total amount of bit volume on each radio link (i.e., V_{ij}s and V_{iB}s). The following algorithm gives an LMM-optimal flow routing solution for time interval $(\mu_{l-1}, \mu_l]$, where $\mu_0 = 0$ and $l = 1, 2, \ldots, n$:

1. Denote $U_l = S - \bigcup_{h=0}^{l-1} S_h$, with $S_0 = \emptyset$. Initialize all flows to zero, i.e., $f_{ij}^{(l)} = 0$, $f_{iB}^{(l)} = 0$ for $1 \leq i, j \leq N$, $j \neq i$.
2. If $U_l = \emptyset$, then stop, else choose a node i from U_l such that
   – node i does not receive data from any other node, or
   – all nodes from which node i receives data are not in U_l.
3. The flow routing at node i during $(\mu_{l-1}, \mu_l]$ is then defined as

$$f_{ij}^{(l)} = \frac{V_{ij}}{V_{iB} + \sum_{m \neq i} V_{im}} \left(\sum_{k \neq i} f_{ki}^{(l)} + r_i \right) \quad (1 \leq j \leq N, j \neq i)$$

$$f_{iB}^{(l)} = \frac{V_{iB}}{V_{iB} + \sum_{m \neq i} V_{im}} \left(\sum_{k \neq i} f_{ki}^{(l)} + r_i \right),$$

where the $f_{ki}^{(l)}$-values, if not zero, have all been defined before calculating the flow routing for node i.
4. Let $U_l = U_l - \{i\}$ and go to Step 2.

Then the out-going flow at node 2 is routed as follows: $f_{24}^{(5)} = 0.4$ Kb/s, $f_{25}^{(5)} = 0.8$ Kb/s, and $f_{2B}^{(5)} = 1.2$ Kb/s.

2.9 Problems

2.1 In an LP problem, suppose that the number of variables and of constraints are N_v and N_c respectively in the standard form. What is the complexity of the solution to this problem?

2.2 For flow routing in a sensor network, we allow an outgoing flow from a node to be split into subflows, each routed to a different next hop node (e.g., (2.5)). Please explain the benefits of flow splitting (1) in terms of traffic flow in the network, and (2) in terms of complexity in mathematical formulation (when compared to single-path routing without flow splitting).

2.3 Why does the MaxCap LP formulation lead to unfairness in rate assignment in r_i, $1 \leq i \leq N$? Give an example of a small network and construct a numerical result to illustrate this point. You may use numbers with normalized units (dimensionless) in this numerical example.

2.4 Suppose that we obtain the following three sorted rate vectors corresponding to three feasible rate allocations (each following a different solution approach):

$a = [0.10, 0.10, 0.10, 0.15, 0.29, 0.29, 0.29, 0.29, 0.29, 0.29]$
$b = [0.10, 0.10, 0.10, 0.10, 0.10, 0.15, 0.15, 0.15, 0.65, 0.65]$
$c = [0.05, 0.06, 0.06, 0.06, 0.12, 0.16, 0.17, 0.26, 0.35, 1.23]$

Which two vectors are definitely not LMM-optimal? Why?

2.5 Discuss the similarity and the difference between the max-min rate allocation problem and the LMM rate allocation problem.

2.6 In this chapter, two incorrect solution approaches for the LMM rate allocation are presented. Explain why these two approaches cannot provide an LMM-optimal rate allocation.

2.7 What is the objective of using parametric analysis (PA) in the solution of the LMM rate allocation problem? Explain the basic idea of PA.

2.8 Is the minimum node set for each rate level under the LMM-optimal rate allocation unique? Is the flow routing solution corresponding to the LMM-optimal rate allocation unique?

2.9 What is the connection between the LMM rate allocation problem and the LMM node lifetime problem in terms of mathematical formulation? What is the mirror relationship between the two problems and when does it occur? What is the application of this mirror relationship?

2.10 For Theorem 2.1, give a proof for the second statement:

Suppose that we have solved Problem $\mathcal{P}_L$ and obtained the LMM-optimal node lifetime t_i for each node i ($i = 1, 2, \ldots, N$). Then under $\mathcal{P}_R$, the LMM rate allocation r_i for node i is

$$r_i = \frac{t_i R}{T}.$$

3

Convex programming and applications

They can because they think they can.

Virgil

3.1 Review of key results in convex optimization

In the previous chapter, we described LP and illustrated its application to solve some interesting problems in wireless networks. In this chapter, we describe convex programming [11], which is a popular tool to solve a wide range of problems in wireless networks. In terms of problem space, LP can be viewed as a special case of convex optimization. To facilitate our description, we define the following terms:

1. **Convex set**:   A set is convex if for any two of its elements z_1 and z_2 and for any $\lambda \in [0, 1]$, $\lambda z_1 + (1 - \lambda)z_2$ is also an element of this set. For example, the set $\{(x, y) : x^2 + y^2 \leq 1\}$ is a convex set but the set $\{(x, y) : 1 \leq x^2 + y^2 \leq 2\}$ is not a convex set.

2. **Convex and concave functions**:   A function $f(x)$ is a convex function if for any x_1 and x_2 and any $\lambda \in [0, 1]$, $f(\lambda x_1 + (1 - \lambda)x_2) \leq \lambda f(x_1) + (1 - \lambda)f(x_2)$, where x can be a single variable or a vector of variables. A function $f(x)$ is a concave function if for any x_1 and x_2 and any $\lambda \in [0, 1]$, $f(\lambda x_1 + (1 - \lambda)x_2) \geq \lambda f(x_1) + (1 - \lambda)f(x_2)$. For example, $f(x) = x^2$ is a convex function, $f(x) = \ln x$ is a concave function, and $f(x) = x^3$ is neither convex nor concave. Observe that if $f(x)$ is a convex function, then $-f(x)$ is a concave function, and vice versa.

3. **Affine function**:   A function $f(x)$ is an affine function if it is both convex and concave. It can be verified that an affine function is a linear function plus some real constant, i.e., $f(x) = ax + b$ for some constants a and b.

A *convex programming problem* usually has the following form:

$$\textbf{CP} \quad \text{Maximize } f(\mathbf{x})$$
$$\text{subject to } \mathbf{g}(\mathbf{x}) \leq \mathbf{0}$$
$$\mathbf{h}(\mathbf{x}) = \mathbf{0}$$
$$\mathbf{x} \in X,$$

where $\mathbf{x}$ is a vector of continuous variables, $f(\mathbf{x})$ is a concave objective function, $\mathbf{g}(\mathbf{x})$ is a vector of convex functions, $\mathbf{h}(\mathbf{x})$ is a vector of affine functions, and X is a convex set. Note that a maximization problem can be readily converted into a minimization problem (and vice verse) by multiplying the coefficients of the objective function by -1, i.e.,

$$\text{Maximum } f(\mathbf{x}) = -\text{Minimum} - f(\mathbf{x}),$$

where $-f(\mathbf{x})$ is a convex function since $f(\mathbf{x})$ is a concave function.

It is easy to verify that any LP is a convex program but a mixed-integer linear programming (MILP) is not. In general, any optimization problem with integer/binary variables is not a convex program because of the discrete nonconvex nature of the set of integer/binary variable values.

A desirable property shared by convex programming problems is that a local optimal solution is also a global optimal solution [11]. Therefore, for a given convex programming problem, we can apply any efficient local optimization approach, e.g., deflected or conjugate (sub)gradient methods or gradient projection methods [11], to improve the current solution iteratively and eventually converge to an optimal solution. In particular, interior point methods can obtain a $(1 - \varepsilon)$-optimal solution in $O(\log_{10}(\frac{1}{\varepsilon}))$ time [114], where ε is the gap to the optimal solution. Due to this low-order complexity, we can obtain a solution very close to the optimal solution with a low complexity, e.g., when $\varepsilon = 10^{-9}$, the complexity is only $O(9)$. Thus, in practice, we can regard a convex programming problem as a problem that can be solved optimally with a low-order complexity.

For the foregoing convex programming CP, its *Lagrangian dual problem* is given by

$$\textbf{LD} \quad \text{Minimize } \Theta(\mathbf{u}, \mathbf{v})$$
$$\text{subject to } \mathbf{u} \geq \mathbf{0},$$

where $\mathbf{u}$ is the Lagrangian multiplier (or dual variable) associated with the constraint $\mathbf{g}(\mathbf{x}) \leq \mathbf{0}$, the unconstrained variable $\mathbf{v}$ is the Lagrangian multiplier associated with the constraint $\mathbf{h}(\mathbf{x}) = \mathbf{0}$, and where

$$\Theta(\mathbf{u}, \mathbf{v}) \equiv \max_{\mathbf{x} \in X}\{f(\mathbf{x}) - \mathbf{u}^T \mathbf{g}(\mathbf{x}) - \mathbf{v}^T \mathbf{h}(\mathbf{x})\}.$$

The function $f(\mathbf{x}) - \mathbf{u}^T \mathbf{g}(\mathbf{x}) - \mathbf{v}^T \mathbf{h}(\mathbf{x})$ is called the *Lagrangian function* and provides an upper bound on $f(\mathbf{x})$ for any feasible solution to CP. Hence,

$$\Theta(\mathbf{u}, \mathbf{v}) \geq \max_{\mathbf{x} \in X, \mathbf{g}(\mathbf{x}) \leq \mathbf{0}, \mathbf{h}(\mathbf{x}) = \mathbf{0}} \{f(\mathbf{x}) - \mathbf{u}^T \mathbf{g}(\mathbf{x}) - \mathbf{v}^T \mathbf{h}(\mathbf{x})\}$$

$$\geq \max_{\mathbf{x} \in X, \mathbf{g}(\mathbf{x}) \leq \mathbf{0}, \mathbf{h}(\mathbf{x}) = \mathbf{0}} \{f(\mathbf{x})\}$$

for any $\mathbf{u} \geq \mathbf{0}$, and therefore $v(LD) \geq v(CP)$, where $v(p)$ denotes the optimal value for any problem p. This latter relationship is known as the *weak duality* property. Under certain constraint qualifications (e.g., Slater's condition, which requires the existence of an $\hat{\mathbf{x}} \in X$ such that $h(\hat{\mathbf{x}}) = 0$, $\mathbf{g}(\hat{\mathbf{x}}) < 0$, and the origin belongs to the interior of the set $\{\mathbf{h}(\mathbf{x}) : \mathbf{x} \in X\}$, convex programming problems have the *strong duality relationship* $v(LD) = v(CP)$, i.e., both the original primal problem and its dual problem have the same optimal objective value [11]. Moreover, given an optimal solution to one problem, the optimal solution to the other can be constructed using a suitable algorithmic procedure [11]. Therefore, a primal problem can be possibly solved in its dual domain.

An excellent reference on convex optimization is [11]. In the rest of this chapter, we offer a case study to show how convex optimization can be applied to solve a problem for a multi-hop MIMO network.

3.2 Case study: Cross-layer optimization for multi-hop MIMO networks

As a case study for convex optimization, we consider a cross-layer optimization problem for multi-hop MIMO networks. Under traditional single transmit/receive antenna node architecture (also known as *single-input single-output* or SISO), channel capacity is limited by the well-known Shannon capacity: $C = \log(1 + \mathsf{SNR})$ per Hz, where SNR represents the signal-to-noise ratio. However, thanks to the work by Winters [169], Foschini and Gans [41], and Telatar [155] on *multiple-input multiple-output* (MIMO), much higher spectrum efficiency and capacity gain can now be achieved. The benefits of substantial improvements in capacity at no additional cost of spectrum have already positioned MIMO as one of the major breakthroughs in modern wireless communications [17; 48; 122]. To date, MIMO has found its way into many wireless network standards, such as wireless LAN (802.11n), WiMAX access networks (802.16e), and 4G systems.

Employing MIMO in a multi-hop network is not trivial. As discussed by Winters [170], a multi-hop MIMO network with each node equipped with M antennas does not necessarily mean that the network throughput is also increased by M-fold. The potential network throughput gain with the use of MIMO depends on the coordinated mechanisms at the physical, link, and network layers. As a result, joint optimization across multiple layers is not only desirable, but also necessary.

Consider a set of user communication sessions in a multi-hop MIMO network. The objective is to maximize some network utility function, which is a function of active sessions' throughput. A number of issues will be considered in this problem, including (i) at the physical layer, how to determine the optimal input covariance matrix at each transmitting node, (ii) at the link layer, how to optimally allocate bandwidth on different outgoing links, (iii) at the network layer, how to perform optimal flow routing in the network, and (iv) at the transport layer, how to optimally establish a session between a source and destination nodes.

We will show that this challenging cross-layer optimization problem can be formulated into a convex programming problem. Just like other convex programming problems, this problem can be solved in its dual domain. But for this particular problem, we will show that it has some special structure that allows us to decompose the problem in the dual domain into two subproblems. That is, the dual problem can be de-coupled in to a transport-network layer subproblem and a link-physical layer subproblem. Both subproblems are also convex programs. Then, we propose a cutting-plane method for the dual problem. We show that our method allows an easy recovery of a primal feasible and optimal solution.

We organize the rest of the sections as follows. In Section 3.3, we describe the network model and problem formulation. Section 3.4 presents a decomposition framework and resulting subproblems. In Section 3.5, we show the cutting-plane method to solve the dual problem. Section 3.6 describes a method to recover an optimal primal solution. Section 3.7 presents numerical results. Section 3.8 summarizes this chapter.

3.3 Network model

We assume orthogonal channels on all links in the network (similar to that in [27; 83; 105]). Note that orthogonal channels do not require as many channels as the number of active links in the network since we can reuse channels on links that are spatially far away from each other. This is called *spatial reuse* and is commonly used in wireless networks to improve channel efficiency. Since efficient channel assignment algorithms have been well studied in the literature, we will assume that channel assignment has been taken care of and thus will not be part of our problem formulation.

Under orthogonal channels, each link in the network is assigned some channel. A node may transmit to different receiving nodes on different links, each with a nonoverlapping channel. Further, a node can simultaneously transmit and receive on different channels and does not cause any self-interference.

In this chapter, we focus on how to jointly optimize input covariance matrices determination at the physical layer, bandwidth allocation at the link layer, session's flow rate assignment at the transport layer and flow routing at the network layer.

We first introduce the notation for matrices, vectors, and complex scalars. We use boldface to denote matrices and vectors. For a matrix $\mathbf{A}$, $\mathbf{A}^{\dagger}$ denotes the conjugate transpose. $\text{Tr}\{\mathbf{A}\}$ denotes the trace of $\mathbf{A}$. $\text{Diag}\{\mathbf{A}_1, \ldots, \mathbf{A}_n\}$ represents the block diagonal matrix with matrices $\mathbf{A}_1, \ldots, \mathbf{A}_n$ on its main diagonal. We denote $\mathbf{I}$ as the identity matrix with dimension determined from the context. $\mathbf{A} \succeq 0$ represents that $\mathbf{A}$ is Hermitian and positive semidefinite (PSD). $\mathbf{1}$ and $\mathbf{0}$ denote vectors whose elements are all ones and zeros, respectively, where their dimensions are determined from the context. $(\mathbf{v})_i$ represents the ith entry of a vector $\mathbf{v}$. For a real vector $\mathbf{v}$ and a real matrix $\mathbf{A}$, $\mathbf{v} \geq \mathbf{0}$ and $\mathbf{A} \geq \mathbf{0}$ mean that all entries in $\mathbf{v}$ and $\mathbf{A}$ are nonnegative, respectively. We let $\mathbf{e}_i$ be the unit column vector where the ith entry is 1 and all the other entries are 0, and where the dimension of $\mathbf{e}_i$ is determined from the context. The operator $\langle \cdot, \cdot \rangle$ represents the inner product of two vectors or matrices. Table 3.1 lists the notations used in this chapter.

3.3.1 MIMO input covariance matrices

Denote x_l^i as the signal sent through the ith antenna at the transmitter of link l. Let matrix $\mathbf{Q}_l$ represent the input covariance matrix of link l, where the (i, j)th element is $\mathbb{E}[x_l^i (x_l^j)^{\dagger}]$. A covariance matrix is Hermitian and PSD. Physically, the diagonal elements of $\mathbf{Q}_l$ represent power allocation at different antennas at the transmitting node of link l. Then we have that

$$\sum_{l \in \mathcal{O}(n)} \text{Tr}\{\mathbf{Q}_l\} \leq P_{\max} \quad (1 \leq n \leq N),$$

where $\mathcal{O}(n)$ is the set of outgoing links that from node n and $P_{\max}$ is the maximum transmit power of a node. We use the matrix $\mathbf{Q} \triangleq \begin{bmatrix} \mathbf{Q}_1 \ \mathbf{Q}_2 \ \ldots \ \mathbf{Q}_L \end{bmatrix} \in \mathbb{C}^{M \times (M \cdot L)}$ to denote the collection of all input covariance matrices in the network, where L is the number of links.

3.3.2 Link capacity and bandwidth allocation

Let the matrix $\mathbf{H}_l \in \mathbb{C}^{M \times M}$ represent the wireless channel gain matrix from the transmitting node to the receiving node of link l, where M is the number of antenna elements at each node. Suppose that $\mathbf{H}_l$ is known at the transmitting node of link l. Although wireless channels in reality are time varying, we consider a "constant" channel model in this chapter, i.e., $\mathbf{H}_l$'s coherence time is larger than the considered transmission period. This simplification is still of much interest as it allows us to find the ergodic capacity for block-wise fading channels [55].

Table 3.1 Notation.

Symbol	Definition
$\mathbf{1}, \mathbf{0}$	Vectors whose elements are all ones or zeros
$\mathbf{A}$	The node-link incidence matrix
$B_{\mathcal{O}(n)}$	The bandwidth allocated at node n for transmission
$B_{\mathcal{I}(n)}$	The bandwidth allocated at node n for reception
$c(W_l, \mathbf{Q}_l)$	The link capacity of a MIMO link l with bandwidth W_l and input covariance matrix $\mathbf{Q}_l$
$d(f)$	The destination node of session f
$\mathbf{e}_i$	Vector where the ith entry is 1 and all the other entries are 0
F	The number of sessions in the network
$\mathbf{H}_l$	The channel gain matrix from the transmitting node to the receiving node of link l
$\mathbf{I}$	The identity matrix with dimension determined from the context
$\mathcal{I}(n)$	The set of incoming links to node n
$\mathcal{L}$	The set of all MIMO links in the network
L	The number of links in the network
M	The number of antenna elements at each node
$\mathcal{N}$	The set of nodes in the network
N	The number of nodes in the network
$\mathcal{O}(n)$	The set of outgoing links from node n
$P_{\max}$	The maximum transmit power of a node
$\mathbf{Q}_l$	The input covariance matrix of link l
$\mathbf{Q}$	$= \begin{bmatrix} \mathbf{Q}_1 & \mathbf{Q}_2 & \dots & \mathbf{Q}_L \end{bmatrix}$, the collection of all input covariance matrices in the network
r_f	The flow rate of session f
$\mathbf{R}_f$	The flow rate vector of session f
$\mathbf{R}$	$= \begin{bmatrix} \mathbf{R}_1 & \mathbf{R}_2 & \dots & \mathbf{R}_F \end{bmatrix}$, the collection of all source-sink vectors
$s(f)$	The source node of session f
u_l	Lagrangian multiplier for link l's capacity constraint (3.10)
$\mathbf{u}$	$= \begin{bmatrix} u_1 & u_2 & \dots & u_L \end{bmatrix}$, the collection of all dual variables
W_l	The communication bandwidth of link l
$\mathbf{W}$	$= \begin{bmatrix} W_1 & W_2 & \dots & W_L \end{bmatrix}^T$, the collection of all bandwidth variables
$y_l^{(f)}$	The amount of flow rate on link l that is attributed to session f
$\mathbf{y}^{(f)}$	$= [\, y_1^{(f)} \; y_2^{(f)} \; \dots \; y_L^{(f)} \,]^T$, the flow rate vector of session f on L links
$\mathbf{Y}$	$= \begin{bmatrix} \mathbf{y}^{(1)} & \mathbf{y}^{(2)} & \dots & \mathbf{y}^{(F)} \end{bmatrix}$, the collection of all flow rate vectors
ρ_l	The ratio of path-loss coefficient and noise power for link l

The link capacity of a MIMO link l in an AWGN channel can be computed as

$$c(W_l, \mathbf{Q}_l) \triangleq W_l \log_2 \det \left(\mathbf{I} + \rho_l \mathbf{H}_l \mathbf{Q}_l \mathbf{H}_l^{\dagger} \right), \qquad (3.1)$$

where W_l represents the channel bandwidth of link l and ρ_l is the ratio of path-loss coefficient and noise power for link l. It can be readily verified that $c(W_l, \mathbf{Q}_l)$ is a monotonically increasing concave function for $W_l > 0$ and $\mathbf{Q}_l \succeq 0$.

We assume the sum of the bandwidth of all outgoing links at a node n cannot exceed $B_{\mathcal{O}(n)}$, the bandwidth at this node that is allocated for transmission. We have

$$\sum_{l \in \mathcal{O}(n)} W_l \leq B_{\mathcal{O}(n)} \quad (1 \leq n \leq N).$$

Similarly, we assume the sum of the bandwidth of all incoming links at node n cannot exceed $B_{\mathcal{I}(n)}$, the bandwidth at this node that is allocated for reception. We have

$$\sum_{l \in \mathcal{I}(n)} W_l \leq B_{\mathcal{I}(n)} \quad (1 \leq n \leq N).$$

We denote the vector $\mathbf{W} = \begin{bmatrix} W_1 & W_2 & \dots & W_L \end{bmatrix}^T \in \mathbb{R}^L$ as the collection of all bandwidth variables.

It is easy to see that the values of W_l and $\mathbf{Q}_l$, i.e., the allocation of bandwidth and power on link l, directly affect the capacity of link l. The determination of these values will be part of our cross-layer optimization problem.

3.3.3 Flow routing

The topology of a multi-hop MIMO network can be represented by a directed graph, denoted by $\mathcal{G} = \{\mathcal{N}, \mathcal{L}\}$, where $\mathcal{N}$ and $\mathcal{L}$ are the set of nodes and all possible MIMO links in the network, respectively. We assume that a link between two nodes exists if the distance between the two is no greater than the transmission range R_T, i.e., $\mathcal{L} = \{(n, m) : D_{nm} \leq R_T, n, m \in \mathcal{N}, n \neq m\}$, where D_{nm} represents the distance between nodes n and m and R_T is determined by a node's maximum transmission power. We assume $\mathcal{G}$ is connected.

The network topology of $\mathcal{G}$ can be mathematically modeled as a *node-link incidence matrix* (NLIM) $\mathbf{A} \in \mathbb{R}^{N \times L}$ [10], with entry a_{nl} being defined as:

$$a_{nl} = \begin{cases} 1 & \text{if } n \text{ is the transmitting node of link } l, \\ -1 & \text{if } n \text{ is the receiving node of link } l, \\ 0 & \text{otherwise.} \end{cases}$$

We assume there are F sessions in the network. We denote the source and destination nodes of session f, $1 \leq f \leq F$, as $s(f)$ and $d(f)$, respectively. We allow flow splitting inside the network so as to offer more flexibility in flow routing and better load balancing. Denote r_f as the flow rate of session f, $1 \leq f \leq F$, which are optimization variables.

Denote $y_l^{(f)} \geq 0$ as the amount of flow on link l attributed to session f. For each session f, we have the following flow balance at node n:

$$\sum_{l \in \mathcal{O}(n)} y_l^{(f)} - \sum_{l \in \mathcal{I}(n)} y_l^{(f)} = \begin{cases} r_f & \text{if } n = s(f), \\ -r_f & \text{if } n = d(f), \\ 0 & \text{otherwise,} \end{cases} \qquad (3.2)$$

where $\mathcal{I}(n)$ is the set of links that are incoming to node n.

To rewrite (3.2) compactly, we use $\mathbf{R}_f \in \mathbb{R}^N$ to denote the right-hand-side (RHS) of constraints for a session f, i.e., $(\mathbf{R}_f)_{s(f)} = r_f$, $(\mathbf{R}_f)_{d(f)} = -r_f$, and $(\mathbf{R}_f)_n = 0$ for $1 \leq n \leq N, n \neq s(f), n \neq d(f)$. We define $\mathbf{y}^{(f)} = [y_1^{(f)} \; y_2^{(f)} \; \ldots \; y_L^{(f)}]^T \in \mathbb{R}^L$ as the *flow vector* on all L links attributed to session f. With NLIM, for any session f, (3.2) can be written as $\mathbf{A}\mathbf{y}^{(f)} = \mathbf{R}_f$.

In addition, denoting the matrix $\mathbf{R} \triangleq [\mathbf{R}_1 \; \mathbf{R}_2 \; \ldots \; \mathbf{R}_F] \in \mathbb{R}^{N \times F}$ as the collection of all source-sink vectors $\mathbf{R}_f$, we have $\mathbf{R}_f = \mathbf{R}\mathbf{e}_f$. The matrix $\mathbf{R}$ needs to satisfy the following constraints:

$$(\mathbf{R}\mathbf{e}_f)_{s(f)} = r_f \quad (1 \leq f \leq F)$$
$$\langle \mathbf{1}, \mathbf{R}\mathbf{e}_f \rangle = 0 \quad (1 \leq f \leq F)$$
$$\mathbf{R}\mathbf{e}_f =_{s(f),d(f)} \mathbf{0} \quad (1 \leq f \leq F),$$

where the notation "$=_{x,y}$" represents the component-wise equality of a vector except at the xth and yth entries. Denote the matrix $\mathbf{Y} \triangleq [\mathbf{y}^{(1)} \; \mathbf{y}^{(2)} \; \ldots \; \mathbf{y}^{(F)}] \in \mathbb{R}^{L \times F}$ as the collection of all flow vectors $\mathbf{y}^{(f)}$. Then, the flow conservation law for all sessions can be written as

$$\mathbf{A}\mathbf{Y} = \mathbf{R}.$$

Since all flows traversing a link cannot exceed this link's capacity limit, we have $\sum_{f=1}^F y_l^{(f)} \leq c(W_l, \mathbf{Q}_l)$ for $1 \leq l \leq L$. Using matrix-vector notation, this can be compactly written as

$$\langle \mathbf{1}, \mathbf{Y}^T \mathbf{e}_l \rangle \leq c(W_l, \mathbf{Q}_l) \quad (1 \leq l \leq L).$$

Note that link capacity $c(W_l, \mathbf{Q}_l)$ depends on bandwidth allocation and the input covariance matrix. Also note that this constraint couples three layers (i.e., network, link, and physical).

3.3.4 Problem formulation

Our cross-layer optimization problem involves joint optimization of session's flow rate assignment at the transport layer, flow routing at the network layer, bandwidth allocation at the link layer, and input covariance matrices at the physical layer. For the objective function, we adopt the proportional fairness utility function, $\ln(r_f)$, for each session f [79], which is to maximize the sum of utilities of all sessions. Putting together the physical layer constraints in Section 3.3.1, the link layer constraints in Section 3.3.2, and the transport-network layer constraints in Section 3.3.3, we have the following problem formulation:

$$\mathcal{P}: \text{Maximize} \sum_{f=1}^{F} \ln(r_f)$$

$$\text{subject to} \sum_{l \in \mathcal{O}(n)} \text{Tr}\{\mathbf{Q}_l\} \leq P_{\max} \qquad (1 \leq n \leq N) \qquad (3.3)$$

$$\sum_{l \in \mathcal{O}(n)} W_l \leq B_{\mathcal{O}(n)} \qquad (1 \leq n \leq N) \qquad (3.4)$$

$$\sum_{l \in \mathcal{I}(n)} W_l \leq B_{\mathcal{I}(n)} \qquad (1 \leq n \leq N) \qquad (3.5)$$

$$\mathbf{Re}_f =_{s(f),d(f)} \mathbf{0} \qquad (1 \leq f \leq F) \qquad (3.6)$$

$$\langle \mathbf{1}, \mathbf{Re}_f \rangle = 0 \qquad (1 \leq f \leq F) \qquad (3.7)$$

$$(\mathbf{Re}_f)_{s(f)} = r_f \qquad (1 \leq f \leq F) \qquad (3.8)$$

$$\mathbf{AY} = \mathbf{R} \qquad (3.9)$$

$$\langle \mathbf{1}, \mathbf{Y}^T \mathbf{e}_l \rangle \leq c(W_l, \mathbf{Q}_l) \qquad (1 \leq l \leq L) \qquad (3.10)$$

$$r_f, \mathbf{Y} \geq \mathbf{0} \qquad (1 \leq f \leq F)$$

$$\mathbf{Q}_l \succeq 0, \ W_l \geq 0 \qquad (1 \leq l \leq L),$$

where $c(W_l, \mathbf{Q}_l)$ is defined in (3.1).

It can be verified that Problem $\mathcal{P}$ is a convex programming by definition. Such a convex problem can be solved by standard approaches in [11]. On the other hand, Problem $\mathcal{P}$ can also be solved in its dual domain. In particular, we will show that the dual problem can be decomposed into subproblems, each of which can be solved separately. Further, such approach is shown to be amenable to distributed implementation. Given the significance of such a dual decomposition approach in the literature [79; 90; 98], we examine this approach in detail in the rest of this chapter.

3.4 Dual problem decomposition

In general, it is always better to decompose a large optimization problem into smaller subproblems and solve each subproblem separately in a suitable coordinated fashion. We find that in Problem $\mathcal{P}$, the network layer variables and the link/physical layer variables are coupled only through the link capacity constraints (3.10). Thus, we may decompose the problem into one network layer problem and one link-physical layer problem. But Problem $\mathcal{P}$ itself cannot be decomposed. However, as we discussed in Section 3.1, we can solve Problem $\mathcal{P}$ in its dual domain. In this section, we show how to decompose the Lagrangian dual formulation of Problem $\mathcal{P}$ and solve the smaller subproblems.

Generally, given a nonlinear programming problem, several different Lagrangian dual problems can be constructed depending on which constraints are chosen to be dualized, i.e., associated with the Lagrangian dual variable [11]. If the strong duality result holds, then all of these dual problems have the same optimal objective value as the original primal problem. For our cross-layer optimization problem, there is a particular dual formulation that can be decomposed. Further, an optimal solution to the original primal problem can be constructed from an optimal solution to the dual problem.

We first map Problem $\mathcal{P}$ to the convex programming problem CP discussed in Section 3.1 and then obtain its dual problem. Note that in Problem $\mathcal{P}$, the constraints (3.3) are power (physical layer) constraints; the constraints (3.4) and (3.5) are bandwidth allocation (link layer) constraints; the constraints (3.6)–(3.8) are flow routing (network layer) constraints; the constraint (3.9) is a coupling constraint that involved both the transport layer and the network layer; and the constraints (3.10) are coupling constraints that involve both the network layer and the physical/link layers. Thus, to obtain a dual problem that can be decomposed into smaller subproblems, we let $\mathbf{g}(\mathbf{x}) = \langle \mathbf{1}, \mathbf{Y}^T \mathbf{e}_l \rangle - c(W_l, \mathbf{Q}_l)$. Therefore, we have $\mathbf{x} = (\mathbf{R}, \mathbf{Y}, \mathbf{Q}, \mathbf{W})$, $f(\mathbf{x}) = \sum_{f=1}^{F} \ln(r_f)$, $\mathbf{g}(\mathbf{x}) = \langle \mathbf{1}, \mathbf{Y}^T \mathbf{e}_l \rangle - c(W_l, \mathbf{Q}_l)$, and

$$
\mathbf{X} = \left\{ (\mathbf{R}, \mathbf{Y}, \mathbf{Q}, \mathbf{W}) \left|
\begin{array}{ll}
\mathbf{AY} = \mathbf{R} & \\
\mathbf{R}\mathbf{e}_f =_{s(f),d(f)} \mathbf{0} & (1 \le f \le F) \\
\langle \mathbf{1}, \mathbf{R}\mathbf{e}_f \rangle = 0 & (1 \le f \le F) \\
(\mathbf{R}\mathbf{e}_f)_{s(f)} = r_f & (1 \le f \le F) \\
\sum_{l \in \mathcal{O}(n)} \mathrm{Tr}\{\mathbf{Q}_l\} \le P_{\max} & (1 \le n \le N) \\
\sum_{l \in \mathcal{O}(n)} W_l \le B_{\mathcal{O}(n)} & (1 \le n \le N) \\
\sum_{l \in \mathcal{I}(n)} W_l \le B_{\mathcal{I}(n)} & (1 \le n \le N) \\
r_f, \mathbf{Y} \ge \mathbf{0} & (1 \le f \le F) \\
\mathbf{Q}_l \succeq 0, W_l \ge 0 & (1 \le l \le L)
\end{array}
\right.
\right\}.
$$

Note that there is no constraint $\mathbf{h}(\mathbf{x}) = \mathbf{0}$ in Problem $\mathcal{P}$. Then its dual problem is given by

$$
\text{Minimize } \Theta(\mathbf{u})
$$
$$
\text{subject to } \mathbf{u} \ge \mathbf{0},
$$

where

$$
\Theta(\mathbf{u}) = \max_{(\mathbf{R}, \mathbf{Y}, \mathbf{Q}, \mathbf{W}) \in \mathbf{X}} \left\{ \sum_{f=1}^{F} \ln(r_f) + \sum_{l=1}^{L} u_l \left(c(W_l, \mathbf{Q}_l) - \langle \mathbf{1}, \mathbf{Y}^T \mathbf{e}_l \rangle \right) \right\}.
$$

$$(3.11)$$

It is easy to recognize that, for any given Lagrangian multiplier $\mathbf{u}$, the Lagrangian can be separated into two terms:

$$
\Theta(\mathbf{u}) = \Theta_{\text{tspt-net}}(\mathbf{u}) + \Theta_{\text{link-phy}}(\mathbf{u}),
$$

where $\Theta_{\text{tspt}-\text{net}}$ and $\Theta_{\text{link}-\text{phy}}$ are two subproblems corresponding to the transport-network layer and the link-physical layer respectively:

$$\Theta_{\text{tspt}-\text{net}}(\mathbf{u}) \triangleq \text{Maximize} \sum_f \ln\left(r_f\right) - \sum_l u_l \langle \mathbf{1}, \mathbf{Y}^T \mathbf{e}_l \rangle$$

$$\text{subject to } \mathbf{AY} = \mathbf{R}$$

$$\mathbf{Re}_f =_{s(f),d(f)} \mathbf{0} \qquad\qquad (1 \le f \le F)$$
$$\langle \mathbf{1}, \mathbf{Re}_f \rangle = 0 \qquad\qquad (1 \le f \le F)$$
$$(\mathbf{Re}_f)_{s(f)} = r_f \qquad\qquad (1 \le f \le F)$$
$$r_f, \mathbf{Y} \ge \mathbf{0} \qquad\qquad (1 \le f \le F)$$

$$\Theta_{\text{link}-\text{phy}}(\mathbf{u}) \triangleq \text{Maximize} \sum_l u_l c(W_l, \mathbf{Q}_l)$$

$$\text{subject to } \sum_{l \in \mathcal{O}(n)} \text{Tr}\{\mathbf{Q}_l\} \le P_{\max} \quad (1 \le n \le N)$$
$$\sum_{l \in \mathcal{O}(n)} W_l \le B_{\mathcal{O}(n)} \qquad (1 \le n \le N)$$
$$\sum_{l \in \mathcal{I}(n)} W_l \le B_{\mathcal{I}(n)} \qquad (1 \le n \le N)$$
$$\mathbf{Q}_l \succeq 0, W_l \ge 0 \qquad\quad (1 \le l \le L).$$

For the transport-network layer subproblem, we can regard $\sum_f \ln\left(r_f\right)$ as the revenue (measured by a rate utility function) and $\sum_l u_l \langle \mathbf{1}, \mathbf{Y}^T \mathbf{e}_l \rangle$ as the cost to achieve session rates. For the link-physical layer subproblem, we can regard $\sum_l u_l c(W_l, \mathbf{Q}_l)$ as the revenue, which is measured by a sum of weighted link capacities. Then, the Lagrangian dual problem of $\mathcal{P}$ can be transformed into the following problem:

$$\mathcal{D}: \quad \text{Minimize } \Theta_{\text{tspt}-\text{net}}(\mathbf{u}) + \Theta_{\text{link}-\text{phy}}(\mathbf{u})$$
$$\text{subject to } \mathbf{u} \ge \mathbf{0}.$$

The Lagrangian dual problem requires us to find the optimal $\mathbf{u}$ such that $\Theta_{\text{tspt}-\text{net}}(\mathbf{u})$ and $\Theta_{\text{link}-\text{phy}}(\mathbf{u})$ can be minimized. Thus, we should be able to evaluate the subproblem functions $\Theta_{\text{tspt}-\text{net}}(\mathbf{u})$ and $\Theta_{\text{link}-\text{phy}}(\mathbf{u})$ for any given $\mathbf{u}$. Note that to evaluate $\Theta_{\text{tspt}-\text{net}}(\mathbf{u})$ in the network layer subproblem, the objective function is concave and all constraints are affine. Hence, this subproblem is a convex program, as described in Section 3.1. Following the same token, we can verify that the problem of evaluating $\Theta_{\text{link}-\text{phy}}(\mathbf{u})$ is also a convex program. Therefore, both $\Theta_{\text{tspt}-\text{net}}(\mathbf{u})$ and $\Theta_{\text{link}-\text{phy}}(\mathbf{u})$ can be readily solved by many efficient convex programming methods. In the next section, we show how to solve Problem $\mathcal{D}$.

3.5 Solving the Lagrangian dual problem

In this section, we show how to solve Problem $\mathcal{D}$ by a cutting-plane method. The basic idea of this method is as follows. We first reformulate Problem $\mathcal{D}$ as a linear program with an infinite number of constraints. Then we build a linear relaxation for Problem $\mathcal{D}$ using a finite number of constraints. We begin with a linear relaxation with only one constraint. During each iteration, we solve the relaxed problem, whose solution provides a lower bound for Problem $\mathcal{D}$. If this solution is not feasible to Problem $\mathcal{D}$, we generate an additional linear constraint to cut the current solution. With this additional linear constraint, we obtain a tighter linear relaxation for the next iteration. Eventually, if we obtain a linear relaxation using a finite number of constraints whose solution to Problem $\mathcal{D}$ is feasible, then we would have derived an optimal solution to Problem $\mathcal{D}$.

We now present the details of the cutting-plane method. Let $z = \Theta(\mathbf{u}) = \max_{(\mathbf{R},\mathbf{Y},\mathbf{Q},\mathbf{W})\in\mathbf{X}} \left\{ \sum_{f=1}^{F} \ln \left(r_f \right) + \sum_{l=1}^{L} u_l \left(c(W_l, \mathbf{Q}_l) - \langle \mathbf{1}, \mathbf{Y}^T \mathbf{e}_l \rangle \right) \right\}$. Thus, for each $\mathbf{u} \geq \mathbf{0}$, the linear inequality $z \geq \sum_{f=1}^{F} \ln \left(r_f \right) + \sum_{l=1}^{L} u_l \cdot \left(c(W_l, \mathbf{Q}_l) - \langle \mathbf{1}, \mathbf{Y}^T \mathbf{e}_l \rangle \right)$ must hold for all $(\mathbf{R}, \mathbf{Y}, \mathbf{Q}, \mathbf{W}) \in \mathbf{X}$. Problem $\mathcal{D}$ is therefore equivalent to

$$
\begin{aligned}
\text{Minimize} \qquad & z \\
\text{subject to } & z \geq \sum_{f} \ln \left(r_f \right) + \sum_{l} u_l \left(c(W_l, \mathbf{Q}_l) - \langle \mathbf{1}, \mathbf{Y}^T \mathbf{e}_l \rangle \right), \qquad (3.12) \\
& \forall (\mathbf{R}, \mathbf{Y}, \mathbf{Q}, \mathbf{W}) \in \mathbf{X} \quad \mathbf{u} \geq \mathbf{0}.
\end{aligned}
$$

Note that each given point $(\mathbf{R}, \mathbf{Y}, \mathbf{Q}, \mathbf{W})$ yields a linear constraint for z and $\mathbf{u}$. Therefore, we can regard (3.12) as an LP problem having an infinite number of constraints, with each constraint corresponding to a point $(\mathbf{R}, \mathbf{Y}, \mathbf{Q}, \mathbf{W}) \in \mathbf{X}$.

Instead of enumerating all (infinite number of) points, we consider only a finite number of points and add more points as needed. This gives us a *relaxed problem* (since we consider only a subset of constraints). In the first iteration, we begin with one point $(\mathbf{R}^{(0)}, \mathbf{Y}^{(0)}, \mathbf{Q}^{(0)}, \mathbf{W}^{(0)}) \in \mathbf{X}$.

In the kth iteration, we have k points $(\mathbf{R}^{(j)}, \mathbf{Y}^{(j)}, \mathbf{Q}^{(j)}, \mathbf{W}^{(j)}) \in \mathbf{X}$, $j = 0, \ldots, k-1$, for which the relaxed problem is given as follows:

$$
\begin{aligned}
\text{Minimize } & z \\
\text{subject to } & z \geq \sum_{f} \ln \left(r_f^{(j)} \right) + \sum_{l} u_l \left(c(W_l^{(j)}, \mathbf{Q}_l^{(j)}) - \langle \mathbf{1}, \mathbf{Y}^{(j)T} \mathbf{e}_l \rangle \right) \\
& (0 \leq j < k) \quad \mathbf{u} \geq \mathbf{0}.
\end{aligned}
$$

$$(3.13)$$

This is an LP problem with k constraints and can be solved efficiently. Denote $(z^{(k)}, \mathbf{u}^{(k)})$ as an optimal solution to this relaxed problem. Then $z^{(k)}$ is a lower bound for the optimal z^* to the Lagrangian dual problem (3.12). If this solution to problem (3.12) is feasible, then $z^{(k)}$ is an upper bound for z^*. Since $z^{(k)}$ is both a lower and upper bound for z^*, we have $z^{(k)} = z^*$, i.e., it is an optimal solution to problem (3.12). To check the feasibility of $(z^{(k)}, \mathbf{u}^{(k)})$ to the Lagrangian dual problem (3.12), we must check whether

$$
z^{(k)} \geq \max_{(\mathbf{R},\mathbf{Y},\mathbf{Q},\mathbf{W})\in\mathbf{X}} \sum_{f} \ln \left(r_f \right) + \sum_{l} u_l^{(k)} \left(c(W_l, \mathbf{Q}_l) - \langle \mathbf{1}, \mathbf{Y}^T \mathbf{e}_l \rangle \right).
$$

To do this, we consider the following problem:

$$\text{Maximize} \sum_f \ln\left(r_f\right) + \sum_l u_l^{(k)}\left(c(W_l, \mathbf{Q}_l) - \langle \mathbf{1}, \mathbf{Y}^T \mathbf{e}_l\rangle\right)$$
$$\text{subject to } (\mathbf{R}, \mathbf{Y}, \mathbf{Q}, \mathbf{W}) \in \mathbf{X}. \tag{3.14}$$

Note that this problem is exactly the problem in (3.11). Therefore, as discussed in Section 3.4, we can decompose this problem into two smaller subproblems and solve each subproblem efficiently.

Algorithm 3.1 Cutting-plane algorithm for solving Problem $\mathcal{D}$

Initialization:
    Find a point $(\mathbf{R}^{(0)}, \mathbf{Y}^{(0)}, \mathbf{Q}^{(0)}, \mathbf{W}^{(0)}) \in \mathbf{X}$.
    Set $k = 1$.
Main Loop:
    1. Solve the relaxed problem in (3.13) and obtain its solution $(z^{(k)}, \mathbf{u}^{(k)})$.
    2. Solve (3.14) and obtain an optimal solution $(\mathbf{R}^{(k)}, \mathbf{Y}^{(k)}, \mathbf{Q}^{(k)}, \mathbf{W}^{(k)})$
       with objective value $\Theta(\mathbf{u}^{(k)})$.
    3. If $z^{(k)} \geq \Theta(\mathbf{u}^{(k)})$, then stop and $(z^{(k)}, \mathbf{u}^{(k)})$ is an optimal dual solution.
    4. Otherwise, add constraint (3.15) to the relaxed problem. Set k to $k + 1$,
       and return to Step 1.

Denote $(\mathbf{R}^{(k)}, \mathbf{Y}^{(k)}, \mathbf{Q}^{(k)}, \mathbf{W}^{(k)})$ as an optimal solution to (3.14) and $\Theta(\mathbf{u}^{(k)})$ as the corresponding optimal objective value. If $z^k \geq \Theta(\mathbf{u}^{(k)})$, then $(z^{(k)}, \mathbf{u}^{(k)})$ is feasible for (3.12) and is an optimal solution to this Lagrangian dual problem. Otherwise, for $(z^{(k)}, \mathbf{u}^{(k)})$, the inequality constraint in (3.12) is not satisfied for $(\mathbf{R}^{(k)}, \mathbf{Y}^{(k)}, \mathbf{Q}^{(k)}, \mathbf{W}^{(k)})$. Thus, we add the constraint:

$$z \geq \sum_f \ln\left(r_f^{(k)}\right) + \sum_l u_l\left(c(W_l^{(k)}, \mathbf{Q}_l^{(k)}) - \langle \mathbf{1}, \mathbf{Y}^{(k)T}\mathbf{e}_l\rangle\right) \tag{3.15}$$

to (3.13), and solve the relaxed linear program again in the $(k + 1)$th iteration (now with $(k + 1)$ points). Obviously, $(z^{(k)}, \mathbf{u}^{(k)})$ will not show up in the $(k + 1)$th iteration because it is cut off by (3.15). The cutting-plane algorithm is summarized in Algorithm 3.1.

3.6 Constructing a primal optimal solution

Thus far, we have solved Problem $\mathcal{D}$. We now show how we can construct an optimal solution to Problem $\mathcal{P}$.

Suppose that Algorithm 3.1 terminates at iteration $k = K$. Then we have $K + 1$ points $(\mathbf{R}^{(j)}, \mathbf{Y}^{(j)}, \mathbf{Q}^{(j)}, \mathbf{W}^{(j)}) \in \mathbf{X}$, $j = 0, \ldots, K$. The last solved LP problem in Algorithm 3.1 is as follows:

Minimize z

subject to $z \geq \sum_f \ln\left(r_f^{(j)}\right) + \sum_l u_l \left(c(W_l^{(j)}, \mathbf{Q}_l^{(j)}) - \langle \mathbf{1}, \mathbf{Y}^{(j)T} \mathbf{e}_l \rangle\right)$

$(0 \leq j \leq K) \, \mathbf{u} \geq 0.$

$$(3.16)$$

As discussed in Section 2.1, when we solve (3.16), we have also solved its dual problem, which is given as follows:

Maximize $\sum_{j=0}^{K} \tau_j \sum_{f=1}^{F} \ln(r_f^{(j)})$

subject to $\sum_{j=0}^{K} \tau_j \left(\langle \mathbf{1}, \mathbf{Y}^{(j)T} \mathbf{e}_l \rangle - c(W_l^{(j)}, \mathbf{Q}_l^{(j)})\right) \leq 0, \forall l$

$\sum_{j=0}^{K} \tau_j = 1$

$\tau_j \geq 0, \forall j.$

$$(3.17)$$

By the LP duality theory, the optimal values τ_j^*, for $j = 0, \ldots, K$, are immediately available after we solve (3.16). Furthermore, (3.16) and (3.17) have the same optimal objective value $z^{(K)}$.

To construct an optimal solution to Problem $\mathcal{P}$, we first need to construct a feasible solution. Such a feasible solution $(\mathbf{R}, \mathbf{Y}, \mathbf{Q}, \mathbf{W})$ should satisfy (3.10) and $(\mathbf{R}, \mathbf{Y}, \mathbf{Q}, \mathbf{W}) \in \mathbf{X}$. Since τ_j^*, for $j = 0, \ldots, K$, is a solution of (3.17), we have $\sum_{j=0}^{K} \tau_j^* (\mathbf{R}^{(j)}, \mathbf{Y}^{(j)}, \mathbf{Q}^{(j)}, \mathbf{W}^{(j)}) \in \mathbf{X}$. Let

$$(\bar{\mathbf{R}}, \bar{\mathbf{Y}}, \bar{\mathbf{Q}}, \bar{\mathbf{W}}) = \sum_{j=0}^{K} \tau_j^* (\mathbf{R}^{(j)}, \mathbf{Y}^{(j)}, \mathbf{Q}^{(j)}, \mathbf{W}^{(j)}).$$

Then $(\bar{\mathbf{R}}, \bar{\mathbf{Y}}, \bar{\mathbf{Q}}, \bar{\mathbf{W}}) \in \mathbf{X}$. If we can verify that $(\bar{\mathbf{R}}, \bar{\mathbf{Y}}, \bar{\mathbf{Q}}, \bar{\mathbf{W}})$ also satisfies (3.10), then it is a feasible solution to Problem $\mathcal{P}$. In fact, we can prove the following theorem for $(\bar{\mathbf{R}}, \bar{\mathbf{Y}}, \bar{\mathbf{Q}}, \bar{\mathbf{W}})$. The proof is given in the Appendix, which is a simplified proof of Theorem 6.5.1 in [11] for our particular problem.

Theorem 3.1
$(\bar{\mathbf{R}}, \bar{\mathbf{Y}}, \bar{\mathbf{Q}}, \bar{\mathbf{W}})$ is a feasible and optimal solution to Problem $\mathcal{P}$.

3.7 Numerical results

In this section, we present some numerical results to gain more quantitative understanding of the problem and solution. We first describe the simulation setting. As shown in Fig. 3.1, we consider 15 nodes uniformly deployed in an 1200 m $\times$ 1200 m square area. Each node in the network is equipped with two antennas. The maximum transmit power for each node n is set to $P_{\max} = 20$ dBm (0.1 W). The path-loss index is 3 and the noise power is $4 \cdot 10^{-21}$ W. The total bandwidth at each node for transmission and reception are 20 MHz and

Figure 3.1 Network topology of
a 15-node network.

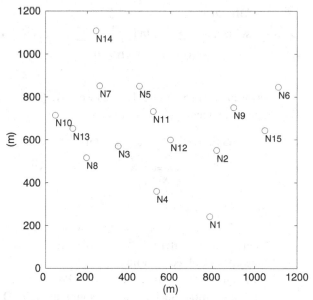

30 MHz, respectively. There are three sessions in the network: node 14 to node 1, node 6 to node 10, and node 5 to node 4. The randomly generated channel gain matrices are shown in Table 3.2.

After executing our solution procedure, we find that the optimal rates for these three sessions are $r_1 = 125$ Mb/s, $r_2 = 190.6$ Mb/s, and $r_3 = 258$ Mb/s. The optimal objective value is 6.64. The routings and flow rates of sessions 1, 2, and 3 are shown in Figs. 3.2, 3.3, and 3.4, respectively. These figures show that flow routings for sessions 1, 2, and 3 are all multi-path and multi-hop. It can be easily verified that the flow rates in Figs. 3.2, 3.3, and 3.4 satisfy flow conservation at each node.

Table 3.3 shows the optimal bandwidth allocation of the network. Table 3.4 shows the optimal input covariance matrices of the network. In Table 3.4, each cell with four entries corresponds to a 2×2 **Q**-matrix, which represents an input covariance matrix. Consider the transmission from N11 to N7 as an example: $\mathbf{Q}_{(11,7)}$ in Table 3.4 says that the power allocations to the two antennas at N11 are 11.83 mW and 12.10 mW. Also, the signals sent through these two antennas, denoted by x_1 and x_2, should also be correlated with power $\mathbb{E}[x_1 x_2^\dagger] = (-0.25 - 0.01i)$ mW and $\mathbb{E}[x_2 x_1^\dagger] = (-0.25 + 0.01i)$ mW.

It can be observed from Tables 3.3 and 3.4 that not every node allocates its full power and total bandwidth for its outgoing links. For example, N14 has only one outgoing link (14, 7). We can see that $W_{(14,7)} = 14.2$ MHz and $\text{Tr}(\mathbf{Q}_{(14,7)}) = 79.47$ mW, which shows N14 does not transmit at full power and does not utilize all of its assigned bandwidth. Such a link is not a bottleneck link. Even if the transmitter of a nonbottleneck link increases its total transmit power and bandwidth, the end-to-end session rate

Table 3.2 Channel gain matrices.

$\mathbf{H}_{(1,4)}$	$\begin{matrix} 0.79 - 0.62i & -0.57 + 1.33i \\ 0.70 + 1.32i & -0.17 - 0.57i \end{matrix}$	$\mathbf{H}_{(8,13)}$	$\begin{matrix} -0.47 - 2.12i & 1.56 + 0.97i \\ 0.07 - 0.71i & 0.42 + 0.18i \end{matrix}$
$\mathbf{H}_{(2,9)}$	$\begin{matrix} 0.14 + 0.35i & -0.17 - 0.19i \\ -1.75 + 0.81i & -0.86 - 0.33i \end{matrix}$	$\mathbf{H}_{(9,2)}$	$\begin{matrix} 0.14 + 0.35i & -1.75 + 0.81i \\ -0.17 - 0.19i & -0.86 - 0.33i \end{matrix}$
$\mathbf{H}_{(2,12)}$	$\begin{matrix} 0.21 + 0.67i & -0.04 + 1.14i \\ -1.08 - 1.70i & 0.03 + 0.68i \end{matrix}$	$\mathbf{H}_{(9,6)}$	$\begin{matrix} 0.20 - 0.15i & 2.00 - 0.52i \\ -0.47 - 0.77i & -0.46 + 0.71i \end{matrix}$
$\mathbf{H}_{(2,15)}$	$\begin{matrix} -0.00 + 0.16i & -1.12 + 0.36i \\ 0.79 - 0.85i & 0.47 + 0.46i \end{matrix}$	$\mathbf{H}_{(9,15)}$	$\begin{matrix} -1.36 - 0.36i & -0.43 - 0.45i \\ -0.37 - 0.37i & 0.24 - 0.73i \end{matrix}$
$\mathbf{H}_{(3,4)}$	$\begin{matrix} 0.73 + 0.58i & 1.06 + 0.58i \\ -0.92 - 0.44i & -0.07 - 1.40i \end{matrix}$	$\mathbf{H}_{(10,7)}$	$\begin{matrix} -0.27 - 0.03i & -0.29 - 0.39i \\ -0.76 + 0.32i & 0.92 - 1.02i \end{matrix}$
$\mathbf{H}_{(3,8)}$	$\begin{matrix} 0.68 - 0.99i & 0.66 - 0.08i \\ -0.20 + 0.99i & -0.53 + 0.54i \end{matrix}$	$\mathbf{H}_{(10,8)}$	$\begin{matrix} -0.15 - 0.83i & 0.76 - 0.59i \\ 0.00 - 0.47i & -0.24 - 0.77i \end{matrix}$
$\mathbf{H}_{(3,11)}$	$\begin{matrix} -1.52 - 0.51i & 0.12 - 0.05i \\ -0.01 - 0.41i & 2.17 - 0.35i \end{matrix}$	$\mathbf{H}_{(10,13)}$	$\begin{matrix} -0.31 - 0.72i & 1.07 - 0.30i \\ 1.12 + 0.73i & -0.92 - 0.47i \end{matrix}$
$\mathbf{H}_{(3,12)}$	$\begin{matrix} -0.23 - 2.93i & 0.29 + 0.70i \\ -0.94 - 0.39i & -0.19 - 1.41i \end{matrix}$	$\mathbf{H}_{(11,3)}$	$\begin{matrix} -1.52 - 0.51i & -0.01 - 0.41i \\ 0.12 - 0.05i & 2.17 - 0.35i \end{matrix}$
$\mathbf{H}_{(3,13)}$	$\begin{matrix} -0.46 + 0.63i & 0.45 + 1.09i \\ 0.70 - 0.42i & -0.11 - 0.23i \end{matrix}$	$\mathbf{H}_{(11,5)}$	$\begin{matrix} 1.02 + 1.43i & -1.79 + 0.84i \\ -0.36 + 0.34i & -1.02 - 0.32i \end{matrix}$
$\mathbf{H}_{(4,1)}$	$\begin{matrix} 0.79 - 0.62i & 0.70 + 1.32i \\ -0.57 + 1.33i & -0.17 - 0.57i \end{matrix}$	$\mathbf{H}_{(11,7)}$	$\begin{matrix} 0.63 - 1.60i & 0.75 - 0.28i \\ 0.25 - 1.17i & -0.75 - 0.02i \end{matrix}$
$\mathbf{H}_{(4,3)}$	$\begin{matrix} 0.73 + 0.58i & -0.92 - 0.44i \\ 1.06 + 0.58i & -0.07 - 1.40i \end{matrix}$	$\mathbf{H}_{(11,12)}$	$\begin{matrix} 1.19 - 0.47i & -0.35 - 0.30i \\ -0.69 + 0.92i & -0.54 + 0.26i \end{matrix}$
$\mathbf{H}_{(4,12)}$	$\begin{matrix} 0.94 + 0.06i & -0.57 - 0.75i \\ 0.01 + 0.65i & 0.28 + 0.35i \end{matrix}$	$\mathbf{H}_{(12,2)}$	$\begin{matrix} 0.21 + 0.67i & -1.08 - 1.70i \\ -0.04 + 1.14i & 0.03 + 0.68i \end{matrix}$
$\mathbf{H}_{(5,7)}$	$\begin{matrix} 0.26 - 1.07i & -1.19 - 0.50i \\ -0.15 + 0.70i & 1.25 - 0.45i \end{matrix}$	$\mathbf{H}_{(12,3)}$	$\begin{matrix} -0.23 - 2.93i & -0.94 - 0.39i \\ 0.29 + 0.70i & -0.19 - 1.41i \end{matrix}$
$\mathbf{H}_{(5,11)}$	$\begin{matrix} 1.02 + 1.43i & -0.36 + 0.34i \\ -1.79 + 0.84i & -1.02 - 0.32i \end{matrix}$	$\mathbf{H}_{(12,4)}$	$\begin{matrix} 0.94 + 0.06i & 0.01 + 0.65i \\ -0.57 - 0.75i & 0.28 + 0.35i \end{matrix}$
$\mathbf{H}_{(6,9)}$	$\begin{matrix} 0.20 - 0.15i & -0.47 - 0.77i \\ 2.00 - 0.52i & -0.46 + 0.71i \end{matrix}$	$\mathbf{H}_{(12,11)}$	$\begin{matrix} 1.19 - 0.47i & -0.69 + 0.92i \\ -0.35 - 0.30i & -0.54 + 0.26i \end{matrix}$
$\mathbf{H}_{(6,15)}$	$\begin{matrix} -1.07 - 0.20i & -1.12 - 0.38i \\ -0.37 - 0.19i & 0.91 - 1.14i \end{matrix}$	$\mathbf{H}_{(13,3)}$	$\begin{matrix} -0.46 + 0.63i & 0.70 - 0.42i \\ 0.45 + 1.09i & -0.11 - 0.23i \end{matrix}$
$\mathbf{H}_{(7,5)}$	$\begin{matrix} 0.26 - 1.07i & -0.15 + 0.70i \\ -1.19 - 0.50i & 1.25 - 0.45i \end{matrix}$	$\mathbf{H}_{(13,7)}$	$\begin{matrix} 0.95 + 1.00i & 0.91 + 0.15i \\ -0.77 - 0.44i & -1.64 - 0.18i \end{matrix}$
$\mathbf{H}_{(7,10)}$	$\begin{matrix} -0.27 - 0.03i & -0.76 + 0.32i \\ -0.29 - 0.39i & 0.92 - 1.02i \end{matrix}$	$\mathbf{H}_{(13,8)}$	$\begin{matrix} -0.47 - 2.12i & 0.07 - 0.71i \\ 1.56 + 0.97i & 0.42 + 0.18i \end{matrix}$
$\mathbf{H}_{(7,11)}$	$\begin{matrix} 0.63 - 1.60i & 0.25 - 1.17i \\ 0.75 - 0.28i & -0.75 - 0.02i \end{matrix}$	$\mathbf{H}_{(13,10)}$	$\begin{matrix} -0.31 - 0.72i & 1.12 + 0.73i \\ 1.07 - 0.30i & -0.92 - 0.47i \end{matrix}$
$\mathbf{H}_{(7,13)}$	$\begin{matrix} 0.95 + 1.00i & -0.77 - 0.44i \\ 0.91 + 0.15i & -1.64 - 0.18i \end{matrix}$	$\mathbf{H}_{(14,7)}$	$\begin{matrix} -0.90 - 0.02i & -0.56 - 0.22i \\ -0.96 + 0.18i & -0.11 - 0.04i \end{matrix}$
$\mathbf{H}_{(7,14)}$	$\begin{matrix} -0.90 - 0.02i & -0.96 + 0.18i \\ -0.56 - 0.22i & -0.11 - 0.04i \end{matrix}$	$\mathbf{H}_{(15,2)}$	$\begin{matrix} -0.00 + 0.16i & 0.79 - 0.85i \\ -1.12 + 0.36i & 0.47 + 0.46i \end{matrix}$
$\mathbf{H}_{(8,3)}$	$\begin{matrix} 0.68 - 0.99i & -0.20 + 0.99i \\ 0.66 - 0.08i & -0.53 + 0.54i \end{matrix}$	$\mathbf{H}_{(15,6)}$	$\begin{matrix} -1.07 - 0.20i & -0.37 - 0.19i \\ -1.12 - 0.38i & 0.91 - 1.14i \end{matrix}$
$\mathbf{H}_{(8,10)}$	$\begin{matrix} -0.15 - 0.83i & 0.00 - 0.47i \\ 0.76 - 0.59i & -0.24 - 0.77i \end{matrix}$	$\mathbf{H}_{(15,9)}$	$\begin{matrix} -1.36 - 0.36i & -0.37 - 0.37i \\ -0.43 - 0.45i & 0.24 - 0.73i \end{matrix}$

Figure 3.2 Routing and flow rates of session 1 (in Mb/s).

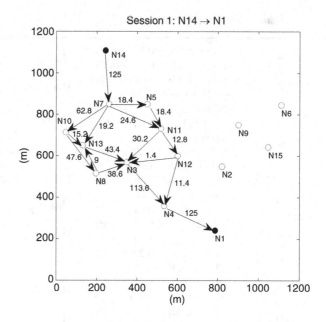

Figure 3.3 Routing and flow rates of session 2 (in Mb/s).

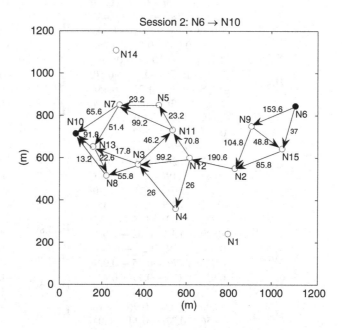

cannot be increased due to the minimum bottleneck link along the path. In this network, it can be verified that N3, N7, N11, and N12 are bottleneck nodes. For example, at N11, $W_{(11,3)} + W_{(11,5)} + W_{(11,7)} + W_{(11,12)} = 20$ MHz and $\mathrm{Tr}(Q_{(11,3)}) + \mathrm{Tr}(\mathbf{Q}_{(11,5)}) + \mathrm{Tr}(\mathbf{Q}_{(11,7)}) + \mathrm{Tr}(\mathbf{Q}_{(11,12)}) = 100$ mW. This means that the power and bandwidth at N11 has been fully utilized and cannot be further increased.

Figure 3.4 Routing and flow rates of session 3 (in Mb/s).

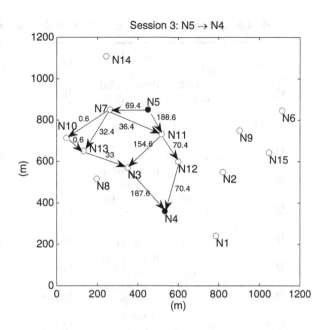

Table 3.3 Bandwidth allocation of the 15-node network (in MHz).

$W_{(2,12)}$	8.20	$W_{(6,9)}$	7.40	$W_{(9,2)}$	7.00	$W_{(12,4)}$	3.60
$W_{(3,4)}$	3.40	$W_{(6,15)}$	9.00	$W_{(9,15)}$	6.80	$W_{(12,11)}$	7.80
$W_{(3,8)}$	2.80	$W_{(7,5)}$	4.00	$W_{(10,8)}$	4.80	$W_{(13,3)}$	4.20
$W_{(3,11)}$	4.60	$W_{(7,10)}$	6.00	$W_{(10,13)}$	4.80	$W_{(13,8)}$	3.60
$W_{(3,13)}$	5.00	$W_{(7,11)}$	4.40	$W_{(11,3)}$	4.40	$W_{(13,10)}$	6.60
$W_{(4,1)}$	8.80	$W_{(7,13)}$	3.00	$W_{(11,5)}$	4.20	$W_{(14,7)}$	14.2
$W_{(4,3)}$	4.80	$W_{(8,3)}$	4.80	$W_{(11,7)}$	4.60	$W_{(15,2)}$	8.00
$W_{(5,7)}$	7.60	$W_{(8,10)}$	5.40	$W_{(11,12)}$	6.80		
$W_{(5,11)}$	10.6	$W_{(8,13)}$	5.20	$W_{(12,3)}$	4.80		

The convergence process of the Lagrangian dual problem is illustrated in Fig. 3.5. In this figure, "Dual UB" denotes the current objective value of the Lagrangian dual function, which can be thought of as an upper bound of the primal objective value. "Primal Feasible Solution" denotes the current primal feasible solution recovered from the Lagrangian dual solution, which can be thought of as providing a lower bound to the optimal primal objective value. During each iteration, the cutting-plane method solves the Lagrangian dual problem. The optimal objective value of the Lagrangian dual problem (or the upper bound to the primal objective value) is nonincreasing with iterations. Meanwhile, the best primal feasible objective value keeps increasing with iterations. As expected, the upper bound and the lower bound converge and give the optimal solution, as shown in this figure. For this 15-node ad hoc network, the cutting-plane algorithm converged in approximately 115 iterations. The optimal value of the network utility function is 56.85 (in ln(b/s)).

Convex programming and applications

Table 3.4 Input covariance matrices in the 15-node network (mW).

	Matrix			Matrix	
$\mathbf{Q}_{(1,4)}$	35.46	$0.00-0.00i$	$\mathbf{Q}_{(8,13)}$	12.80	$0.03-0.04i$
	$0.00+0.00i$	35.46		$0.03+0.04i$	12.82
$\mathbf{Q}_{(2,9)}$	11.82	$0.00-0.00i$	$\mathbf{Q}_{(9,2)}$	17.23	$-0.09+0.10i$
	$0.00+0.00i$	11.82		$-0.09-0.10i$	17.47
$\mathbf{Q}_{(2,12)}$	24.51	$-0.08-0.08i$	$\mathbf{Q}_{(9,6)}$	14.78	$-0.10-0.08i$
	$-0.08+0.08i$	23.98		$-0.10+0.08i$	14.55
$\mathbf{Q}_{(2,15)}$	11.82	$0.00-0.00i$	$\mathbf{Q}_{(9,15)}$	17.21	$0.32+0.28i$
	$0.00+0.00i$	11.82		$0.32-0.28i$	16.77
$\mathbf{Q}_{(3,4)}$	8.90	$0.01+0.00i$	$\mathbf{Q}_{(10,7)}$	11.82	$0.00-0.00i$
	$0.01-0.00i$	8.91		$0.00+0.00i$	11.82
$\mathbf{Q}_{(3,8)}$	7.36	$0.10+0.01i$	$\mathbf{Q}_{(10,8)}$	11.82	$0.0-0.00i$
	$0.10-0.01i$	7.23		$0.016+0.00i$	11.82
$\mathbf{Q}_{(3,11)}$	11.37	$0.00+0.33i$	$\mathbf{Q}_{(10,13)}$	11.82	$0.00-0.00i$
	$0.00-0.33i$	12.04		$0.00+0.00i$	11.82
$\mathbf{Q}_{(3,12)}$	9.60	$0.13-0.18i$	$\mathbf{Q}_{(11,3)}$	10.76	$0.00+0.00i$
	$0.13+0.18i$	10.27		$0.00-0.00i$	10.70
$\mathbf{Q}_{(3,13)}$	12.22	$0.35+0.69i$	$\mathbf{Q}_{(11,5)}$	10.42	$0.01-0.12i$
	$0.35-0.69i$	12.09		$0.01+0.12i$	10.39
$\mathbf{Q}_{(4,1)}$	21.84	$-0.09+0.02i$	$\mathbf{Q}_{(11,7)}$	11.83	$-0.25-0.01i$
	$-0.09-0.02i$	21.87		$-0.25+0.01i$	12.10
$\mathbf{Q}_{(4,3)}$	11.82	$0.000-0.000i$	$\mathbf{Q}_{(11,12)}$	16.76	$-0.04+0.02i$
	$0.00+0.00i$	11.82		$-0.04-0.02i$	17.04
$\mathbf{Q}_{(4,12)}$	11.82	$0.00-0.00i$	$\mathbf{Q}_{(12,2)}$	8.87	$0.00-0.00i$
	$0.00+0.00i$	11.82		$0.00+0.00i$	8.87
$\mathbf{Q}_{(5,7)}$	19.62	$-0.01-0.08i$	$\mathbf{Q}_{(12,3)}$	11.75	$0.00-0.00i$
	$-0.01+0.08i$	19.68		$0.00+0.00i$	11.73
$\mathbf{Q}_{(5,11)}$	26.56	$-0.10-0.14i$	$\mathbf{Q}_{(12,4)}$	9.22	$0.04-0.09i$
	$-0.10+0.14i$	26.90		$0.04+0.09i$	9.25
$\mathbf{Q}_{(6,9)}$	21.22	$0.00-0.00i$	$\mathbf{Q}_{(12,11)}$	20.85	$0.20+0.11i$
	$0.00+0.00i$	21.22		$0.20-0.11i$	19.47
$\mathbf{Q}_{(6,15)}$	22.84	$0.21+0.14i$	$\mathbf{Q}_{(13,3)}$	11.09	$-0.03-0.05i$
	$0.21-0.14i$	23.24		$-0.03+0.05i$	11.08
$\mathbf{Q}_{(7,5)}$	9.54	$0.00-0.04i$	$\mathbf{Q}_{(13,7)}$	8.87	$0.00-0.00i$
	$0.00+0.04i$	9.51		$0.00+0.00i$	8.87
$\mathbf{Q}_{(7,10)}$	14.19	$-0.05-0.09i$	$\mathbf{Q}_{(13,8)}$	10.00	$0.54-0.59i$
	$-0.05+0.09i$	13.82		$0.54+0.59i$	9.69
$\mathbf{Q}_{(7,11)}$	12.48	$1.15-0.43i$	$\mathbf{Q}_{(13,10)}$	15.70	$0.15+0.01i$
	$1.15+0.43i$	11.21		$0.15-0.01i$	15.71
$\mathbf{Q}_{(7,13)}$	7.54	$0.12-0.02i$	$\mathbf{Q}_{(14,7)}$	39.76	$0.19+0.04i$
	$0.12+0.02i$	7.51		$0.19-0.04i$	39.71
$\mathbf{Q}_{(7,14)}$	7.10	$0.00-0.00i$	$\mathbf{Q}_{(15,2)}$	20.58	$-0.01+0.16i$
	$0.00+0.00i$	7.10		$-0.01-0.16i$	20.50
$\mathbf{Q}_{(8,3)}$	12.18	$0.00-0.00i$	$\mathbf{Q}_{(15,6)}$	11.82	$0.00-0.00i$
	$0.00+0.00i$	12.15		$0.00+0.00i$	11.82
$\mathbf{Q}_{(8,10)}$	15.00	$0.05-0.05i$	$\mathbf{Q}_{(15,9)}$	13.09	$0.10-0.08i$
	$0.05+0.05i$	14.95		$0.10+0.08i$	12.95

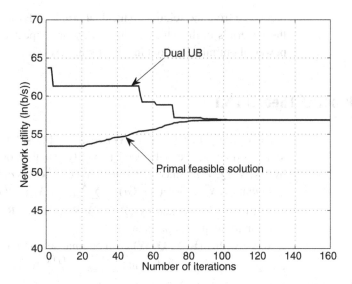

Figure 3.5 Convergence behavior of primal and dual solutions for the 15-node network.

3.8 Chapter summary

Convex programming is a popular and powerful tool to study nonlinear optimization problems. Once a problem is shown to be a convex program, then there are standard solution techniques and we may even apply directly a solver to obtain an optimal solution. For many cross-layer convex optimization problems, the research community is more interested in exploring a solution in the dual domain. There are two reasons for this approach. First, many cross-layer problems, once properly formulated in the dual domain, can be decomposed into subproblems, each of which may be de-coupled from the other layers. Such a layering-based decomposition offers better insights and interpretations for the underlying problems. Second, once a problem is decomposed in its dual domain, the solution may be implemented in a distributed fashion, which is a highly desirable feature for networking researchers. Although this second point was not well discussed in this chapter, it was discussed extensively in the literature and readers are encouraged to explore this on their own (perhaps by first studying related work [79; 90; 98]).

In this chapter, we illustrated the above points by studying a cross-layer optimization problem for a multi-hop MIMO network. We show that the problem can be formulated as a convex program. The problem is a cross-layer optimization problem as it involves variables at the network, link, and physical layers. By studying the problem in its dual domain, we showed that the dual problem can be decomposed into two subproblems: one subproblem solely involving variables at the network layer and the other problem involving variables at the link and physical layers. We described how the dual problem can be solved by a cutting-plane method and how the solution to the primal problem can be recovered from the solution to the dual problem. By thoroughly understanding the materials in this chapter, the readers are expected to be equipped with a powerful optimization tool in networking research.

Appendix: Proof of Theorem 3.1

We first verify that $(\bar{\mathbf{R}}, \bar{\mathbf{Y}}, \bar{\mathbf{Q}}, \bar{\mathbf{W}})$ is a feasible solution to the primal problem. To do this, we only need to verify whether or not (3.10) is satisfied by $(\bar{\mathbf{R}}, \bar{\mathbf{Y}}, \bar{\mathbf{Q}}, \bar{\mathbf{W}})$. Since the function $\langle \mathbf{1}, \mathbf{Y}^T \mathbf{e}_l \rangle - c(W_l, \mathbf{Q}_l)$ is convex, we have that $\langle \mathbf{1}, \bar{\mathbf{Y}}^T \mathbf{e}_l \rangle - c(\bar{W}_l, \bar{\mathbf{Q}}_l) \leq \sum_{j=0}^{K} \tau_j^* (\langle \mathbf{1}, \mathbf{Y}^{(j)T} \mathbf{e}_l \rangle - c(W_l^{(j)}, \mathbf{Q}_l^{(j)})) \leq 0$, i.e., (3.10) is satisfied by $(\bar{\mathbf{R}}, \bar{\mathbf{Y}}, \bar{\mathbf{Q}}, \bar{\mathbf{W}})$. Thus, $(\bar{\mathbf{R}}, \bar{\mathbf{Y}}, \bar{\mathbf{Q}}, \bar{\mathbf{W}})$ is a feasible solution to the primal problem.

To show that $(\bar{\mathbf{R}}, \bar{\mathbf{Y}}, \bar{\mathbf{Q}}, \bar{\mathbf{W}})$ is an optimal solution to the primal problem, let $\bar{z} = \sum_{f=1}^{F} \ln(\bar{r}_f) = \sum_{f=1}^{F} \ln\left(\sum_{j=0}^{K} \tau_j^* r_f^{(j)}\right)$ be the primal objective value corresponding to $(\bar{\mathbf{R}}, \bar{\mathbf{Y}}, \bar{\mathbf{Q}}, \bar{\mathbf{W}})$ and let z^* be the optimal objective value of both the primal and dual problems. Then, we have

$$\bar{z} = \sum_{f=1}^{F} \ln\left(\sum_{j=0}^{K} \tau_j^* r_f^{(j)}\right) \geq \sum_{f=1}^{F} \sum_{j=0}^{K} \tau_j^* \ln(r_f^{(j)}) = \sum_{j=0}^{K} \tau_j^* \sum_{f=1}^{F} \ln(r_f^{(j)})$$
$$= z^{(K)} \geq \Theta(\mathbf{u}^{(K)}) \geq z^*.$$

The second inequality holds by the concavity of $\sum \ln(\cdot)$. The fourth equality holds because τ_j^*, for $0 \leq j \leq K$, is an optimal solution to (3.17) and $z^{(K)}$ is the optimal objective value of (3.17). The fifth inequality holds by the terminating condition of Algorithm 3.1. The last inequality holds because $\Theta(\mathbf{u}^{(K)})$ is the objective value of the dual problem under a given $\mathbf{u}^{(K)}$, while z^* is the optimal objective value of the dual problem among all $\mathbf{u}$ values.

On the other hand, $\bar{z}$ is the objective value of the primal problem achieved by a feasible solution $(\bar{\mathbf{R}}, \bar{\mathbf{Y}}, \bar{\mathbf{Q}}, \bar{\mathbf{W}})$, while z^* is the optimal objective value of the primal problem. Thus, we have $\bar{z} \leq z^*$.

Combining both results, we have $\bar{z} = z^*$, i.e., $(\bar{\mathbf{R}}, \bar{\mathbf{Y}}, \bar{\mathbf{Q}}, \bar{\mathbf{W}})$ is an optimal solution to the primal problem.

3.9 Problems

3.1 Which of the following sets are convex and which are not? Why?
(a) The set in Fig. 3.6.
(b) The set in Fig. 3.7.
(c) $\{(x_1, x_2) : x_1^2 + x_2^2 \geq 4\}$.
(d) $\{(x_1, x_2) : x_1 - 2x_2^2 \geq 0\}$.
(e) $\{(x_1, x_2) : 1 \leq x_1^2 + x_2^2 \leq 4\}$.

Figure 3.6 Set 1.

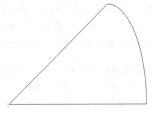

Figure 3.7 Set 2.

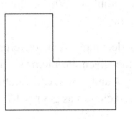

Figure 3.8 A simple network.

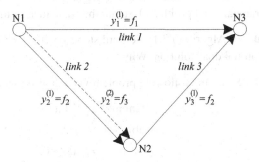

3.2 Which of the following functions are convex, concave, affine, or neither? Why?

(a) $f(x) = x^2 + x^4$, $x \in \mathcal{R}$.

(b) $f(x) = x + x^3$, $x \in \mathcal{R}$.

(c) $f(x) = B \log_2(1 + x)$, B is a positive constant and $x \in \mathcal{R}^+$.

(d) $f(x_1, x_2) = x_1 x_2$, for $x_1, x_2 \in \mathcal{R}^+$.

(e) $f(x_1, x_2) = 5x_1 - 10x_2$, for $x_1, x_2 \in \mathcal{R}$.

3.3 Compared to a nonconvex program, why is it generally easier to obtain an optimal solution for a convex programming problem?

3.4 Consider a small network shown in Fig. 3.8. Suppose there are two sessions in this network. The first session is from N1 to N3 and includes two routing paths: one from N1 to N3 directly and the other from N1 to N3 via N2. The second session is from N1 to N2 and includes the direct routing path only. Suppose $y_1^{(1)} = f_1$, $y_2^{(1)} = y_3^{(1)} = f_2$, $y_2^{(2)} = f_3$. Give the expressions for **A** (the node-link incidence matrix), **Y** (the collection of all flow vectors), and **R** (the collection of all source-sink vectors) in the problem formulation.

3.5 In the networking research community, why is the sum of the log function of session rates a preferred objective function than the sum of rates?

3.6 In the formulation of Problem $\mathcal{P}$ in this chapter, which constraints are associated with the transport, network, link, and physical layers, respectively? Which constraints are associated with multiple layers?

3.7 The Problem $\mathcal{P}$ discussed in this chapter is a convex program. Instead of solving the problem directly using standard convex optimization approaches, in this chapter we solve the Lagrangian dual problem and then recover an optimal solution to the original problem. Why is this approach appealing to the networking research community?

3.8 In the dual problem decomposition, different Lagrangian dual problems can be devised for the given Problem $\mathcal{P}$, depending on which constraints are treated as $\mathbf{g(x)} \leq \mathbf{0}$ and how set $\mathbf{X}$ is defined. Why is the constraint $< \mathbf{1}, \mathbf{Y}^T e_l > -c(W_l, \mathbf{Q}_l) \leq 0$ chosen as $\mathbf{g(x)} \leq \mathbf{0}$?

3.9 In this chapter, a cutting-plane method is employed to solve the Lagrangian dual problem. Explain the basic idea of this method.

3.10 In Algorithm 3.1, the third step says if $z^k \geq \Theta(\mathbf{u}^k)$, then $(z^{(k)}, \mathbf{u}^{(k)})$ is an optimal dual solution. Why?

3.11 Solve the following problem via a dual-based approach:

$$\text{Minimize } x_1^2 - \ln(x_2)$$
$$\text{subject to } x_1 + 2x_2 - 5 \geq 0$$
$$x_1, x_2 \geq 0.$$

4 Design of polynomial-time exact algorithm

When one door of happiness closes, another opens; but often we look so long at the closed door that we do not see the one which has opened for us.

Helen Keller

4.1 Problem complexity vs. solution complexity

The previous two chapters focus on developing optimal polynomial-time solutions following formal optimization methods from operations research (OR). For some problems, such an approach may not be always effective, and could lead to nonpolynomial-time solutions. For these problems, a customized approach following algorithm design from computer science (CS) could be more effective and lead to a polynomial-time solution.

It is important to distinguish a (solution) algorithm's complexity from the underlying problem's complexity. A problem's complexity determines the potential complexity of *any* algorithm that is designed to solve this problem. That is, for a problem not in P, unless $P = NP$, any algorithm that can find an optimal solution to this problem must have nonpolynomial-time complexity. In contrast, if an algorithm (design to solve the problem) has a nonpolynomial-time complexity, we cannot claim that this problem is not in P. Another algorithm designed by someone else may well solve the problem with a polynomial-time complexity.

In this chapter, we illustrate the above approaches and ideas with a case study. This case study is concerned with an optimal relay node assignment problem that arises in cooperative communications (CC) [139]. We first formulate the problem as a mixed-integer linear programming (MILP), following OR's optimization approach. An MILP can be solved optimally, but with a nonpolynomial-time complexity in general. However, this problem is in fact in P, as we can design an exact algorithm with polynomial-time complexity following a CS approach.

4.2 Case study: Optimal cooperative relay node assignment

Wireless channels are considered unreliable due to signal fading. Spatial diversity, in the form of employing multiple antennas (i.e., MIMO), has proved to be very effective in increasing network capacity and reliability. However, equipping a wireless node with multiple antennas may not always be practical, as the footprint of multiple antennas may not fit on a wireless node (particularly a handheld device). In order to achieve spatial diversity without requiring multiple antennas on the same node, the so-called cooperative communications (CC) have been introduced [87; 136; 137]. Under CC, each node is equipped with only a single antenna, and spatial diversity is achieved by exploiting the antennas on other nodes in the network through cooperative relaying. In Section 4.3, we give a brief overview of CC so as to set the stage of our case study.

As expected, the choice of a relay node plays a critical role in the performance of CC [18; 22; 180]. An improperly chosen relay node may offer a smaller data rate for a source–destination pair than that under direct transmission. In Section 4.4, we describe the relay node assignment problem in a network environment, where multiple source–destination pairs compete for the same pool of relay nodes in the network. Our objective is to assign the available relay nodes to different source–destination pairs so as to maximize the minimum data rate among all pairs.

In Section 4.5, we will show that this problem can be formulated as a mixed-integer linear programming (MILP). In general, an MILP has a nonpolynomial complexity. In Section 4.6, instead of studying this MILP, we develop an optimal polynomial-time algorithm, called ORA, that directly solves the original problem. A novel idea in this algorithm is a "linear marking" mechanism, which yields a low complexity for each iteration. In Section 4.7, we give a proof of the optimality of ORA and Section 4.8 presents numerical results. Section 4.9 summarizes this chapter. The case study in this chapter provides an interesting instance to show that a complex problem formulation (seemingly corresponding to a nonpolynomial-time solution) cannot be used as a way to show that the problem is not in P.

4.3 Cooperative communications: a primer

The essence of CC is best explained by a three-node example in Fig. 4.1. In this figure, node s is the source node, node d is the destination node, and node

Figure 4.1 A three-node relay channel for CC.

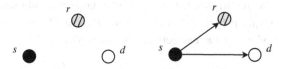

(a) One source, one destination, and one relay node.

(b) Source node transmits in the first time slot.

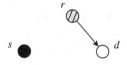

(c) Relay node transmits in the second time slot.

r is a relay node. Transmission from s to d is done on a frame-by-frame basis. Within a frame, there are two time slots. In the first time slot, source node s makes a transmission to the destination node d. Due to the broadcast nature of wireless communications, this transmission is also overheard by the relay node r. In the second time slot, node r forwards the data received in the first time slot to node d. Note that such a two-slot structure is necessary for CC due to the half-duplex nature of most wireless transceivers.

In this section, we provide expressions for the achievable data rate under CC and direct transmissions (i.e., no cooperation). For CC, we consider both AF and DF coding schemes at the relay node [87].

Amplify-and-forward (AF)   Let h_{sd}, h_{sr}, and h_{rd} capture the effect of path-loss, shadowing, and fading between nodes s and d, s and r, and r and d, respectively. Denote $z_d[t]$ and $z_r[t]$ as the zero-mean background noise at nodes d and r in the tth time slot, respectively, with variance σ_d^2 and σ_r^2.

Denote x_s as the signal transmitted by source node s in the first time slot. Then the received signal at destination node d, y_{sd}, can be expressed as

$$y_{sd} = h_{sd}x_s + z_d[1], \tag{4.1}$$

and the received signal at the relay node r, y_{sr}, is

$$y_{sr} = h_{sr}x_s + z_r[1]. \tag{4.2}$$

In the second time slot, relay node r transmits to destination node d. The received signal at d, y_{rd}, can be expressed as

$$y_{rd} = h_{rd}\alpha_r y_{sr} + z_d[2],$$

where α_r is the amplifying factor at relay node r and y_{sr} is given in (4.2). Thus, we have

$$y_{rd} = h_{rd}\alpha_r(h_{sr}x_s + z_r[1]) + z_d[2]. \tag{4.3}$$

The amplifying factor α_r at relay node r should satisfy the power constraint $\alpha_r^2(|h_{sr}|^2 P_s + \sigma_r^2) = P_r$, where P_s and P_r are the transmission powers at nodes s and r, respectively. So, α_r is given by

$$\alpha_r^2 = \frac{P_r}{|h_{sr}|^2 P_s + \sigma_r^2}.$$

We can rewrite (4.1), (4.2), and (4.3) in the following compact matrix form:

$$\mathbf{Y} = \mathbf{H}x_s + \mathbf{B}\mathbf{Z},$$

where

$$\mathbf{Y} = \begin{bmatrix} y_{sd} \\ y_{rd} \end{bmatrix}, \quad \mathbf{H} = \begin{bmatrix} h_{sd} \\ \alpha_r h_{rd} h_{sr} \end{bmatrix}, \quad \mathbf{B} = \begin{bmatrix} 0 & 1 & 0 \\ \alpha_r h_{rd} & 0 & 1 \end{bmatrix}, \quad \text{and } \mathbf{Z} = \begin{bmatrix} z_r[1] \\ z_d[1] \\ z_d[2] \end{bmatrix}.$$

(4.4)

It has been shown in [87] that the above channel, which combines both the direct path (s to d) and the relay path (s to r to d), can be modeled as a one-input, two-output complex Gaussian noise channel. The achievable data rate $C_{\text{AF}}(s, r, d)$ from s to d is given by

$$C_{\text{AF}}(s, r, d) = \frac{W}{2} \log_2 \left[\det \left(\mathbf{I} + (P_s \mathbf{H}\mathbf{H}^\dagger)(\mathbf{B}E[\mathbf{Z}\mathbf{Z}^\dagger]\mathbf{B}^\dagger)^{-1} \right) \right], \quad (4.5)$$

where W is the bandwidth, $\det(\cdot)$ is the determinant function, $\mathbf{I}$ is the identity matrix, the superscript "$\dagger$" represents the complex conjugate transpose, and $E[\cdot]$ is the expectation function.

Substituting (4.4) into (4.5) and performing algebraic manipulations, we have

$$C_{\text{AF}}(s, r, d) = \frac{W}{2} \log_2 \left(1 + \frac{P_s}{\sigma_d^2}|h_{sd}|^2 + \frac{P_s|h_{sr}|^2 P_r|h_{rd}|^2}{P_s\sigma_d^2|h_{sr}|^2 + P_r\sigma_r^2|h_{rd}|^2 + \sigma_r^2\sigma_d^2} \right).$$

(4.6)

Denote

$$\text{SNR}_{sd} = \frac{P_s}{\sigma_d^2}|h_{sd}|^2, \quad \text{SNR}_{sr} = \frac{P_s}{\sigma_r^2}|h_{sr}|^2, \quad \text{and SNR}_{rd} = \frac{P_r}{\sigma_d^2}|h_{rd}|^2.$$

We have

$$C_{\text{AF}}(s, r, d) = W \cdot I_{\text{AF}}(\text{SNR}_{sd}, \text{SNR}_{sr}, \text{SNR}_{rd}),$$

where

$$I_{\text{AF}}(\text{SNR}_{sd}, \text{SNR}_{sr}, \text{SNR}_{rd}) = \frac{1}{2} \log_2 \left(1 + \text{SNR}_{sd} + \frac{\text{SNR}_{sr} \cdot \text{SNR}_{rd}}{\text{SNR}_{sr} + \text{SNR}_{rd} + 1} \right).$$

Decode-and-forward (DF)　　Under this mode, relay node r decodes and estimates the received signal from source node s in the first time slot, then transmits the estimated data to destination node d in the second time slot. The achievable data rate for DF under the two time-slot structure is given by [87]:

$$C_{\mathrm{DF}}(s, r, d) = W \cdot I_{\mathrm{DF}}(\mathrm{SNR}_{sd}, \mathrm{SNR}_{sr}, \mathrm{SNR}_{rd}),$$

where

$$I_{\mathrm{DF}}(\mathrm{SNR}_{sd}, \mathrm{SNR}_{sr}, \mathrm{SNR}_{rd})$$
$$= \frac{1}{2} \min\{\log_2(1 + \mathrm{SNR}_{sr}), \log_2(1 + \mathrm{SNR}_{sd} + \mathrm{SNR}_{rd})\}.$$

Note that $I_{\mathrm{AF}}(\cdot)$ and $I_{\mathrm{DF}}(\cdot)$ are increasing functions of P_s and P_r, respectively. This suggests that, in order to achieve the maximum data rate under either mode, both source node and relay node should transmit at their maximum powers. In this chapter, we let $P_s = P_r = P$.

Direct transmission   When CC is not used, source node s transmits to destination node d in both time slots. The achievable data rate from node s to node d is

$$C_{\mathrm{D}}(s, d) = W \log_2(1 + \mathrm{SNR}_{sd}).$$

Based on the above results, we have two observations. First, comparing C_{AF} (or C_{DF}) to C_{D}, we can see that CC may not always be better than direct transmissions. For example, a poor choice of relay node could make the achievable data rate under CC lower than that under direct transmissions. This fact underlines the significance of relay node selection in CC. Second, although AF and DF are different mechanisms, the data rates for both of them have the same form, i.e., a function of SNR_{sd}, SNR_{sr}, and SNR_{rd}. Therefore, we only need to develop one optimal relay node assignment algorithm, which should work for both AF and DF. Table 4.1 lists the notations used in this chapter.

4.4 The relay node assignment problem

Based on the background in the last section, we consider a relay node assignment problem in a network setting. There are N nodes in an ad hoc network, with each node being either a source node, a destination node, or a potential relay node (see Fig. 4.2). In order to avoid interference, we assume that orthogonal channels are available in the network (e.g., using OFDMA) for CC [87]. The path-loss between nodes u and v is captured in h_{uv} and is given *a priori*. Denote $\mathcal{N}_s = \{s_1, s_2, \ldots, s_{N_s}\}$ as the set of source nodes, $\mathcal{N}_d = \{d_1, d_2, \ldots, d_{N_d}\}$ as the set of destination nodes, and $\mathcal{N}_r = \{r_1, r_2, \ldots, r_{N_r}\}$ as the set of relays (see Fig. 4.2). We consider unicast where every source node s_i is paired with a destination node d_i, i.e., $N_d = N_s$. Each node is equipped with a single transceiver and can transmit/receive within one channel at a time.

Table 4.1 Notation.

Symbol	Definition
A_{s_i,r_j}	Decision variable for relay node assignment
$C_R(s_i, r_j)$	Achievable rate for s_i–d_i pair when relay node r_j is used
$C_R(s_i, \emptyset)$	Achievable rate for s_i–d_i pair under direct transmission
$C_{\min}$	The minimum rate among all source–destination pairs
h_{uv}	Effect of path-loss, shadowing, and fading from node u to node v
$\mathcal{N}_r$	Set of relay nodes in the network
$\mathcal{N}_s$	Set of source nodes in the network
N_r	$= |\mathcal{N}_r|$, number of relay nodes in the network
N_s	$= |\mathcal{N}_s|$, number of source nodes in the network
N	Number of all the nodes in the network
P	Maximum transmission power
r_j	The jth relay node, $r_j \in \mathcal{N}_r$
$\mathcal{R}_\psi(s_i)$	The relay node assigned to s_i under ψ
s_i	The ith source node, $s_i \in \mathcal{N}_s$
$S_\psi(r_j)$	The source node that uses r_j under ψ
SNR_{uv}	The signal-to-noise ratio between nodes u and v
W	Channel bandwidth
x_s	Signal transmitted by node s
y_{uv}	Received signal at node v (form node u)
$z_v[t]$	Background noise at node v in the tth time slot
α_r	Amplifying factor at relay r
σ_v^2	Variance of background noise at node v
ψ	A solution for relay node assignment

Figure 4.2 An ad hoc network consisting of source nodes (senders), destination nodes (receivers), and relay nodes.

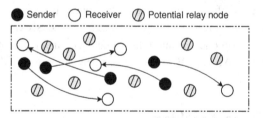

Further, we assume that each node can only serve a unique role of source, destination, or relay. That is, $N_r = N - 2N_s$.

Note that a source node may not always get a relay node. There are two possible scenarios in which this may happen. First, there may not be a sufficient number of relay nodes in the network (e.g., $N_r < N_s$). In this case, some source nodes will not have relay nodes. Second, even if there are enough relay nodes, a sender may choose not to use a relay node if it leads to a lower data rate than direct transmission.

We now discuss the objective function of our problem. Although different objectives can be used, a widely used objective for CC is the achievable data rate. For the multi-session network environment considered in this chapter (see Fig. 4.2), each source–destination pair will have a different achievable data rate after we apply a relay node assignment algorithm. So a plausible objective is to maximize the minimum achievable data rate among all the source–destination pairs.

More formally, denote $\mathcal{R}(s_i)$ as the relay node assigned to s_i, and $\mathcal{S}(r_j)$ as the source node that uses r_j. For both AF and DF, its achievable data rate can be written as (see Section 4.3)

$$W \cdot I_R(\text{SNR}_{s_i,d_i}, \text{SNR}_{s_i,\mathcal{R}(s_i)}, \text{SNR}_{\mathcal{R}(s_i),d_i}),$$

with $I_R(\cdot) = I_{AF}(\cdot)$ when AF is employed, and $I_R(\cdot) = I_{DF}(\cdot)$ when DF is employed. In case s_i does not use a relay, we denote $\mathcal{R}(s_i) = \emptyset$ and the achievable data rate is the direct transmission capacity, i.e.,

$$C_R(s_i, \emptyset) = C_D(s_i, d_i).$$

Combining both cases, we have

$$C_R(s_i, \mathcal{R}(s_i)) = \begin{cases} W \cdot I_R(\text{SNR}_{s_i,d_i}, \text{SNR}_{s_i,\mathcal{R}(s_i)}, \text{SNR}_{\mathcal{R}(s_i),d_i}) & \text{if } \mathcal{R}(s_i) \neq \emptyset, \\ W \log(1 + \text{SNR}_{s_i,d_i}) & \text{if } \mathcal{R}(s_i) = \emptyset. \end{cases}$$
(4.7)

Note that d_i is not listed in function $C_R(s_i, \mathcal{R}(s_i))$ since for each source node s_i, its corresponding destination node d_i is deterministic.

Denote by $C_{\min}$ the objective function, which is the minimum achievable data rate among all source nodes. We have

$$C_{\min} = \min\{C_R(s_i, \mathcal{R}(s_i)) : s_i \in \mathcal{N}_s\}.$$

Our goal is to find an optimal relay node assignment for all the source–destination pairs such that $C_{\min}$ is maximized.

4.5 An optimization-based formulation

In this section, we present a formulation for the relay node assignment problem based on an optimization approach from OR. To describe relay node assignment mathematically, we denote a binary variable A_{s_i,r_j} as

$$A_{s_i,r_j} = \begin{cases} 1 \text{ if relay } r_j \text{ is assigned to source } s_i, \\ 0 \text{ otherwise.} \end{cases}$$

Since a source node s_i will be assigned at most one relay node, we have

$$\sum_{j=1}^{N_r} A_{s_i,r_j} \leq 1.$$

Table 4.2 A mathematical formulation for relay node assignment problem.

Maximize $C_{\min}$
subject to $C_{\min} \le \left(1 - \displaystyle\sum_{j=1}^{N_r} A_{s_i, r_j} \right) C_R(s_i, \emptyset)$
$\qquad\qquad + \displaystyle\sum_{j=1}^{N_r} A_{s_i, r_j} C_R(s_i, r_j)$ $(s_i \in \mathcal{N}_s)$
$\displaystyle\sum_{i=1}^{N_s} A_{s_i, r_j} \le 1$ $(r_j \in \mathcal{N}_r)$
$\displaystyle\sum_{j=1}^{N_r} A_{s_i, r_j} \le 1$ $(s_i \in \mathcal{N}_s)$
$A_{s_i, r_j} \in \{0, 1\}, C_{\min} \ge 0$ $(s_i \in \mathcal{N}_s, r_j \in \mathcal{N}_r)$

Due to the unique role of each node, a relay node r_j can be assigned to at most one source node. That is

$$\sum_{i=1}^{N_s} A_{s_i, r_j} \le 1.$$

Note that if any $A_{s_i, r_j} = 1$, then a relay r_j is assigned to source s_i. By (4.7), the achievable data rate from source s_i to destination d_i by using relay r_j is $C_R(s_i, r_j)$. On the other hand, if $\sum_{j=1}^{N_r} A_{s_i, r_j} = 0$, then there is no relay assigned to source s_i. By (4.7), the achievable data rate from source s_i to destination d_i is $C_R(s_i, \emptyset)$. Combining these two cases, the achievable data rate from source s_i to destination d_i is

$$\left(1 - \sum_{j=1}^{N_r} A_{s_i, r_j} \right) C_R(s_i, \emptyset) + \sum_{j=1}^{N_r} A_{s_i, r_j} C_R(s_i, r_j).$$

Based on the above discussion, one formulation for the optimal relay node assignment problem is given in Table 4.2. This formulation is a mixed-integer linear programming (MILP) and is NP-hard in general [46]. As discussed in Section 4.1, this does not mean that our optimal relay node assignment problem is NP-hard because an NP-hard formulation is not a proof of a problem's NP-hardness. In fact, this problem can be viewed as a maximin linear assignment problem, for which there exists polynomial-time algorithms (see [3], [25], [89], [126], and [161] for example, although these papers address differently structured problems, the proposed algorithms can be adapted to solve our problem in polynomial time as well).

Instead of applying a general algorithm for the maximin linear assignment problem, we design in the next section a specialized polynomial-time algorithm (along with an optimality proof) for our relay node assignment problem.

The existence of such a polynomial-time algorithm implies that the problem belongs to the class P of polynomially solvable problems.

4.6 An exact algorithm

4.6.1 Basic idea

The polynomial-time algorithm that we will describe is called the *Optimal Relay Assignment* (ORA) algorithm. Fig. 4.3 shows the flow chart of the ORA algorithm.

Initially, the ORA algorithm starts with a random but feasible relay node assignment. By "feasible," we mean that each source–destination pair can be assigned at most one relay node and that a relay node can be assigned only once. Such an initial feasible assignment is easy to construct, e.g., a direct transmission between each source–destination pair (without the use of a relay) is a special case of a feasible assignment.

Starting with this initial assignment, ORA adjusts the assignment during each iteration, with the goal of increasing the objective function $C_{\min}$. Specifically, during each iteration, ORA identifies the source node that corresponds to the current $C_{\min}$. Then, ORA helps this source node search for a better relay such that this "bottleneck" data rate can be increased. In the case that the selected relay is already assigned to another source node, a further adjustment of the relay node assignment for that source node is necessary (so that its current relay can be released). Such an adjustment might have a chain effect on a number of source nodes in the network. It is important that for any adjustment made on a relay node assignment, the affected source node still maintains a data rate larger than $C_{\min}$. There are only two outcomes from such an iteration: (i) a better assignment is found, in which case ORA moves on to the next iteration; or (ii) a better assignment cannot be found, in which case ORA terminates.

There are two key issues that we must address in the design of this algorithm. First, for any nonoptimal solution, the algorithm should be able to find a better solution. As a result, upon termination the final assignment is optimal. Second, its running time must be polynomial. We will show that ORA addresses both issues successfully. Specifically, we show that the complexity of the ORA algorithm is polynomial in Section 4.6.4. We will also give a correctness proof of its optimality in Section 4.7.

4.6.2 Algorithmic details

To begin with, the ORA algorithm performs a "preprocessing" step. In this step, for each source–destination pair, the source node s_i considers each relay

Figure 4.3 A flow chart of the ORA algorithm.

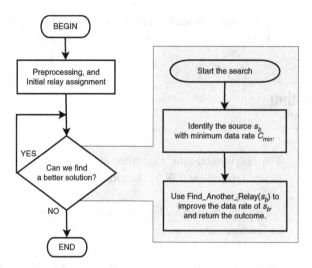

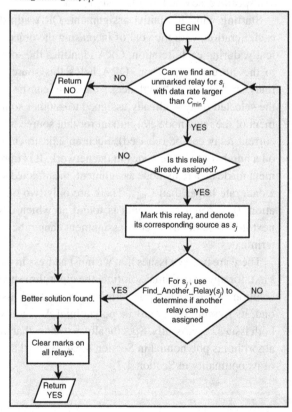

Figure 4.4 An example tree topology in the ORA algorithm for finding a better solution.

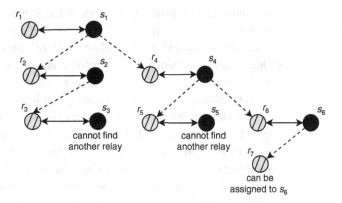

node r_j in the network and computes the corresponding data rate $C_R(s_i, r_j)$ by (4.7). Each source node s_i also computes the achievable data rate $C_R(s_i, \emptyset)$ by (4.7) under direct transmissions (i.e., without the use of a relay node). After these computations, each source node s_i can identify those relay nodes that can offer an increase in its data rate compared to direct transmissions, i.e., those relays with $C_R(s_i, r_j) > C_R(s_i, \emptyset)$. Obviously, it only makes sense to consider these relays for CC. That is, for a source node s_i, those relay nodes with rates not greater than direct transmissions can be removed from further consideration.

After the preprocessing step, we enter the initial assignment step. The objective of this step is to obtain an initial feasible solution for the ORA algorithm so that it can start its iteration. In the preprocessing step, we have already identified the list of relay nodes for each source node that can increase its data rate compared to direct transmission. We can randomly assign a relay node from this list to a source node. Note that once a relay node is assigned to a source node, it cannot be assigned again to another source node. If there is no relay node available for a source node, then this source node will simply employ direct transmission. Upon the completion of this assignment, each source node will have a data rate that is no less than that under direct transmission.

The next step in the ORA algorithm is to find a better assignment, which is an iterative process. This is the key step in the ORA algorithm. The detail of this step is shown in the bottom portion of Fig. 4.3. As a starting point of this step, the ORA algorithm identifies the smallest data rate C_{min} among all sources. The ORA algorithm aims to increase this minimum data rate for the corresponding source node, while having all other source nodes maintain their data rates above C_{min}. Without loss of generality, we use Fig. 4.4 to illustrate the search process.

- Suppose ORA identifies that s_1 has the smallest rate C_{min} under the current assignment (with relay node r_1). In case of a tie, i.e., when two or more source nodes have the same smallest data rate, the tie is broken by choosing the source node with the highest node index. Source s_1 examines other relays with a rate larger than $C_{\min}$. If it cannot find such a relay, then no better solution is found and the ORA algorithm terminates.

- Otherwise, i.e., when there exist better relays, we consider these relays in a *nonincreasing* order of data rate (should it be assigned to s_1). That is, we try the relay that can offer the *maximum* possible increase in data rate first. In case of a tie, i.e., when two or more relay nodes offer the same maximum data rate, the tie is broken by choosing the relay node with the highest node index.

- Suppose that source node s_1 considers relay node r_2. If this relay node is not yet assigned to any other source node, then r_2 can be immediately assigned to s_1. In this simple case, we find a better solution and the current iteration is completed.

- Otherwise, i.e., if r_2 is already assigned to a source node, say s_2, we mark r_2 to indicate that r_2 is "under consideration" and check whether r_2 can be released by s_2.

- To release r_2, source node s_2 needs to find another relay (or use direct transmission) while making sure that such a new assignment still keeps its data rate larger than the current $C_{\min}$. This process is identical to that for s_1, with the only (but important) difference being that s_2 will not consider a relay that has already been "marked," because that relay node has already been considered by a source node encountered earlier in this iteration.

- Suppose that source node s_2 now considers relay r_3. We consider the following three cases. If this relay node is not yet assigned to any source node, then r_3 can be assigned to s_2, r_2 can be assigned to s_1, and the current iteration is completed. Moreover, if the relay under consideration by s_2 is the one that is being used by the source node that initiated the iteration, i.e., relay r_1, then it is easy to see that r_1 can be taken away from s_1. A better solution (r_1 is assigned to s_2 and r_2 is assigned to s_1) is found and the current iteration is completed. Otherwise, we mark r_3 and check further to see whether r_3 can be released by its corresponding source node, say s_3. We also note that s_2 can consider direct transmission if it offers a data rate larger than $C_{\min}$.

- Suppose that s_3 cannot find any "unmarked" relay that offers a data rate larger than $C_{\min}$ and its achievable data rate under direct transmission is not larger than $C_{\min}$. Then s_2 cannot use r_3 as its relay.

- If any "unmarked" relay that offers a data rate larger than $C_{\min}$ cannot be assigned to s_2, then s_1 cannot use r_2 and the algorithm will move on to consider the next relay on its nonincreasing rate list, say r_4.

- The search continues, with relay nodes being marked along the way, until a better solution is found or no better solution can be found. For example, in Fig. 4.4, s_6 finds a new relay r_7. As a result, we have a new assignment, where r_7 is assigned to s_6, r_6 is assigned to s_4, and r_4 is assigned to s_1.

Table 4.3 An example illustrating the operation of the ORA algorithm.

	$\emptyset$	r_1	r_2	r_3	r_4	r_5	r_6	r_7
(a) After initial assignment. The start of the first iteration.								
s_1	14	7	24	5	14	_15_	17	9
s_2	9	8	10	11	_20_	10	12	11
$\rightarrow s_3$	11	10	_13_	17	21	8	9	19
s_4	12	8	9	12	11	10	9	_18_
s_5	10	9	18	_19_	24	9	13	23
s_6	7	18	12	6	11	11	_17_	20
s_7	_16_	1	9	4	14	19	8	12
(b) The start of the second iteration.								
$\rightarrow s_1$	14	7	24	5	14	_15_	17	9
s_2	9	8	10	11	_20_	10	12	11
s_3	11	10	13	_17_	21	8	9	19
s_4	12	8	9	12	11	10	9	_18_
s_5	10	9	_18_	19	24	9	13	23
s_6	7	18	12	6	11	11	_17_	20
s_7	_16_	1	9	4	14	19	8	12
(c) The start of the third iteration.								
s_1	14	7	24	5	14	15	_17_	9
s_2	9	8	10	11	_20_	10	12	11
s_3	11	10	13	_17_	21	8	9	19
s_4	12	8	9	12	11	10	9	_18_
s_5	10	9	_18_	19	24	9	13	23
s_6	7	_18_	12	6	11	11	17	20
$\rightarrow s_7$	16	1	9	4	14	_19_	8	12
(d) Final assignment upon ORA termination.								
s_1	14	7	24	5	14	15	_17_	9
s_2	9	8	10	11	_20_	10	12	11
$\rightarrow s_3$	11	10	13	_17_	21	8	9	19
s_4	12	8	9	12	11	10	9	_18_
s_5	10	9	_18_	19	24	9	13	23
s_6	7	_18_	12	6	11	11	17	20
s_7	16	1	9	4	14	_19_	8	12

Note that the "mark" on a relay node will not be cleared throughout the search process in the same iteration. We call this a "linear marking" mechanism. These marks will only be cleared when the current iteration terminates and before the start of the next iteration. A pseudocode for the ORA algorithm is shown in Algorithm 4.1. Note that ORA works regardless of whether $N_r \geq N_s$ or $N_r < N_s$.

We now use an example to illustrate the operation of the ORA algorithm, in particular, its "linear marking" mechanism.

Algorithm 4.1 The ORA algorithm

Main algorithm

1. Calculate $C_R(s_i, \emptyset)$ and $C_R(s_i, r_j)$ for each source node s_i and relay node r_j.

2. For each source node s_i, build a list of relay nodes following nonincreasing order of rates (including $C_R(s_i, \emptyset)$).*

3. Find an initial relay node assignment.

4. while (true) {

5.   Set all relay nodes in the network as "unmarked."

6.   Denote s_b as the bottleneck source node with rate $C_{\min}$.* Its relay node is $\mathcal{R}(s_b)$.

7.   Find_Another_Relay($s_b, \mathcal{R}(s_b), C_{\min}, s_b$).

8.   If s_b cannot find another relay

9.   break }

Subroutines

Find_Another_Relay($s_i, r_j, C_{\min}, s_b$) {

10. Search s_i's relay node list {

11.   Suppose the current considered relay node is r_k.

12.   If $C_R(s_i, r_k) \leq C_{\min}$

13.   break.

14.   If relay r_k is "marked"

15.   continue.

16.   Check_Relay_Availability($r_k, C_{\min}, s_b$)

17.   If r_k is available {

18.   Remove relay node r_j's assignment to s_i.

19.   Assign relay node r_k to s_i.

20.   return "s_i finds another relay." } }

21. return "s_i cannot find another relay" }

Check_Relay_Availability($r_j, C_{\min}, s_b$) {

22. If $r_j = \emptyset$, $r_j = \mathcal{R}(s_b)$, or r_j is not assigned to any source node

23.   return "r_j is available."

24. Set r_j as "marked."

25. Find_Another_Relay($\mathcal{S}(r_j), r_j, C_{\min}, s_b$)

26. If $\mathcal{S}(r_j)$ can find another relay

27.   return "r_j is available."

28. else

29.   return "r_j is unavailable." }

Note: * A tie is broken by choosing the node with the highest node index.

Example 4.1

Suppose that there are seven source–destination pairs and seven relay nodes in the network.

Table 4.3(a) shows the data rate for each source node s_i when relay node r_j is assigned to it. The symbol Ø indicates direct transmission. Also shown in Table 4.3(a) is an initial relay node assignment, which is indicated by an underscore on the intersecting row (s_i) and column (r_j). That is, the initial assignment is r_5 for s_1, r_4 for s_2, r_2 for s_3, r_7 for s_4, r_3 for s_5, r_6 for s_6, and Ø for s_7. Note that the initial assignment step ensures that the achievable data rate for each source–destination pair is no less than that under direct transmission.

Given the initial relay node assignment in Table 4.3(a), source s_3 is identified as the bottleneck source node s_b, with the current $C_{min} = 13$. Since consideration of relay nodes is given in the order of nonincreasing (from largest to smallest) data rate, r_4 is therefore considered for s_3 first. Since r_4 is already assigned to source node s_2, we "mark" r_4 now. Next we check whether or not s_2 can find another relay. But any other relay (or direct transmission) will result in a data rate no greater than the current $C_{min} = 13$. This means that r_4 cannot be taken away from s_2. Therefore, s_3 will then consider the next relay node that offers the second largest rate value, i.e., r_7. Since r_7 is already assigned to sender s_4, we "mark" r_7. Next, the ORA algorithm will check whether or not s_4 can find another relay. But none of the relay nodes except r_7 can offer a data rate larger than the current C_{min} to s_4. So r_7 cannot be taken away from s_4. Therefore, node s_3 will now check for the relay node that offers the next highest rate, i.e., r_3. Since r_3 is already assigned to s_5, we "mark" r_3 now. Next, the ORA algorithm checks to see if s_5 can find another relay; s_5 checks relay nodes in nonincreasing order of data rate values. Since r_4 (with largest rate) and r_7 (with the second largest rate) are both marked, they will not be considered (this is the essence of the linear marking mechanism of the ORA algorithm, which ensures that the linear complexity of examining relay nodes in each iteration). The next relay to be considered is r_2, which offers a rate of $18 > C_{min} = 13$. Moreover, r_2 is the relay node assigned to $s_b = s_3$. Thus, r_2 can be reassigned to s_5 and r_3 can be reassigned to s_3. The new assignment after the first iteration is shown in Table 4.3(b). Now the objective value, C_{min}, is updated to 15, which corresponds to s_1. Before the second iteration, all markings done in the first iteration are cleared.

In the second iteration, the ORA algorithm will identify s_1 as the source node with the minimum data rate in the network. The algorithm will then perform a new search for a better relay node for source s_1. Similar to the first iteration, the assignments for other source nodes can change during this search process, but all assignments should result in data rates larger than 15. The details of this iteration are similar to those in the first iteration and we leave them as a homework exercise.

Similar to the previous two iterations, in the third iteration, the ORA algorithm will identify the source node with minimum data rate, which is now s_7. The algorithm will then perform a new search to improve the data rate of s_7.

Finally, in the fourth and final iteration, the algorithm will try to improve the data rate of s_3 and will be unsuccessful. As a result, the ORA algorithm will terminate.

The final assignment upon termination of the ORA algorithm is shown in Table 4.3(d), with the optimal (maximum) value of C_{min} being 17.

4.6.3   Caveat on the marking mechanism

We now re-visit the marking mechanism in the ORA algorithm. Although different marking mechanisms may be designed to achieve an optimal solution, their complexities may differ significantly. In this subsection, we first present a naive marking mechanism, which appears to be a natural approach but unfortunately leads to an exponential complexity for each iteration. Then we discuss our marking mechanism and show that it leads to a linear complexity of examining relay nodes in each iteration.

A natural and naive approach is to perform unmarking during an iteration. This approach is best explained with an example. Again, let's look at Fig. 4.4. Source node s_1 first considers r_2. Since r_2 is now being considered by s_1 and is used by s_2 in the current solution, r_2 is marked. Source node s_2 considers r_3, which is already assigned to s_3. Since s_3 cannot release r_3 without reducing its data rate below the current C_{min}, this branch of search is futile and s_1 moves on to consider its next candidate relay node r_4. Since r_4 is currently assigned to s_4, we mark r_4 and try to find a new relay for s_4. Now the question is: should those marks on r_2 and r_3 that we put earlier in this iteration be removed so that both of them can be considered by s_4? Under this naive approach, r_2 and r_3 should be unmarked so that they can be considered as candidate relay nodes for s_4 in its search. Similarly, when we try to find a relay for s_6, relay nodes r_2, r_3, r_4, and r_5 should be unmarked so that they can be considered as candidate relay nodes for s_6, in addition to r_7. In summary, under this approach, each relay node that has been considered earlier in the search process by a source node should be unmarked when this source node moves on to consider the next relay node, so that this relay node can be in the pool of candidate relay nodes to be considered in the search process. It is not hard to show that such a marking/unmarking mechanism will consider all possible assignments and can guarantee to find an optimal solution upon termination. However, it can be shown that the complexity of this approach is exponential for each iteration. We leave it as a homework exercise.

In contrast, under the ORA algorithm, unmarking of marked nodes is only performed at the end of an iteration (so that there is a clean start for the next iteration). That is, marked relay nodes in the search process will remain marked throughout the iteration. For example, in Fig. 4.4, when s_4 tries to find another relay, it will no longer consider r_2 and r_3 that have already been marked. Similarly, when s_6 tries to find another relay, it will not consider r_2, r_3, r_4, and r_5. As a result, any relay node will be considered at most once in the search process and the pool of candidate relay nodes shrinks linearly as the search continues. It is not hard to see that this marking mechanism leads to a linear complexity of examining relay nodes in each iteration of the ORA algorithm.

Although such a linear marking mechanism is attractive in its complexity, a natural question we would ask is: "How could such a 'linear marking' lead to an optimal solution, as it may overlook many possible assignments that potentially increase $C_{\min}$?" This is precisely the question that we will address in Section 4.7, where we will prove that ORA can guarantee that its final solution is optimal (Theorem 4.1).

Remark 4.1
We comment here that the linear marking mechanism can be alternatively viewed as a flow augmentation step within a specially defined max-flow problem (see Problem 4.12 in the exercise).

4.6.4 Complexity analysis

We now analyze the computational complexity of the ORA algorithm in Algorithm 4.1.

Before first iteration The complexity of Step 1 is $O(N_s(N_r+1)) = O(N_sN_r)$. The complexity of Step 2 is $O(N_s(N_r+1)\log_2(N_r+1)) = O(N_sN_r\log_2 N_r)$. The complexity of Step 3 is $O(N_s)$.

Each iteration The dominant component in complexity is the recursive process of searching an unmarked relay node and checking its availability. This process may involve all source nodes, each of which checks its relay nodes' marking status and availability. Since a source node needs to check at most N_r relay nodes and the number of source nodes is N_s, the complexity of checking relay nodes' marking status and availability in an iteration is $O(N_sN_r)$. Given this is the dominant complexity in an iteration, we have that the complexity of an iteration is $O(N_sN_r)$.

Number of iterations We examine the maximum number of iterations that the ORA can execute. Note that the number of different $C_{\min}$ values that we may encounter is no more than the number of source–relay pairs and that the $C_{\min}$ value for the current bottleneck node s_b is increased after each improving iteration. As a result, the number of iterations that the algorithm can go through is $O(N_sN_r)$.

Complexity of all iterations   The complexity of all iterations is $O(N_s N_r) \cdot O(N_s N_r) = O(N_s^2 N_r^2)$.

Overall complexity   Thus, the overall complexity of ORA algorithm can be summed up as $O(N_s N_r) + O(N_s N_r \log_2 N_r) + O(N_s) + O(N_s^2 N_r^2) = O(N_s^2 N_r^2)$.

4.7 Proof of optimality

In this section, we give a correctness proof of the ORA algorithm. That is, upon termination of the ORA algorithm, the solution (i.e., objective value and the corresponding relay node assignment) is optimal.

Our proof is based on contradiction. Denote ψ as the final solution obtained by the ORA algorithm, with the objective value being $C_{\min}$. For ψ, denote the relay node assigned to source node s_i as $\mathcal{R}_\psi(s_i)$. Conversely, for ψ, denote the source node that uses relay node r_j as $\mathcal{S}_\psi(r_j)$.

Assume that there exists a better solution $\hat{\psi}$ than ψ. That is, the objective value for $\hat{\psi}$, denoted as $\hat{C}_{\min}$, is greater than that for ψ, i.e., $\hat{C}_{\min} > C_{\min}$. For $\hat{\psi}$, denote the relay node assigned to source node s_i as $\mathcal{R}_{\hat{\psi}}(s_i)$, the source node that uses relay node r_j as $\mathcal{S}_{\hat{\psi}}(r_j)$.

The key idea in the proof is to exploit the marking status of the ORA algorithm at the end of the last iteration, which is a nonimproving iteration. Specifically, in the beginning of this last iteration, ORA will select a "bottleneck" source node, which we denote as s_b. ORA will then try to improve the solution by searching for a better relay node for this bottleneck source node. Since the last iteration is a nonimproving iteration, ORA will not find a better solution, and thus will terminate. We will show that $\mathcal{R}_\psi(s_b)$ is not marked at the end of the last iteration of ORA. On the other hand, if there exists a better solution $\hat{\psi}$ than ψ, we will show that $\mathcal{R}_\psi(s_b)$ is marked at the end of the last iteration of ORA. This leads to a contradiction and thus the assumption of the existence of a better solution $\hat{\psi}$ cannot hold. We begin our proof with the following fact:

Fact 4.1

For the bottleneck source node s_b under ψ, its relay node $\mathcal{R}_\psi(s_b)$ is not marked at the end of the last iteration of the ORA algorithm.

Proof. In ORA algorithm, a relay node r_j is marked only if $r_j \neq \emptyset$, $r_j \neq \mathcal{R}_\psi(s_b)$, and r_j is assigned to some source node (see Check_Relay_Availability() in Algorithm 4.1). Thus, $\mathcal{R}_\psi(s_b)$ cannot be marked at the end of the last iteration of the ORA algorithm.   □

Fact 4.1 will be the basis for contradiction in our proof of Theorem 4.1, the main result of this section. Now we present the following three claims, which

Figure 4.5 The sequence of
nodes under analysis in the
proof of optimality.

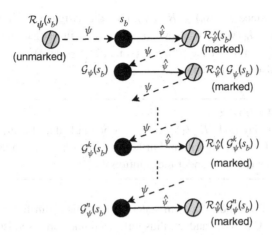

Figure 4.5 The sequence of nodes under analysis in the proof of optimality.

recursively examine relay node assignments under $\hat{\psi}$. First, for the relay node assigned to s_b in $\hat{\psi}$, i.e., $\mathcal{R}_{\hat{\psi}}(s_b)$, we have:

Claim 4.1

Relay node $\mathcal{R}_{\hat{\psi}}(s_b)$ must be marked at the end of the last iteration of the ORA algorithm. Further, it cannot be $\emptyset$ and must be assigned to some source node under solution ψ.

Proof. Since $\hat{\psi}$ is a better solution than ψ, we have $C_R(s_b, \mathcal{R}_{\hat{\psi}}(s_b)) \geq \hat{C}_{\min} > C_{\min}$. Thus, by construction, ORA will consider the availability of relay node $\mathcal{R}_{\hat{\psi}}(s_b)$ for s_b in its last iteration of ψ. Since the ORA algorithm cannot find a better solution in this last iteration, $\mathcal{R}_{\hat{\psi}}(s_b)$ must be marked and then the outcome of checking the availability of $\mathcal{R}_{\hat{\psi}}(s_b)$ shows that it is unavailable. By the "linear marking" process, the mark on $\mathcal{R}_{\hat{\psi}}(s_b)$ will not be cleared throughout the search process in the last iteration. Thus, $\mathcal{R}_{\hat{\psi}}(s_b)$ is marked at the end of the last iteration of the ORA algorithm.

We now prove the second statement by contradiction. If $\mathcal{R}_{\hat{\psi}}(s_b)$ were $\emptyset$, then s_b would choose $\emptyset$ in the last iteration of ORA since it can offer $C_R(s_b, \mathcal{R}_{\hat{\psi}}(s_b)) > C_{\min}$. But this contradicts the fact that we are now in the last iteration of ORA, which is a nonimproving iteration. So $\mathcal{R}_{\hat{\psi}}(s_b)$ cannot be $\emptyset$. Further, since we proved that $\mathcal{R}_{\hat{\psi}}(s_b)$ is marked at the end of the last iteration of the ORA algorithm, it must be assigned to some source node already. $\qquad\square$

By the definition of $\mathcal{S}_\psi(\cdot)$, we have that $\mathcal{R}_{\hat{\psi}}(s_b)$ is assigned to source node $\mathcal{S}_\psi(\mathcal{R}_{\hat{\psi}}(s_b))$ in solution ψ. To simplify the notation, define the function $\mathcal{G}_\psi(\cdot)$ as

$$\mathcal{G}_\psi(\cdot) = \mathcal{S}_\psi(\mathcal{R}_{\hat{\psi}}(\cdot)). \tag{4.8}$$

Thus, relay node $\mathcal{R}_{\hat{\psi}}(s_b)$ is assigned to source node $\mathcal{G}_\psi(s_b)$ in ψ (see Fig. 4.5).

Since $\mathcal{R}_{\hat{\psi}}(s_b) \neq \mathcal{R}_{\psi}(s_b)$, they are assigned to different source nodes in ψ, i.e., $\mathcal{G}_{\psi}(s_b) \neq s_b$. Now we recursively investigate the relay node assigned to $\mathcal{G}_{\psi}(s_b)$ under the solution $\hat{\psi}$, i.e., $\mathcal{R}_{\hat{\psi}}(\mathcal{G}_{\psi}(s_b))$. We have the following claim (also see Fig. 4.5):

Claim 4.2

Relay node $\mathcal{R}_{\hat{\psi}}(\mathcal{G}_{\psi}(s_b))$ must be marked at the end of the last iteration of the ORA algorithm. Further, it cannot be $\emptyset$ and must be assigned to some source node under the solution ψ.

The proofs for both statements in this claim follow the same token as that for Claim 4.1 and are thus left as a homework exercise.

Again, by the definition of $\mathcal{S}_{\psi}(\cdot)$, we have that the relay node $\mathcal{R}_{\hat{\psi}}(\mathcal{G}_{\psi}(s_b))$ is assigned to source node $\mathcal{S}_{\psi}(\mathcal{R}_{\hat{\psi}}(\mathcal{G}_{\psi}(s_b)))$ in solution ψ. By (4.8), we have source node $\mathcal{S}_{\psi}(\mathcal{R}_{\hat{\psi}}(\mathcal{G}_{\psi}(s_b))) = \mathcal{G}_{\psi}(\mathcal{G}_{\psi}(s_b))$. To simplify the notation, we define the function $\mathcal{G}_{\psi}^2(\cdot)$ as

$$\mathcal{G}_{\psi}^2(\cdot) = \mathcal{G}_{\psi}(\mathcal{G}_{\psi}(\cdot)).$$

Thus, the relay node $\mathcal{R}_{\hat{\psi}}(\mathcal{G}_{\psi}(s_b))$ is assigned to source node $\mathcal{G}_{\psi}^2(s_b)$ in ψ. Now we have two cases: the source node $\mathcal{G}_{\psi}^2(s_b)$ may or may not be a node in $\{s_b, \mathcal{G}_{\psi}(s_b)\}$. If the source node $\mathcal{G}_{\psi}^2(s_b)$ is a node in $\{s_b, \mathcal{G}_{\psi}(s_b)\}$, then we terminate our recursive procedure. Otherwise, we continue to consider its relay node in $\hat{\psi}$.

In general, we use the following notation:

$$\mathcal{G}_{\psi}^0(s_b) = s_b$$

$$\mathcal{G}_{\psi}^k(s_b) = \mathcal{G}_{\psi}(\mathcal{G}_{\psi}^{k-1}(s_b)) \quad (k = 1, 2, \ldots). \tag{4.9}$$

Since the number of source nodes is finite, our recursive procedure will terminate in a finite number of steps. Suppose that we terminate at $k = n$.

Following the same token for Claims 4.1 and 4.2, we can obtain a similar claim for each of the relay nodes $\mathcal{R}_{\hat{\psi}}(\mathcal{G}_{\psi}^2(s_b))$, $\mathcal{R}_{\hat{\psi}}(\mathcal{G}_{\psi}^3(s_b))$, $\ldots, \mathcal{R}_{\hat{\psi}}(\mathcal{G}_{\psi}^k(s_b)), \ldots, \mathcal{R}_{\hat{\psi}}(\mathcal{G}_{\psi}^n(s_b))$ (see Fig. 4.5). Thus, we can generalize the statements in Claims 4.1 and 4.2 for relay node $\mathcal{R}_{\hat{\psi}}(\mathcal{G}_{\psi}^k(s_b))$ and have the following claim:

Claim 4.3

Relay node $\mathcal{R}_{\hat{\psi}}(\mathcal{G}_{\psi}^k(s_b))$ must be marked at the end of the last iteration of the ORA algorithm. Further, it cannot be $\emptyset$ and must be assigned to some source node under solution ψ, $k = 0, 1, 2, \ldots, n$.

Proof. Since $\hat{\psi}$ is a better solution than ψ, we have $C_R(\mathcal{G}_\psi^k(s_b),$ $\mathcal{R}_{\hat{\psi}}(\mathcal{G}_\psi^k(s_b))) \geq \hat{C}_{\min} > C_{\min}$. Note that $\mathcal{G}_\psi^k(s_b)$ is some source node in the solution ψ obtained by ORA, whereas $\mathcal{R}_{\hat{\psi}}(\mathcal{G}_\psi^k(s_b))$ is the relay node assigned to this source node in the hypothetical better solution $\hat{\psi}$. Our goal is to show that ORA should have marked this relay node in its last iteration.

Since $C_R(\mathcal{G}_\psi^k(s_b), \mathcal{R}_{\hat{\psi}}(\mathcal{G}_\psi^k(s_b))) > C_{\min}$ and $\mathcal{R}_{\hat{\psi}}(\mathcal{G}_\psi^k(s_b))$ is not assigned to $\mathcal{G}_\psi^k(s_b)$ in the last iteration of ORA, then by construction of ORA, ORA must have checked the availability of $\mathcal{R}_{\hat{\psi}}(\mathcal{G}_\psi^k(s_b))$ for $\mathcal{G}_\psi^k(s_b)$ during the last iteration, then marked it, and then determined that it is unavailable for $\mathcal{G}_\psi^k(s_b)$. Moreover, due to the "linear marking" process, this mark on $\mathcal{R}_{\hat{\psi}}(\mathcal{G}_\psi^k(s_b))$ should remain after the last iteration of ORA. Thus, we conclude that $\mathcal{R}_{\hat{\psi}}(\mathcal{G}_\psi^k(s_b))$ is marked at the end of the last iteration of the ORA algorithm.

We now prove the second statement by contradiction. If $\mathcal{R}_{\hat{\psi}}(\mathcal{G}_\psi^k(s_b))$ is $\emptyset$, then $\mathcal{G}_\psi^k(s_b)$ will choose $\emptyset$ in the last iteration since it can offer $C_R(\mathcal{G}_\psi^k(s_b), \mathcal{R}_{\hat{\psi}}(\mathcal{G}_\psi^k(s_b)) > C_{\min}$, and finally s_b will be able to get a better relay node. But this contradicts the fact that this last iteration is a nonimproving iteration. So, $\mathcal{R}_{\hat{\psi}}(\mathcal{G}_\psi^k(s_b))$ cannot be $\emptyset$. Further, since we proved that $\mathcal{R}_{\hat{\psi}}(\mathcal{G}_\psi^k(s_b))$ is marked at the end of the last iteration of the ORA algorithm, it must be assigned to some source node already. $\square$

Referring to Fig. 4.5, we have Claim 4.3 for a set of relay nodes $\mathcal{R}_{\hat{\psi}}(s_b), \mathcal{R}_{\hat{\psi}}(\mathcal{G}_\psi(s_b)), \ldots, \mathcal{R}_{\hat{\psi}}(\mathcal{G}_\psi^n(s_b))$. Our recursive procedure terminates at $\mathcal{R}_{\hat{\psi}}(\mathcal{G}_\psi^n(s_b))$ because its assigned source node in solution ψ is a node in $\{s_b, \mathcal{G}_\psi(s_b), \ldots, \mathcal{G}_\psi^n(s_b)\}$. We are now ready to prove the following theorem, which is the main result of this section:

Theorem 4.1

Upon termination of the ORA algorithm, the obtained solution ψ is optimal.

Proof. Suppose that the solution ψ obtained by the ORA algorithm is not optimal. Then, there exists a better solution $\hat{\psi}$. We have Fact 4.1 and Claim 4.3. We will show a contradiction.

Under Claim 4.3, we proved that the relay node $\mathcal{R}_{\hat{\psi}}(\mathcal{G}_\psi^n(s_b))$ is assigned to a source node in the solution ψ obtained by ORA. Since our recursive procedure terminates at $\mathcal{R}_{\hat{\psi}}(\mathcal{G}_\psi^n(s_b))$, its assigned source node in solution ψ is a node in $\{s_b, \mathcal{G}_\psi(s_b), \ldots, \mathcal{G}_\psi^n(s_b)\}$. But we also know that under ψ, the source nodes $\mathcal{G}_\psi(s_b), \mathcal{G}_\psi^2(s_b), \mathcal{G}_\psi^3(s_b), \ldots, \mathcal{G}_\psi^n(s_b)$ have relay nodes $\mathcal{R}_{\hat{\psi}}(s_b), \mathcal{R}_{\hat{\psi}}(\mathcal{G}_\psi(s_b)),$ $\mathcal{R}_{\hat{\psi}}(\mathcal{G}_\psi^2(s_b)), \ldots, \mathcal{R}_{\hat{\psi}}(\mathcal{G}_\psi^{n-1}(s_b))$, respectively. Thus, $\mathcal{R}_{\hat{\psi}}(\mathcal{G}_\psi^n(s_b))$ can only be assigned to s_b in solution ψ. On the other hand, the relay node assigned to s_b in the solution ψ is denoted by $\mathcal{R}_\psi(s_b)$. Therefore, we must have $\mathcal{R}_{\hat{\psi}}(\mathcal{G}_\psi^n(s_b)) = \mathcal{R}_\psi(s_b)$.

Now, Claim 4.3 states that $\mathcal{R}_{\hat{\psi}}(\mathcal{G}_{\psi}^n(s_b))$ must be marked after the last iteration, whereas Fact 4.1 states that the relay node assigned to the bottleneck source node, i.e., $\mathcal{R}_{\psi}(s_b)$, cannot be marked. Since both $\mathcal{R}_{\psi}(s_b)$ and $\mathcal{R}_{\hat{\psi}}(\mathcal{G}_{\psi}^n(s_b))$ are the same relay node, we have a contradiction. Thus, our assumption that there exists a better solution $\hat{\psi}$ than ψ does not hold. $\square$

Note that the proof of Theorem 4.1 does not depend on the initial assignment in the ORA algorithm. So we have the following corollary for the ORA algorithm:

Corollary 4.1
Under any feasible initial relay node assignment, the ORA algorithm can find an optimal relay node assignment.

4.8 Numerical examples

In this section, we present some numerical results to illustrate some properties of the ORA algorithm.

4.8.1 Simulation setting

We consider a 100-node ad hoc network. The location of each node is given in Table 4.4. For this network, we consider both the cases of $N_r \geq N_s$ and $N_r < N_s$. In the first case, we have 30 source–destination pairs and 40 relay nodes. While in the second case, we have 40 source–destination pairs and 20 relay nodes. The role of each node (either as a source, destination, or relay) for each case is shown in Figs. 4.6 and 4.10, respectively, with details given in Table 4.4.

For the simulations, we assume $W = 10$ MHz bandwidth for each channel. The maximum transmission power at each node is set to 1 W. Each relay node employs AF for CC. We assume that h_{sd} only includes the path-loss component between nodes s and d and is given by $|h_{sd}|^2 = ||s - d||^{-4}$, where $||s - d||$ is the distance (in meters) between these two nodes and 4 is the path-loss index. For the AWGN channel, we assume the variance of noise is 10^{-10} W at all nodes.

4.8.2 Results

Case 1: $N_r \geq N_s$. In this case (see Fig. 4.6), we have 30 source–destination pairs and 40 relay nodes.

Under ORA, after preprocessing, we start with an initial relay node assignment in the first iteration. Such an initial assignment is not unique. But regardless of the initial relay node assignment, we expect the objective value

Table 4.4 Locations and roles of all the nodes in the network.

Location	Role Case 1	Case 2	Location	Role Case 1	Case 2	Location	Role Case 1	Case 2
$(75, 500)$	s_1	s_1	$(220, 190)$	d_4	d_4	$(380, 370)$	r_7	s_{31}
$(170, 430)$	s_2	s_2	$(660, 190)$	d_5	d_5	$(300, 350)$	r_8	r_8
$(170, 500)$	s_3	s_3	$(430, 630)$	d_6	d_6	$(410, 650)$	r_9	s_{33}
$(250, 650)$	s_4	s_4	$(180, 620)$	d_7	d_7	$(470, 500)$	r_{10}	d_{40}
$(400, 550)$	s_5	s_5	$(750, 625)$	d_8	d_8	$(660, 525)$	r_{11}	s_{39}
$(340, 230)$	s_6	s_6	$(310, 480)$	d_9	d_9	$(600, 425)$	r_{12}	s_{40}
$(390, 150)$	s_7	s_7	$(1100, 180)$	d_{10}	d_{10}	$(510, 200)$	r_{13}	s_{38}
$(460, 280)$	s_8	s_8	$(1110, 360)$	d_{11}	d_{11}	$(575, 325)$	r_{14}	r_{14}
$(700, 500)$	s_9	s_9	$(875, 600)$	d_{12}	d_{12}	$(750, 560)$	r_{15}	r_{15}
$(750, 360)$	s_{10}	s_{10}	$(700, 300)$	d_{13}	d_{13}	$(800, 360)$	r_{16}	r_{16}
$(800, 90)$	s_{11}	s_{11}	$(650, 550)$	d_{14}	d_{14}	$(860, 260)$	r_{17}	r_{17}
$(900, 160)$	s_{12}	s_{12}	$(740, 170)$	d_{15}	d_{15}	$(980, 450)$	r_{18}	r_{18}
$(1125, 300)$	s_{13}	s_{13}	$(410, 810)$	d_{16}	d_{16}	$(950, 310)$	r_{19}	r_{19}
$(1000, 340)$	s_{14}	s_{14}	$(550, 1100)$	d_{17}	d_{17}	$(950, 200)$	r_{20}	d_{37}
$(1025, 540)$	s_{15}	s_{15}	$(150, 790)$	d_{18}	d_{18}	$(100, 1000)$	r_{21}	s_{32}
$(100, 1120)$	s_{16}	s_{16}	$(210, 1110)$	d_{19}	d_{19}	$(310, 980)$	r_{22}	r_{12}
$(150, 920)$	s_{17}	s_{17}	$(530, 720)$	d_{20}	d_{20}	$(250, 800)$	r_{23}	d_{32}
$(330, 1110)$	s_{18}	s_{18}	$(800, 1140)$	d_{21}	d_{21}	$(460, 1010)$	r_{24}	r_{13}
$(450, 890)$	s_{19}	s_{19}	$(1080, 1100)$	d_{22}	d_{22}	$(610, 930)$	r_{25}	d_{34}
$(650, 1050)$	s_{20}	s_{20}	$(940, 790)$	d_{23}	d_{23}	$(680, 760)$	r_{26}	s_{34}
$(700, 640)$	s_{21}	s_{21}	$(1360, 640)$	d_{24}	d_{24}	$(700, 900)$	r_{27}	r_{20}
$(820, 880)$	s_{22}	s_{22}	$(1280, 1120)$	d_{25}	d_{25}	$(910, 1120)$	r_{28}	d_{35}
$(1150, 1060)$	s_{23}	s_{23}	$(1260, 350)$	d_{26}	d_{26}	$(970, 970)$	r_{29}	s_{35}
$(1480, 1120)$	s_{24}	s_{24}	$(1500, 50)$	d_{27}	d_{27}	$(1360, 910)$	r_{30}	r_9
$(1160, 720)$	s_{25}	s_{25}	$(1450, 605)$	d_{28}	d_{28}	$(1200, 920)$	r_{31}	r_{11}
$(1050, 50)$	s_{26}	s_{26}	$(1030, 910)$	d_{29}	d_{29}	$(1250, 690)$	r_{32}	d_{36}
$(1350, 450)$	s_{27}	s_{27}	$(1150, 230)$	d_{30}	d_{30}	$(1290, 180)$	r_{33}	r_{10}
$(1380, 110)$	s_{28}	s_{28}	$(80, 370)$	r_1	d_{31}	$(150, 360)$	r_{34}	r_5
$(1500, 800)$	s_{29}	s_{29}	$(110, 280)$	r_2	r_2	$(1380, 380)$	r_{35}	r_7
$(1500, 300)$	s_{30}	s_{30}	$(160, 300)$	r_3	r_3	$(1220, 60)$	r_{36}	s_{37}
$(200, 50)$	d_1	d_1	$(280, 520)$	r_4	r_4	$(1190, 510)$	r_{37}	s_{36}
$(520, 240)$	d_2	d_2	$(375, 580)$	r_5	d_{39}	$(500, 40)$	r_{38}	d_{38}
$(40, 100)$	d_3	d_3	$(385, 450)$	r_6	r_6	$(50, 805)$	r_{39}	d_{33}
						$(1510, 920)$	r_{40}	r_1

$C_{\min}$ to converge to the optimum (Corollary 4.1). To validate this result, we show the results of running the ORA algorithm under two different initial relay node assignments, denoted as I and II (see Table 4.5).

In Table 4.5, the second column shows the data rate for each source–destination pair under direct transmissions. Note that the minimum rate among all pairs is 1.83 Mbps, which is associated with s_7. The third to fifth columns are results under the initial relay node assignment I and the sixth to eighth columns are results under the initial relay node assignment II. The symbol $\emptyset$ denotes direct transmission. Note that initial relay node assignments I and II are different. As a result, the final assignment is different under I and II.

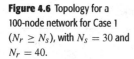

Figure 4.6 Topology for a 100-node network for Case 1 ($N_r \geq N_s$), with $N_s = 30$ and $N_r = 40$.

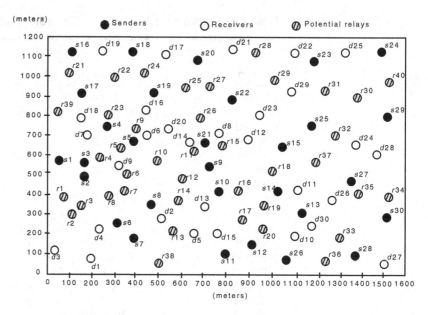

However, the final objective value (i.e., C_{min}) under I and II is identical (4.43 Mbps).

Fig. 4.7 shows the objective value C_{min} at each iteration under initial relay node assignments I and II. Under either initial assignments, C_{min} is a nondecreasing function of iteration number. Note that a higher initial value of C_{min} does not mean that ORA will converge faster.

Case 2: $N_r < N_s$. In this case (see Fig. 4.8), we have 40 source–destination pairs and 20 relay nodes.

Table 4.6 shows the results of this case under two different initial relay node assignments I and II. The second column in Table 4.6 lists the data rate under direct transmissions for each source–destination pair. As discussed at the end of Section 4.6.2, for the case of $N_r < N_s$, it is only necessary to consider relay node assignments for $N_r = 20$ source nodes corresponding to the 20 smallest data rates, i.e., nodes s_1, s_3, s_4, s_5, s_7, s_8, s_{11}, s_{12}, s_{13}, s_{14}, s_{15}, s_{16}, s_{17}, s_{21}, s_{24}, s_{25}, s_{27}, s_{28}, and s_{29}.

Again in Table 4.6, the objective value C_{min} is identical (3.80 Mbps) regardless of different initial relay node assignments (I and II). Note that despite the different final relay node assignments under I and II, the objective value C_{min} is identical.

Fig. 4.9 shows the objective value C_{min} at each iteration under initial relay node assignments I and II. Again, we observe that in Fig. 4.9, C_{min} is a nondecreasing function of iteration number under both initial relay node assignments I and II.

Significance of preprocessing   Now we use a set of numerical results to show the significance of preprocessing in our ORA algorithm. We consider the same network in Fig. 4.6 with 30 source–destination pairs and 40 relay

Table 4.5 Optimal assignments for Case 1 ($N_r \geq N_s$) under two different initial relay node assignments.

Session	C_D (Mbps)	Relay assignment I			Relay assignment II		
		Initial	Final	Final rate (Mbps)	Initial	Final	Final rate (Mbps)
s_1	2.62	$\emptyset$	r_3	6.54	r_3	r_3	6.54
s_2	4.60	r_8	r_7	9.46	r_8	r_7	9.46
s_3	3.81	$\emptyset$	r_2	8.73	r_1	r_1	7.21
s_4	2.75	$\emptyset$	r_4	4.66	r_4	r_4	4.66
s_5	3.15	$\emptyset$	r_{14}	6.47	r_7	r_{14}	6.47
s_6	4.17	$\emptyset$	r_6	9.25	r_{10}	r_6	9.25
s_7	$\boxed{1.83}$	r_6	r_8	4.76	r_6	r_8	4.76
s_8	2.99	$\emptyset$	r_{12}	7.22	r_{16}	r_{12}	7.22
s_9	4.92	r_{12}	r_{10}	9.81	r_{12}	r_{10}	9.81
s_{10}	4.80	r_{18}	$\emptyset$	4.80	$\emptyset$	$\emptyset$	4.80
s_{11}	4.13	r_{16}	r_{20}	9.13	r_{17}	r_{20}	9.13
s_{12}	3.23	$\emptyset$	r_{19}	5.89	r_{18}	r_{18}	5.55
s_{13}	3.68	$\emptyset$	r_{18}	4.84	r_{19}	r_{17}	7.32
s_{14}	4.23	$\emptyset$	r_{16}	7.87	r_{15}	r_{15}	5.29
s_{15}	2.62	r_{17}	r_{17}	4.86	r_{20}	r_{19}	5.84
s_{16}	3.30	$\emptyset$	r_{22}	7.29	r_{22}	r_{22}	7.29
s_{17}	4.17	$\emptyset$	r_{24}	5.62	r_{24}	r_{24}	5.62
s_{18}	6.03	r_{21}	r_{21}	7.37	r_{23}	r_{23}	6.26
s_{19}	8.76	$\emptyset$	$\emptyset$	8.76	$\emptyset$	$\emptyset$	8.76
s_{20}	6.95	$\emptyset$	$\emptyset$	6.95	$\emptyset$	$\emptyset$	6.95
s_{21}	1.90	r_{27}	r_{27}	4.90	r_{27}	r_{27}	4.90
s_{22}	7.65	r_{28}	r_{28}	8.71	r_{28}	r_{28}	8.71
s_{23}	7.55	r_{29}	r_{29}	11.26	r_{29}	r_{28}	11.26
s_{24}	2.12	r_{40}	r_{40}	$\boxed{4.43}$	$\emptyset$	r_{40}	$\boxed{4.43}$
s_{25}	3.90	$\emptyset$	r_{30}	5.87	$\emptyset$	r_{30}	5.87
s_{26}	6.08	r_{36}	r_{36}	6.81	r_{36}	r_{36}	6.81
s_{27}	3.61	$\emptyset$	r_{34}	5.44	$\emptyset$	r_{34}	5.44
s_{28}	2.04	r_{35}	r_{35}	5.29	$\emptyset$	r_{35}	5.29
s_{29}	2.32	r_{30}	r_{31}	4.68	$\emptyset$	r_{31}	4.68
s_{30}	6.60	r_{34}	r_{33}	9.65	r_{34}	r_{33}	9.65

nodes, but we remove the preprocessing step in the ORA algorithm. As an example, the third column of Table 4.7 shows an initial assignment without first going through the preprocessing step. Although the objective value C_{min} also reaches the same optimal value (4.43 Mbps) as that in Table 4.5, the final data rate for some nonbottleneck source nodes could be worse than under direct transmissions. For example, for s_{19}, its final rate is 4.81 Mbps, which is less than the direct transmission rate (8.76 Mbps). Such an event is undetectable without the preprocessing step, as 4.81 Mbps is still greater than the optimal objective value (4.43 Mbps).

Figure 4.7 Case 1 ($N_r \geq N_s$): The objective value $C_{\min}$ at each iteration of the ORA algorithm under two different initial relay node assignments.

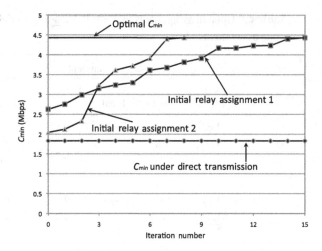

Figure 4.8 Topology for a 100-node network for Case 2 ($N_r < N_s$), with $N_s = 40$ and $N_r = 20$.

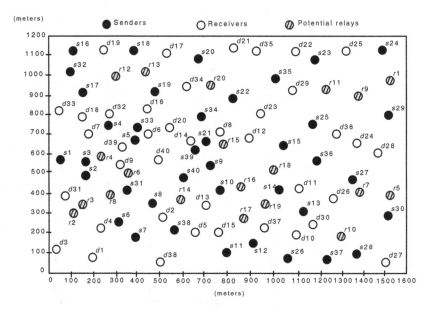

On the other hand, when the preprocessing step is employed, ORA can ensure that the final rate for each source–destination pair is no less than that under direct transmission, as shown in Table 4.5.

4.9 Chapter summary

This chapter illustrates how an optimization problem can be solved following algorithmic techniques from computer science (CS). For certain problems, general optimization methods from operations research (OR) may not always be the best approach. In fact, a formulation following OR's optimiza-

Table 4.6 Optimal assignments for Case 2 (N_r ¡ N_s) under two different initial relay node assignments.

Session	C_D (Mbps)	Relay assignment I			Relay assignment II		
		Initial	Final	Final rate (Mbps)	Initial	Final	Final rate (Mbps)
s_1	2.62	$\emptyset$	r_2	6.62	r_3	r_3	6.54
s_2	2.60	$\emptyset$	$\emptyset$	4.60	$\emptyset$	$\emptyset$	4.60
s_3	3.81	$\emptyset$	$\emptyset$	3.81	$\emptyset$	$\emptyset$	3.81
s_4	2.75	$\emptyset$	r_5	4.66	r_8	r_5	5.20
s_5	3.15	$\emptyset$	r_6	3.80	r_{14}	r_6	3.80
s_6	4.17	$\emptyset$	$\emptyset$	4.17	$\emptyset$	$\emptyset$	4.16
s_7	1.83	$\emptyset$	r_8	4.76	r_6	r_8	4.76
s_8	2.99	$\emptyset$	r_{14}	4.43	$\emptyset$	r_{14}	4.43
s_9	4.92	$\emptyset$	$\emptyset$	4.92	$\emptyset$	$\emptyset$	4.92
s_{10}	4.80	$\emptyset$	$\emptyset$	4.80	$\emptyset$	$\emptyset$	4.80
s_{11}	4.13	$\emptyset$	$\emptyset$	4.13	$\emptyset$	$\emptyset$	4.13
s_{12}	3.23	$\emptyset$	r_{18}	5.55	$\emptyset$	r_{18}	5.55
s_{13}	3.68	$\emptyset$	r_{19}	8.04	$\emptyset$	r_{19}	8.04
s_{14}	4.23	$\emptyset$	$\emptyset$	4.23	$\emptyset$	$\emptyset$	4.23
s_{15}	2.62	$\emptyset$	r_{16}	5.60	$\emptyset$	r_{16}	5.60
s_{16}	3.30	$\emptyset$	r_{12}	7.30	$\emptyset$	r_{12}	7.30
s_{17}	4.17	$\emptyset$	$\emptyset$	4.17	$\emptyset$	$\emptyset$	4.17
s_{18}	6.03	$\emptyset$	$\emptyset$	6.03	r_{13}	$\emptyset$	6.03
s_{19}	8.76	$\emptyset$	$\emptyset$	8.76	r_{12}	r_{13}	8.97
s_{20}	6.95	$\emptyset$	$\emptyset$	6.95	r_{20}	$\emptyset$	6.95
s_{21}	1.90	$\emptyset$	r_{20}	4.90	$\emptyset$	r_{20}	4.90
s_{22}	7.65	$\emptyset$	$\emptyset$	7.65	$\emptyset$	$\emptyset$	7.65
s_{23}	7.55	$\emptyset$	$\emptyset$	7.55	r_{11}	$\emptyset$	7.55
s_{24}	2.12	$\emptyset$	r_9	5.15	$\emptyset$	r_9	5.15
s_{25}	3.91	$\emptyset$	$\emptyset$	3.91	$\emptyset$	$\emptyset$	3.91
s_{26}	6.08	$\emptyset$	$\emptyset$	6.08	r_{10}	$\emptyset$	6.08
s_{27}	3.61	$\emptyset$	r_{10}	5.27	$\emptyset$	r_{10}	5.27
s_{28}	2.04	$\emptyset$	r_7	5.29	$\emptyset$	r_7	5.29
s_{29}	2.32	$\emptyset$	r_{11}	4.68	$\emptyset$	r_{11}	4.68
s_{30}	6.60	$\emptyset$	$\emptyset$	6.60	$\emptyset$	$\emptyset$	6.60
s_{31}	11.06	$\emptyset$	$\emptyset$	11.06	$\emptyset$	$\emptyset$	11.06
s_{32}	17.47	$\emptyset$	$\emptyset$	17.47	$\emptyset$	$\emptyset$	17.47
s_{33}	4.86	$\emptyset$	$\emptyset$	4.86	$\emptyset$	$\emptyset$	4.86
s_{34}	31.34	$\emptyset$	$\emptyset$	31.34	$\emptyset$	$\emptyset$	31.34
s_{35}	37.87	$\emptyset$	$\emptyset$	37.87	$\emptyset$	$\emptyset$	37.87
s_{36}	29.79	$\emptyset$	$\emptyset$	29.79	$\emptyset$	$\emptyset$	29.79
s_{37}	10.65	$\emptyset$	$\emptyset$	10.65	$\emptyset$	$\emptyset$	10.65
s_{38}	38.27	$\emptyset$	$\emptyset$	38.27	$\emptyset$	$\emptyset$	38.27
s_{39}	12.10	$\emptyset$	$\emptyset$	12.10	$\emptyset$	$\emptyset$	12.10
s_{40}	41.70	$\emptyset$	$\emptyset$	41.70	$\emptyset$	$\emptyset$	41.70

Figure 4.9 Case 2 ($N_r < N_s$): The objective value C_{min} at each iteration of the ORA algorithm under two different initial node assignments.

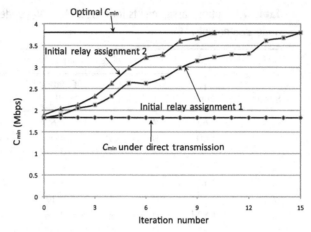

Table 4.7 An example illustrating the importance of preprocessing.

Sender	C_D (Mbps)	Without preprocessing		Final rate (Mbps)
		Initial	Final	
s_1	2.62	r_3	r_3	6.54
s_2	4.60	$\emptyset$	$\emptyset$	4.60
s_3	3.81	$\emptyset$	r_2	8.73
s_4	2.75	r_8	r_4	4.66
s_5	3.15	r_{14}	r_{14}	6.47
s_6	4.17	$\emptyset$	r_6	9.25
s_7	1.83	r_6	r_8	4.76
s_8	2.99	$\emptyset$	r_{12}	7.22
s_9	4.92	$\emptyset$	$\emptyset$	4.92
s_{10}	4.80	$\emptyset$	$\emptyset$	4.80
s_{11}	4.13	$\emptyset$	r_{20}	9.13
s_{12}	3.24	$\emptyset$	r_{18}	5.55
s_{13}	3.68	$\emptyset$	r_{17}	7.32
s_{14}	4.23	$\emptyset$	r_{16}	7.87
s_{15}	2.62	$\emptyset$	r_{19}	5.84
s_{16}	3.30	$\emptyset$	r_{22}	7.30
s_{17}	4.17	$\emptyset$	r_{24}	5.62
s_{18}	6.03	r_{23}	r_{21}	7.37
s_{19}	**8.76**	r_{39}	r_{39}	**4.81**
s_{20}	6.95	r_{26}	r_{26}	7.25
s_{21}	1.90	$\emptyset$	r_{27}	4.90
s_{22}	7.65	r_{28}	r_{28}	8.71
s_{23}	7.55	r_{29}	r_{29}	11.26
s_{24}	2.12	$\emptyset$	r_{40}	4.43
s_{25}	3.91	$\emptyset$	r_{30}	5.87
s_{26}	6.08	r_{33}	r_{33}	7.55
s_{27}	3.61	$\emptyset$	r_{34}	5.45
s_{28}	2.04	$\emptyset$	r_{35}	5.29
s_{29}	2.33	$\emptyset$	r_{31}	4.68
s_{30}	6.60	$\emptyset$	$\emptyset$	6.60

tion approach may lead to a solution with nonpolynomial-time complexity. But a solution with non-polynomial time complexity does not mean that the problem is not in P. In fact, we may well develop a different algorithm to solve the problem with polynomial-time complexity. This is what we have illustrated in this chapter, where we developed a polynomial-time algorithm.

In our case study, we considered a relay node assignment problem in CC. Our objective is to assign a set of available relay nodes to different source–destination pairs so as to maximize the minimum data rate among all the pairs. Following the OR optimization approach, we showed that the problem can be formulated as an MILP, which is NP-hard in general. But this does not mean that the problem is NP-hard. Instead, by following a CS algorithm design approach, we developed a polynomial-time exact algorithm for this problem. A novel idea in this algorithm is a linear marking mechanism, which is able to achieve polynomial-time complexity at each iteration. We gave a formal proof of the optimality of the algorithm.

This chapter concludes the first part of this book, which is on developing optimal solutions.

4.10 Problems

4.1   Can a generally NP-hard formulation for a problem be used to claim that the problem is NP-hard? Why?

4.2   Verify Eq. (4.6) in this chapter.

4.3   Show that $I_{AF}(SNR_{sd}, SNR_{sr}, SNR_{rd})$ is an increasing function of P_s and P_r, respectively. Do the same for $I_{DF}(SNR_{sd}, SNR_{sr}, SNR_{rd})$.

4.4   Consider the simple three-node relay channel in Fig. 4.1. The locations of the source node s and the destination node d are at (0, 0) and (100, 0), respectively, all in meters. The maximum transmit power of both source and relay nodes is 0.1 W. The channel coefficient between two nodes is $h = 2d^{-2}$, where d is the distance between the two nodes. The noise power spectrum density (PSD) at all nodes is 2×10^{-16} W/Hz and the channel bandwidth is 10 MHz.
(a) Calculate the achievable rate under direct transmission between s and d.
(b) If the location of the relay node r is at (50, 33) (in meters), calculate the achievable rates under AF and DF, respectively.
(c) If the location of the relay node r is at (80, 50) (in meter), calculate the achievable rates under AF and DF, respectively.

4.5   If DF is employed for CC, how will it affect the operation of ORA? Justify your answer.

Table 4.8 Initial relay node assignment for a CC network with four sessions and six relay nodes.

	Ø	r_1	r_2	r_3	r_4	r_5	r_6
s_1	4	17	8	15	5	<u>8</u>	15
s_2	6	<u>12</u>	8	16	14	8	13
s_3	3	15	<u>6</u>	21	13	14	13
s_4	11	19	15	<u>12</u>	16	9	15

Table 4.9 Initial relay node assignment for a CC network with seven sessions and seven relay nodes.

	Ø	r_1	r_2	r_3	r_4	r_5	r_6	r_7
s_1	1	17	6	<u>15</u>	10	20	11	15
s_2	6	11	6	17	<u>9</u>	18	10	13
s_3	6	5	18	20	16	18	15	<u>12</u>
s_4	4	18	12	9	6	15	<u>19</u>	11
s_5	<u>15</u>	19	4	15	16	19	8	15
s_6	9	6	<u>11</u>	16	15	18	11	15
s_7	6	11	19	5	7	<u>9</u>	18	15

4.6 In practice, the number of relay nodes may or may not be less than the number of source nodes, i.e., $N_r < N_s$ or $N_r \geq N_s$. How will this affect the operation of the ORA algorithm?

4.7 Describe the marking mechanism in the ORA algorithm. Why do we call it "linear"? Explain its role in the algorithm's complexity.

4.8 For the naive marking mechanism described in Section 4.6.3, show that the number of checks for relay nodes within an iteration is $O(N_r \cdot N_r!)$ if $N_r < N_s$ and $O(\frac{N_r \cdot N_r!}{(N_r - N_s + 1)!})$ if $N_r \geq N_s$.

4.9 For Example 4.1 in Section 4.6.2, give the details of how ORA and linear marking work in the second and third iterations, respectively.

4.10 Apply the ORA algorithm to find an optimal solution for the data given in Table 4.8. Show your work for each iteration.

4.11 Apply the ORA algorithm to find an optimal solution for the data given in Table 4.9. Show your work for each iteration.

4.12 Give a complete proof of Claim 4.2.

4.13 For any given relay node assignment with an objective value $C_{\min}$, we can define a directional graph as follows (see Fig. 4.10 as an example). The set of nodes in the graph is $\mathcal{N}_s \bigcup \mathcal{N}_r \bigcup \{S, D, \emptyset\}$, where S, D, and $\emptyset$ are dummy source, destination, and relay nodes, respectively. The graph has the following

Figure 4.10 A sample for the
construct graph in Problem 4.13.

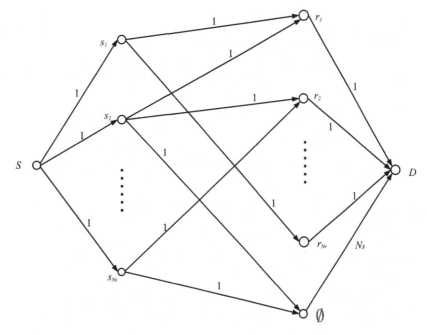

edges: (i) edges (S, s_i) with unit capacity, for each $s_i \in \mathcal{N}_s$; (ii) edges (s_i, r_j) with unit capacity, for each $s_i \in \mathcal{N}_s, r_j \in \mathcal{N}_r, C_R(s_i, r_j) > C_{\min}$, and edges $(s_i, \emptyset)$ with unit capacity, for each $s_i \in \mathcal{N}_s, C_R(s_i, \emptyset) > C_{\min}$; and (iii) edges (r_j, D) with unit capacity, for each $r_j \in \mathcal{N}_r$, and an edge $(\emptyset, D)$ with capacity N_s. Show how such a max-flow problem for the defined graph is related to the problem of identifying an improving solution.

4.14   In each iteration of the ORA algorithm, we try to find a better solution. Denote $\mathcal{C}$ as the set of all possible values for the maximum $C_{\min}$ value in an optimal solution. Then in the worst case, ORA decreases $|\mathcal{C}|$ by one in an iteration. We can improve the ORA algorithm as follows. In each iteration, instead of just finding a better solution, we identify the median value C_{med} for the current set $\mathcal{C}$ and try to find a solution with an objective value of at least C_{med}. We may or may not find such a solution. But for either case, we can decrease $|\mathcal{C}|$ to $\frac{1}{2}|\mathcal{C}|$ and thus the number of iterations can be significantly decreased. Give a detailed description of this enhanced ORA (EORA) algorithm and analyze its complexity. (Hint: The complexity of one iteration in EORA is higher than that for ORA, since it may be necessary to find better relay nodes for multiple sources in an iteration of EORA.)

II

Methods for Near-optimal and Approximation Solutions

5 Branch-and-bound framework and application

Vision is not enough, it must be combined with venture. It is not enough to stare up the steps, we must step up the stairs.

Vaclav Havel

5.1 Review of branch-and-bound framework

In this chapter, we present a general and effective approach to solve NP-hard or NP-complete problems, despite their potential exponential worst-case complexity. Although the complexity of this approach is nonpolynomial, when designed appropriately, this approach may still be a viable approach to solve many problems that arise in practice, particularly when their problem sizes are not terribly large.

The approach that we describe is called *branch-and-bound* [113]. It is a powerful general-purpose framework to solve nonconvex programming problems. Such a framework aims to obtain a $(1 - \varepsilon)$-optimal solution for a small given $\varepsilon \geq 0$. Obviously, the smaller ε is, the higher is the complexity. Thus, we need to select a suitable ε based on the optimality requirement. The efficiency of the algorithm depends on its ability to correctly remove large portions of the solution space during each iterative step within the process of identifying the best solution. As a result, we progressively focus on smaller and smaller portions of the solution space and thus we are able to find a $(1 - \varepsilon)$-optimal solution much faster than a brute-force exhaustive search.

We now describe the branch-and-bound framework for a minimization problem (the procedure for a maximization problem is similar). Initially, we need to determine the solution space for our problem, including the set of feasible values for each of the optimization variables. The framework consists of two steps, namely, the *bounding* step and the *branching* step.

Bounding step   The goal of this step is to develop a lower bound and an upper bound for a particular problem. First of all, by using some *relaxation* techniques, branch-and-bound obtains a relaxation for the original problem and its solution provides a *lower bound* (LB) for the objective function. Such a relaxation is usually in the form of an LP or a convex programming problem, which we can solve in polynomial time (given that the relaxation itself is of polynomial size). It is clear that we should make this relaxation as tight as possible, i.e., the obtained lower bound should be as close to the optimal value as possible. The particular approach to develop a tight and polynomial-time solvable relaxation needs to be carefully designed by the user.

Using the relaxation solution as a starting point, branch-and-bound employs a *local search* algorithm to find a feasible solution to the original problem, which provides an *upper bound* (UB) for the objective function (see Fig. 5.1(a) for an example). Again, the local search algorithm needs to be carefully designed by the user. This computation of lower and upper bounds constitutes the *bounding* step of the branch-and-bound framework.

If the obtained lower and upper bounds are within an ε-tolerance of each other, i.e., $LB \geq (1 - \varepsilon)UB$, we are done, i.e., the current feasible solution is $(1 - \varepsilon)$-optimal. If the relaxation is not tight, then the lower bound LB could be significantly lower than the upper bound UB. To close this gap, we should produce a tighter relaxation that yields a smaller relaxation error. This can be achieved by a *branching* (or called a *partitioning*) step.

Branching step   The branching step includes partitioning problem, updating lower and upper bounds, and/or fathoming a subproblem.

- *Partitioning problem.*   To have a small relaxation error, we can narrow down the value-sets of variables that affect this error. We call these variables partitioning variables. In general, branch-and-bound selects a partitioning variable that evidently contributes most toward the relaxation error and splits its value-set into two (or sometimes more) sets based on its value in the relaxation solution (see Fig. 5.1(b)). For example, if a variable x with a value-set $\{0, 1, \ldots, 10\}$ is selected and its value in the relaxation solution is 3.7, then its value-set can be partitioned into two sets $\{0, 1, 2, 3\}$ (the set of values no more than 3.7) and $\{4, 5, \ldots, 10\}$ (the set of values more than 3.7). Note that if x's value in the relaxation solution is very close to lower bound x_L or upper bound x_U, i.e., $\min\{x - x_L, x_U - x\} < 0.1(x_U - x_L)$, then we should use the bisection rule. That is, if x's value in the relaxation solution is 0.7 (we have $\min\{0.7 - 0, 10 - 0.7\} < 0.1(10 - 0)$), then its value-set can be partitioned into two sets $\{0, 1, \ldots, 4\}$ and $\{5, 6, \ldots, 10\}$. This is called *partitioning* or *branching* (*dichotomous branching* in this case, where an original value-set is split into two, which we assume to be the case in the sequel). Then the original problem (denoted as Problem 1) is divided into two new problems (denoted as Problem 2 and Problem 3).

Figure 5.1 Illustration of the branch-and-bound framework.

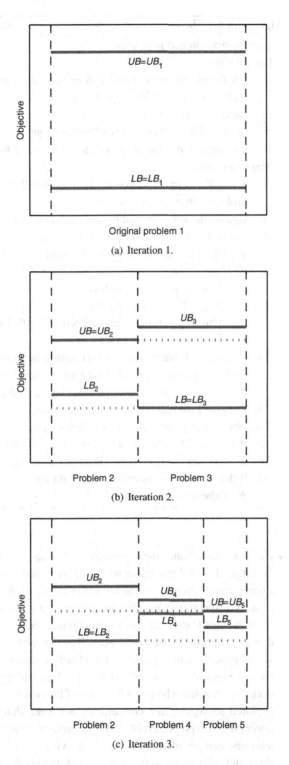

(a) Iteration 1.

(b) Iteration 2.

(c) Iteration 3.

Algorithm 5.1 The branch-and-bound framework for a minimization problem

---

Branch-and-bound framework

Initialization

1. Let the initial upper bound $UB = \infty$ and the initial problem list include only the original problem, denoted as Problem 1.
2. Determine the initial value-set for each partitioning variable.
3. Build a linear relaxation for Problem 1 and obtain its solution $\hat{\psi}_1$.
4. The objective value of $\hat{\psi}_1$ is a lower bound LB_1 to Problem 1.

Main iteration

5. Select Problem z that has the minimum LB_z-value among all problems in the problem list.
6. Update the lower bound by setting $LB = LB_z$.
7. Find a feasible solution ψ_z starting from $\hat{\psi}_z$ via a local search algorithm and denote its objective value as UB_z.
8. If $(UB_z < UB)$ {
9.   Update $\psi_\varepsilon = \psi_z$ and $UB = UB_z$.
10.   If $LB \geq (1 - \varepsilon)UB$, stop with the $(1 - \varepsilon)$-optimal solution ψ_ε. Otherwise, remove all Problems z' with $LB_{z'} \geq (1 - \varepsilon)UB$ from the problem list. }
11. Select a variable that appears to contribute most toward the relaxation error and partition its value-set into two sets based on its value in $\hat{\psi}_z$.
12. Build two new Problems z_1 and z_2 based on the two sets in Step 11.
13. Remove Problem z from the problem list.
14. Obtain LB_{z1} and LB_{z2} for Problems z_1 and z_2 via their relaxations.
15. If $LB_{z1} < (1 - \varepsilon)UB$, add Problem z_1 into the problem list.
16. If $LB_{z2} < (1 - \varepsilon)UB$, add Problem z_2 into the problem list.
17. If the problem list is empty, stop with the $(1 - \varepsilon)$-optimal solution ψ_ε. Otherwise, go to Step 5.

- *Updating lower and upper bounds.* For each of two new problems, the branch-and-bound procedure solves a corresponding relaxation and performs a local search. Hence, we derive lower bounds LB_2 and LB_3 for Problems 2 and 3, respectively. We also possibly obtain feasible solutions that provide upper bounds UB_2 and UB_3 for Problems 2 and 3, respectively. It is possible that a new problem does not have a feasible solution or the (heuristic) local search procedure might fail to find a feasible solution even if one exists. But since the feasible solution ψ_1 to Problem 1, assuming that such a solution has been detected, must be feasible to one of the two new problems (unless some explored subregion that contains ψ_1 has been deleted from consideration during the analysis of Problem 1), we have at hand a feasible solution for at least one new problem by using the previous feasible solution if necessary. Since the relaxations in Problems 2 and 3 are both more tightly constrained than that of Problem 1, we have $\min\{LB_2, LB_3\} \geq LB_1$. The lower bound of the original problem is updated from $LB = LB_1$ to $LB = \min\{LB_2, LB_3\}$. The upper bound of the original problem is updated from $UB = UB_1$ to $UB = \min\{UB, UB_2, UB_3\}$, since the best feasible solution to the original problem is the solution with the smallest UB_i-value.[1] As a result, we now have a smaller gap between LB and UB. If $LB \geq (1 - \varepsilon)UB$, we are done. Otherwise, we choose a problem with the minimum lower bound (Problem 3 in Fig. 5.1(b)) and perform partitioning on this problem.

- *Fathoming a subproblem.* Note that during the branch-and-bound process, if we find a Problem z with $LB_z \geq (1 - \varepsilon)UB$ (see Problem 4 in Fig. 5.1(c)), then we have the following two possible cases:

  Case 1: The global optimal solution is not in Problem z. In this case, the removal of Problem z will not cause the loss of the optimal solution in future iterations.

  Case 2: The global optimal solution is in Problem z. In this case, the optimal (feasible) solution must have an objective value greater than or equal to LB_z, which means that it is also greater than or equal to $(1 - \varepsilon)UB$ (since $LB_z \geq (1 - \varepsilon)UB$). Thus, the current best feasible solution with objective value UB is already a $(1 - \varepsilon)$-optimal solution. Therefore, we can still guarantee $(1 - \varepsilon)$-optimality if we remove this problem from the list for further consideration.

  The foregoing cases highlight the main advantage of the branch-and-bound framework, i.e., we can remove a problem before we completely solve it. This removal of a subproblem from consideration in the active list of problems is called *fathoming* the subproblem. Eventually, when we obtain $LB \geq (1 - \varepsilon)UB$, the branch-and-bound procedure terminates.

It has been shown that, under very general conditions, a branch-and-bound solution procedure always converges [146]. Moreover, although the worst-case complexity of such a procedure is exponential, the actual running time could be fast when all partitioning variables are reasonably bounded integers (e.g., in the case study problem considered in this chapter as well as many other problems in wireless networks), and the relaxations derived are tight.

Algorithm 5.1 presents a pseudocode for the branch-and-bound framework. As discussed, several key components in this framework are problem specific and need to be carefully designed. In this chapter, we will demonstrate how to design these key components in a case study.

---

[1] Note that we always keep the best detected (incumbent) feasible solution in the process.

5.2 Case study: Power control problem for multi-hop cognitive radio networks

The case study that we present in this chapter is on a power control problem for a multi-hop cognitive radio network (CRN). We will show how a branch-and-bound framework can be applied to solve this problem. Cognitive radio (CR) is an enabling physical layer technology to enhance the efficiency of radio spectrum [173]. A CR is a frequency-agile data communication device that has a rich control and monitoring (spectrum sensing) interface [65; 112]. It capitalizes advances in signal processing and radio technology, as well as recent advancements in spectrum policy [130]. A frequency-agile radio module is capable of sensing the available bands [45; 47; 111; 160], reconfiguring RF, and switching to newly selected frequency bands. Thus, a CR can be programmed to tune and operate on specific frequency bands over a wide spectrum range [130]. From an application perspective, CR allows a single radio to provide a wide variety of functions, acting as a cell phone, broadcast receiver, GPS receiver, wireless data terminal, etc.

A fundamental problem for a wireless network is power control. The benefits of per-node-based power control for CRNs were discussed in [150]. In a multi-hop wireless network, power control is challenging since it directly affects the upper layers (scheduling and routing). When each node is allowed to perform power control (which we call per-node-based power control), the problem becomes even more difficult due to its large optimization space. Due to such difficulty, some related work, e.g., [13; 86], assumed synchronized power control, where transmission power at each node in the network is adjustable but is synchronized to be identical. Needless to say, such simplification can hardly reflect real problems in practice.

The main difficulty associated with the per-node-based power control problem is its tight coupling to scheduling and routing. Therefore, a joint formulation of multiple layers is needed, which inevitably leads to a very complex problem. Although there has been some success in the context of *asymptotic* scaling laws (e.g., [60; 86]), theoretically optimal results for a given *finite-sized* network remain limited. For example, in [16], Bhatia and Kodialam optimized power control and routing, but assumed some frequency hopping mechanism was in place for scheduling, which helped simplify the joint consideration of scheduling. In [37], Elbatt and Ephremides optimized power control and scheduling, but assumed routing was given *a priori*. Although [26; 32] aimed to investigate joint power control, scheduling, and routing problems, both of the proposed procedures followed a "de-coupled" approach, where the final solution was obtained one layer at a time (instead of solving a joint optimization problem, as we will do in this chapter). Due to such de-coupling, the final solution can only be suboptimal at best.

In this chapter, we study the per-node-based power control problem for a multi-hop CRN. This problem is both challenging and interesting as it inherits not only all the difficulties associates with per-node-based power control, but also all the unique characteristics associated with a CRN. We develop a formal mathematical model for a joint per-node-based power control, scheduling, and flow routing problem. This joint formulation leads to a *mixed-integer non-linear programming* (MINLP) problem. Subsequently, we develop a unified (instead of a de-coupled) solution procedure based on the branch-and-bound framework and a convex hull relaxation. This solution procedure guarantees a $(1 - \varepsilon)$-optimal solution, where $\varepsilon \geq 0$ is a small prespecified error tolerance parameter.

The remainder of this chapter is organized as follows. In Section 5.3, we develop a unified mathematical model for per-node-based power control, scheduling, and flow routing. In Section 5.4, we formulate the cross-layer optimization problem. Section 5.5 applies the branch-and-bound framework to design a solution procedure. In Section 5.6, we use numerical results to illustrate the solution procedure. Section 5.7 summarizes this chapter.

5.3 Mathematical modeling

Consider a multi-hop CRN consisting of a set of $\mathcal{N}$ nodes. The set of available frequency bands at each node depends on its location and may not be the same. For example, at node i, its available frequency bands may consist of bands I, III, and V, while at a different node j, its available frequency bands may consist of bands I, IV, and VI, and so forth. More formally, denote by $\mathcal{M}_i$ the set of available frequency bands at node i. For simplicity, we assume that the bandwidth of each frequency band is W. Denote $\mathcal{M}$ as the set of all frequency bands present in the network, i.e., $\mathcal{M} = \bigcup_{i \in \mathcal{N}} \mathcal{M}_i$. Table 5.1 lists the notation used in this chapter.

5.3.1 Necessary and sufficient condition for successful transmission

Scheduling for transmission at each node in the network can be done either in-time or frequency domain. In this chapter, we consider scheduling in the frequency domain in the form of frequency bands. Similar modeling can also be done in-time domain following the same token.

Consider a transmission from node i to node j. Suppose that band m is available at both node i and node j, i.e., $m \in \mathcal{M}_{ij}$, where $\mathcal{M}_{ij} = \mathcal{M}_i \bigcap \mathcal{M}_j$. Denote p_{ij}^m as the transmission power from node i to node j in frequency band m. For transmission from node i to node j, a simple model for path attenuation loss g_{ij} is

Table 5.1 Notation.

Symbol	Definition
c_{ij}^m	Link capacity of link $i \to j$ under p_{ij}^m
$d(l)$	Destination node of session $l \in \mathcal{L}$
d_{ij}	Distance between nodes i and j
$f_{ij}(l)$	Data rate that is attributed to session l on link $i \to j$
$\mathbf{f}$	The vector of variables $f_{ij}(l), l \in \mathcal{L}, i \in \mathcal{N}, i \neq d(l), j \in \mathcal{T}_i, j \neq s(l)$
g_{ij}	Path attenuation loss from node i to node j
$\mathcal{I}_j^m$	The set of nodes that can interfere with node j on band m (under full transmission power P)
$\mathcal{L}$	The set of user communication sessions
$\mathcal{M}_i$	The set of available bands at node $i \in \mathcal{N}$
$\mathcal{M}$	$= \sum_{i \in \mathcal{N}} \mathcal{M}_i$, the set of bands in the network
$\mathcal{M}_{ij}$	$= \mathcal{M}_i \bigcap \mathcal{M}_j$, the set of available bands for link $i \to j$
$\mathcal{N}$	The set of nodes in the network
p_{ij}^m	Node i's transmission power to node j on band m
P	The maximum transmission power at a transmitter
P_T	The threshold of minimum receiving power that can be decoded
P_I	The threshold of maximum interference power that is negligible
q_{ij}^m	Transmission power level from node i to node j on band m
$\mathbf{q}$	The vector of variables $q_{ij}^m, i \in \mathcal{N}, m \in \mathcal{M}_i, j \in \mathcal{T}_i^m$
Q	Number of transmission power levels
$r(l)$	Rate requirement of session l
$R_T(p), R_I(p)$	The transmission and interference ranges under power p
R_T^{max}, R_I^{max}	The maximum transmission and interference ranges under full transmission power P
$s(l)$	Source nodes of session $l \in \mathcal{L}$
$\mathcal{T}_i^m$	The set of nodes to which node i can transmit on band m (under full transmission power P)
$\mathcal{T}_i$	$= \bigcup_{m \in \mathcal{M}_i} \mathcal{T}_i^m$, the set of nodes to which node i can transmit on all bands (under full transmission power P)
W	Bandwidth of each frequency band
x_{ij}^m	A binary indicator to denote whether or not band m is used by link $i \to j$
$\mathbf{x}$	The vector of variables $x_{ij}^m, i \in \mathcal{N}, m \in \mathcal{M}_i, j \in \mathcal{T}_i^m$
α	Path-loss index
η	Ambient Gaussian noise density

Notation for the branch-and-bound framework

LB_z, LB	The lower bounds for Problem z and the original problem
UB_z, UB	The upper bounds for Problem z and the original problem
ε	The desired accuracy in the final solution
ψ_z	The solution obtained by local search for Problem z
ψ_ε	A $(1 - \varepsilon)$-optimal solution
Ω_z	The set of all possible values of $(\mathbf{x}, \mathbf{q})$ in Problem z

$$g_{ij} = d_{ij}^{-\alpha}, \tag{5.1}$$

where d_{ij} is the physical distance between nodes i and j and α is the path-loss index. In this context, we assume data transmission from node i to node j is successful only if the received power at node j exceeds a threshold P_T. Denote the transmission range of node i under p_{ij}^m as $R_T(p_{ij}^m)$. Then, based on $g_{ij} \cdot p_{ij}^m \geq P_T$ and (5.1), we can calculate the transmission range of this node as follows:

$$R_T(p_{ij}^m) = \left(\frac{p_{ij}^m}{P_T} \right)^{1/\alpha}. \tag{5.2}$$

Since the receiving node j must be physically within the transmission range of node i, we have

(C-1)
$$d_{ij} \leq \left(\frac{p_{ij}^m}{P_T} \right)^{1/\alpha}.$$

Similarly, we assume that an interference is nonnegligible only if it exceeds a threshold, say P_I, at a receiver. Denote the interference range of node k ($k \in \mathcal{N}$, $k \neq i$) under p_{kh}^m as $R_I(p_{kh}^m)$, where h is the intended receiving node of transmitting node k. Then following the same token as the derivation for the transmission range, we can obtain the interference range of node k as $R_I(p_{kh}^m) = \left(\frac{p_{kh}^m}{P_I} \right)^{1/\alpha}$. Since the receiving node j must not fall in the interference range of any other node k that is transmitting in the same band, we have

(C-2)
$$d_{jk} \geq \left(\frac{p_{kh}^m}{P_I} \right)^{1/\alpha}.$$

As an example, Fig. 5.2(a) shows a network with three links ($1 \to 2$, $3 \to 4$, and $5 \to 6$). For each transmitting node (1, 3, and 5), the inner circle (dashed) represents the transmission range and the outer circle (solid) represents the interference range. Clearly, each receiving node falls in the transmission range of its respective transmitting node. Further, we can see that both receiving nodes 2 and 6 fall in the interference range of node 3. Thus, when link $3 \to 4$ is using a frequency band m for transmission, links $1 \to 2$ and $5 \to 6$ should not use the same band. It should also be noted that when link $3 \to 4$ is not using a frequency band m, both links $1 \to 2$ and $5 \to 6$ may use band m. This is because receiving node 2 does not fall in node 5's interference range and receiving node 6 does not fall in node 1's interference range. Now, consider that each node can adjust its transmission power. In this setting, nodes 1, 3, and 5 can reduce their transmission powers while still maintaining data transmission to the corresponding receiving nodes (see Fig. 5.2(b)). Then receiving nodes 2 and 6 are no longer in node 3's interference range. As a result, both transmitting nodes 1 and 5 can also transmit on band m as node 3 does.

Figure 5.2 A three-link
network.

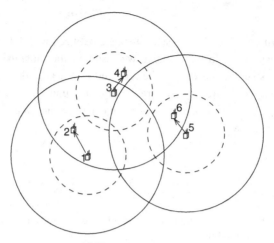

(a) No power control case.

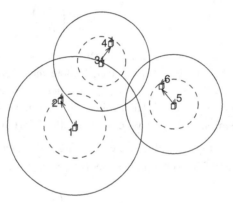

(b) Per-node-based power control case.

5.3.2 Per-node-based power control and scheduling

In this section, we formulate a mathematical model for the joint relationship
between per-node-based power control and scheduling. Suppose $m \in \mathcal{M}_{ij}$.
Denote

$$x_{ij}^m = \begin{cases} 1 & \text{if node } i \text{ transmits data to node } j \text{ on frequency band } m, \\ 0 & \text{otherwise.} \end{cases}$$

As mentioned earlier, we consider scheduling in the frequency domain and
assume unicast communication. Thus once a band m is used by node i for
transmission to node j, this band cannot be used again by node i to transmit to
a different node. That is,

(C-3)
$$\sum_{j \in \mathcal{T}_i^m} x_{ij}^m \leq 1,$$

where $\mathcal{T}_i^m$ is the set of nodes to which node i can transmit on band m under full power P.

Denote by R_T^{max} the maximum transmission range of a node when it transmits at full power P. Then based on (5.2), we have $R_T^{max} = R_T(P) = \left(\frac{P}{P_T}\right)^{1/\alpha}$. Thus, we have $P_T = \frac{P}{(R_T^{max})^\alpha}$. Then, for a node transmitting at a power $p \in [0, P]$, its transmission range is given by

$$R_T(p) = \left(\frac{p}{P_T}\right)^{\frac{1}{\alpha}} = \left[\frac{p(R_T^{max})^\alpha}{P}\right]^{\frac{1}{\alpha}} = \left(\frac{p}{P}\right)^{\frac{1}{\alpha}} \cdot R_T^{max}. \quad (5.3)$$

Similarly, denote R_I^{max} as the maximum interference range of a node when it transmits at full power P. Then, following the same token, we have $R_I^{max} = R_I(P) = \left(\frac{P}{P_I}\right)^{1/\alpha}$ and $P_I = \frac{P}{(R_I^{max})^\alpha}$. For a node transmitting at a power $p \in [0, P]$, its interference range is given by

$$R_I(p) = \left(\frac{p}{P}\right)^{\frac{1}{\alpha}} \cdot R_I^{max}. \quad (5.4)$$

Recall that $\mathcal{T}_i^m$ denotes the set of nodes to which node i can transmit on band m under full power P. More formally, we have

$$\mathcal{T}_i^m = \{j : d_{ij} \le R_T^{max}, j \ne i, m \in \mathcal{M}_j\}.$$

Similarly, denote by $\mathcal{I}_j^m$ the set of nodes that can interfere with node j on band m under full power P. Then,

$$\mathcal{I}_j^m = \{k : d_{jk} \le R_I^{max}, m \in \mathcal{M}_k\}.$$

Note that the definitions of $\mathcal{T}_i^m$ and $\mathcal{I}_j^m$ are both based on full transmission power P. When the power level p is below P, the corresponding transmission and interference ranges will be smaller. Nevertheless, the set of nodes that fall in the transmission range and the set of nodes that can produce interference can be upper bounded by those sets under full transmission power.

By (C-1) and (C-2) (i.e., constraints for successful transmission from node i to node j), (5.3) and (5.4), we have

$$d_{ij} \le R_T(p_{ij}^m) = \left(\frac{p_{ij}^m}{P}\right)^{\frac{1}{\alpha}} \cdot R_T^{max},$$

$$d_{jk} \ge R_I(p_{kh}^m) = \left(\frac{p_{kh}^m}{P}\right)^{\frac{1}{\alpha}} \cdot R_I^{max} \quad (k \in \mathcal{I}_j^m, k \ne i, h \in \mathcal{T}_k^m).$$

Based on the above two constraints, we have the following requirements for the transmission link $i \to j$ and the interfering link $k \to h$:

$$p_{ij}^m \begin{cases} \in \left[\left(\frac{d_{ij}}{R_T^{max}} \right)^\alpha P, P \right] & \text{if } x_{ij}^m = 1, \\ = 0 & \text{if } x_{ij}^m = 0. \end{cases}$$

$$p_{kh}^m \leq \begin{cases} \left(\frac{d_{kj}}{R_I^{max}} \right)^\alpha P & \text{if } x_{ij}^m = 1, \\ P & \text{if } x_{ij}^m = 0. \end{cases} (k \in \mathcal{I}_j^m, k \neq i, h \in \mathcal{T}_k^m).$$

Mathematically, these requirements can be rewritten as follows:

$$\text{(C-1')} \quad p_{ij}^m \in \left[\left(\frac{d_{ij}}{R_T^{max}} \right)^\alpha \cdot P x_{ij}^m, \ P x_{ij}^m \right],$$

$$\text{(C-2')} \quad p_{kh}^m \leq P - \left[1 - \left(\frac{d_{kj}}{R_I^{max}} \right)^\alpha \right] P x_{ij}^m \qquad (k \in \mathcal{I}_j^m, k \neq i, h \in \mathcal{T}_k^m).$$

In addition, for a successful scheduling in the frequency domain, the following two constraints must also hold:

(C-4) For a band $m \in \mathcal{M}_j$ that is available at node j, this band cannot be used for both transmission and receiving. That is, if band m is used at node j for transmission (or receiving), then it cannot be used for receiving (or transmission).

(C-5) Similar to constraint (C-3) on transmission, node j cannot use the same band $m \in \mathcal{M}_j$ to receive from two different nodes.

Note that (C-4) can be viewed as a "self-interference" avoidance constraint where, at the same node j, its transmission to another node h on band m interferes its reception from node i on the same band. It turns out that the above two constraints are mathematically *embedded* in (C-1') and (C-2'). That is, once (C-1') and (C-2') are satisfied, then both the constraints (C-4) and (C-5) are also satisfied. This result is formally stated in the following lemma:

Lemma 5.1

If transmission powers on every transmission link and interference link satisfy (C-1') and (C-2') in the network, then (C-4) and (C-5) are also satisfied.

Proof. We first prove that (C-1') and (C-2') lead to (C-4). To do this, we let $k = j$ in (C-2'). Then (C-2') degenerates to $p_{jh}^m \leq P - P x_{ij}^m$ since $d_{jj} = 0$.

- Suppose that node j is receiving from node i on band m, i.e., $x_{ij}^m = 1$. Then, $p_{jh}^m \leq P - P x_{ij}^m = 0$. Since $p_{jh}^m \geq \left(\frac{d_{jh}}{R_T^{max}} \right)^\alpha P x_{jh}^m$ from (C-1'), we have that x_{jh}^m must be 0. That is, if node j is receiving from node i on band m, then node j cannot transmit to node h in the same band.

- Now, suppose that node j is transmitting to node h on band m, i.e., $x_{jh}^m = 1$. We will show that this implies $x_{ij}^m = 0$. We show this by contradiction. That is, if $x_{ij}^m = 1$, then we have just proved in the foregoing case that $x_{jh}^m = 0$. But this contradicts our initial assumption that $x_{jh}^m = 1$. Therefore, x_{ij}^m must be 0. That is, if node j is transmitting to node h on band m, then node j cannot use the same band for receiving from a node i.

Combining the above two results, we have shown that (C-4) holds.

We now prove that (C-1$'$) and (C-2$'$) also imply (C-5). Again, the proof is based on contradiction. Suppose that (C-5) does not hold. Then node j can receive from two different nodes i and k on the same band m, i.e., $x_{ij}^m = 1$ and $x_{kj}^m = 1$. Note that link $k \to j$ can be viewed as an interfering link with respect to link $i \to j$. This corresponds to letting $h = j$ in (C-2$'$). Then from (C-2$'$), since $x_{ij}^m = 1$, we have $p_{kj}^m \le \left(\frac{d_{kj}}{R_I^{max}} \right)^\alpha P$. Now consider node k is transmitting to node j. Then by (C-1$'$), we have $p_{kj}^m \ge \left(\frac{d_{kj}}{R_T^{max}} \right)^\alpha P$. However, the above two inequalities cannot hold at the same time since we have $R_I^{max} > R_T^{max}$. This gives us a contradiction. Thus, the initial assumption that (C-5) does not hold is incorrect.   □

The significance of Lemma 5.1 is that, since (C-4) and (C-5) are embedded in (C-1$'$) and (C-2$'$), it is sufficient to consider only (C-1$'$), (C-2$'$), and (C-3) in the problem formulation. This helps reduce the number of constraints.

5.3.3 Flow routing and link capacity constraints

We assume that there is a set of $\mathcal{L}$ active user communication (unicast) sessions in the CRN. Denote $s(l)$ and $d(l)$ as respectively the source and destination nodes of session $l \in \mathcal{L}$, and $r(l)$ as the rate requirement (in b/s) of session l. Again, allow flow splitting between a source node and its destination node. The benefits of flow splitting were left as a homework exercise in Chapter 2 (Problem 2.2).

Mathematically, this can be easily modeled based on flow balance at each node. Denote by $f_{ij}(l)$ the data rate on link (i, j) that is attributed to session l, where $i \in \mathcal{N}, i \ne d(l), j \in \mathcal{T}_i = \bigcup_{m \in \mathcal{M}_i} \mathcal{T}_i^m, j \ne s(l)$. Note that for $f_{ij}(l)$, we set $i \ne d(l)$ to ensure that the destination node $d(l)$ will be a sink node and will not transmit data to a relay node i. We also set $j \ne s(l)$ to ensure that a relay node i will not transmit data back to the source node $s(l)$. If node i is the source node of session l, i.e., $i = s(l)$, then

$$\sum_{j \in \mathcal{T}_i} f_{ij}(l) = r(l). \tag{5.5}$$

If node i is an intermediate relay node for session l, i.e., $i \ne s(l)$ and $i \ne d(l)$, then

$$\sum_{\substack{j \in \mathcal{T}_i}}^{j \neq s(l)} f_{ij}(l) = \sum_{\substack{k \in \mathcal{T}_i}}^{k \neq d(l)} f_{ki}(l). \tag{5.6}$$

If node i is the destination node of session l, i.e., $i = d(l)$, then

$$\sum_{k \in \mathcal{T}_i} f_{ki}(l) = r(l). \tag{5.7}$$

It can be easily verified that once (5.5) and (5.6) are satisfied, (5.7) must also be satisfied. As a result, it is sufficient to just include (5.5) and (5.6) in the formulation.

Denote c_{ij}^m as link capacity of link $i \rightarrow j$ under p_{ij}^m. In addition to the above flow balance equations at each node $i \in \mathcal{N}$ for session $l \in \mathcal{L}$, the aggregated flow rates on each radio link cannot exceed this link's capacity. We have

$$\sum_{l \in \mathcal{L}}^{s(l) \neq j, d(l) \neq i} f_{ij}(l) \leq \sum_{m \in \mathcal{M}_{ij}} c_{ij}^m = \sum_{m \in \mathcal{M}_{ij}} W \log_2 \left(1 + \frac{g_{ij}}{\eta W} p_{ij}^m \right), \tag{5.8}$$

where η is the ambient Gaussian noise density.

5.4 Problem formulation

Objective function   In this chapter, we consider how to minimize network resource usage to support a set of user sessions. Network resource usage can be defined in a number of ways, which typically includes bandwidth usage. However, bandwidth usage can only quantify resource usage in spectrum, but cannot take into account of the impact (i.e., interference) of radio transmission in space. For example, a node transmitting with the same channel bandwidth but with different power levels will produce different interference "footprint" areas. To account for a CR's impact on both bandwidth usage in spectrum and interference footprint in space, we introduce the so-called *bandwidth-footprint-product* (BFP) metric in this chapter (also known as space-bandwidth product in [96]).

Since each node in the network will use a number of bands for transmission and each band will have a certain footprint corresponding to its transmission power, our objective is to minimize network-wide BFP, which is the sum of BFPs among all the nodes in the network, i.e.,

$$\min \sum_{i \in \mathcal{N}} \sum_{m \in \mathcal{M}_i} \sum_{j \in \mathcal{T}_i^m} W \cdot \pi (R_I(p_{ij}^m))^2. \tag{5.9}$$

In some sense, minimizing network-wide BFP can be viewed as minimizing a "weighted" version of bandwidth usage, where the weight is the interference footprint area. By (5.4), (5.9) is equivalent to minimizing $W\pi(R_I^{max})^2$ $\sum_{i\in\mathcal{N}}\sum_{m\in\mathcal{M}_i}\sum_{j\in\mathcal{T}_i^m}\left(\frac{p_{ij}^m}{P}\right)^{2/\alpha}$. Since $W\pi(R_I^{max})^2$ is a constant factor, we can remove it from the objective function.

Discretization of transmission powers   For power control, we assume that the transmission power can only be tuned into a finite number of discrete levels between 0 and P. This discretization helps the branch-and-bound process run much faster due to reduced optimization space. This discretization is also consistent to power control in the real world, where in many cases a radio's transmission power can only be tuned into a finite number of discrete levels. To model this discrete version of power control, we introduce an integer parameter Q that represents the total number of power levels to which a transmitter can be adjusted, i.e., $0, \frac{1}{Q}P, \frac{2}{Q}P, \dots, P$. Denote $q_{ij}^m \in \{0, 1, 2, \dots, Q\}$ as the integer power level for p_{ij}^m, i.e., $p_{ij}^m = \frac{q_{ij}^m}{Q}P$. Then (C-1'), (C-2'), and (5.8) can be rewritten as follows:

$$q_{ij}^m \in \left[\left(\frac{d_{ij}}{R_T^{max}}\right)^\alpha Qx_{ij}^m,\ Qx_{ij}^m\right] \tag{5.10}$$

$$q_{kh}^m \le Q - \left[1 - \left(\frac{d_{kj}}{R_I^{max}}\right)^\alpha\right]Qx_{ij}^m \quad (k\in\mathcal{I}_j^m, k\neq i, h\in\mathcal{T}_k^m) \tag{5.11}$$

$$\sum_{l\in\mathcal{L}}^{s(l)\neq j, d(l)\neq i} f_{ij}(l) \le \sum_{m\in\mathcal{M}_{ij}} W\log_2\left(1 + \frac{g_{ij}P}{\eta WQ}q_{ij}^m\right).$$

We can re-formulate (5.11) as follows. Note that by (C-3), there is at most one $h \in \mathcal{T}_k^m$ such that $x_{kh}^m = 1$. As a result, based on (5.11), there is at most one $q_{kh}^m > 0$ in $\sum_{h\in\mathcal{T}_k^m}q_{kh}^m$. Thus, (5.11) can be rewritten as

$$\sum_{h\in\mathcal{T}_k^m}q_{kh}^m \le Q - \left[1 - \left(\frac{d_{kj}}{R_I^{max}}\right)^\alpha\right]Qx_{ij}^m \quad (k\in\mathcal{I}_j^m, k\neq i).$$

This reformulation will help reduce the number of constraints associated with (5.11).

Mathematical formulation   Putting together the objective function and all the constraints for power control, scheduling, and flow routing, we have the following problem formulation:

$$\text{Minimize} \quad \sum_{i\in\mathcal{N}}\sum_{m\in\mathcal{M}_i}\sum_{j\in\mathcal{T}_i^m}\left(\frac{q_{ij}^m}{Q}\right)^{\frac{2}{\alpha}}$$

$$\text{subject to} \quad \sum_{j\in\mathcal{T}_i^m}x_{ij}^m \le 1 \quad\quad (i\in\mathcal{N}, m\in\mathcal{M}_i) \tag{5.12}$$

$$q_{ij}^m - \left(\frac{d_{ij}}{R_T^{max}}\right)^\alpha Q x_{ij}^m \geq 0 \qquad (i \in \mathcal{N}, m \in \mathcal{M}_i, j \in \mathcal{T}_i^m) \qquad (5.13)$$

$$q_{ij}^m - Q x_{ij}^m \leq 0 \qquad (i \in \mathcal{N}, m \in \mathcal{M}_i, j \in \mathcal{T}_i^m) \qquad (5.14)$$

$$\sum_{h \in \mathcal{T}_k^m} q_{kh}^m + \left[1 - \left(\frac{d_{kj}}{R_I^{max}}\right)^\alpha\right] Q x_{ij}^m \leq Q$$

$$(i \in \mathcal{N}, m \in \mathcal{M}_i, j \in \mathcal{T}_i^m, k \in \mathcal{I}_j^m, k \neq i) \qquad (5.15)$$

$$\sum_{l \in \mathcal{L}}^{s(l) \neq j, d(l) \neq i} f_{ij}(l) - \sum_{m \in \mathcal{M}_{ij}} W \log_2 \left(1 + \frac{g_{ij} P}{\eta W Q} q_{ij}^m\right) \leq 0 \quad (i \in \mathcal{N}, j \in \mathcal{T}_i) \qquad (5.16)$$

$$\sum_{j \in \mathcal{T}_i} f_{ij}(l) = r(l) \qquad (l \in \mathcal{L}, i = s(l)) \qquad (5.17)$$

$$\sum_{\substack{j \in \mathcal{T}_i \\ j \neq s(l)}} f_{lj}(l) - \sum_{\substack{k \in \mathcal{T}_i \\ k \neq d(l)}} f_{ki}(l) = 0 \quad (l \in \mathcal{L}, i \in \mathcal{N}, i \neq s(l), d(l)) \qquad (5.18)$$

$$x_{ij}^m \in \{0, 1\}, q_{ij}^m \in \{0, 1, 2, \dots, Q\} \qquad (i \in \mathcal{N}, m \in \mathcal{M}_i, j \in \mathcal{T}_i^m)$$

$$f_{ij}(l) \geq 0 \qquad (l \in \mathcal{L}, i \in \mathcal{N}, i \neq d(l), j \in \mathcal{T}_i, j \neq s(l)),$$

where (5.13) and (5.14) come from (5.10). In this problem formulation, W, g_{ij}, R_T^{max}, R_I^{max}, P, η, $r(l)$, and Q are constants; q_{ij}^m, x_{ij}^m, and $f_{ij}(l)$ are optimization variables.

This optimization problem is in the form of a *mixed-integer nonlinear programming* (MINLP), which is NP-hard in general [46]. In the next section, we develop a solution procedure based on the branch-and-bound framework that we described in Section 5.1.

5.5 A solution procedure

As discussed in Section 5.1, with the branch-and-bound framework, there are several key components that are problem specific and need to be carefully designed. These components are listed as follows and will be designed in Sections 5.5.1 and 5.5.3:

- How to obtain a tight relaxation as well as a tight lower bound?
- How to design a local search algorithm to find a feasible solution and an upper bound?
- How to select a partitioning variable?

5.5.1 Linear relaxation

During each iteration of the branch-and-bound process, we need a relaxation technique to derive a lower bound for the optimal objective function value

Figure 5.3 Illustration of convex hull for a discrete term.

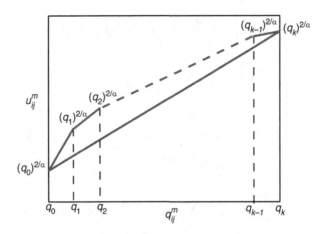

(see Steps 3 and 4 in Algorithm 5.1). Since an LP problem can be solved in polynomial time, we aim to obtain a polynomially sized linear relaxation for the underlying problem. For a nonlinear discrete term, we propose to use a convex hull relaxation. This is done by introducing a new variable u_{ij}^m for the nonlinear discrete term $(q_{ij}^m)^{2/\alpha}$. Suppose that $q_{ij}^m \in \{q_0, q_1, \ldots, q_K\}$, where $q_0 \equiv (q_{ij}^m)_L < q_1 < \cdots < q_K \equiv (q_{ij}^m)_U$. The convex hull of feasible (q_{ij}^m, u_{ij}^m)-coordinates (see Fig. 5.3) can be formulated as follows:

$$u_{ij}^m - \frac{(q_K)^{2/\alpha} - (q_0)^{2/\alpha}}{q_K - q_0} q_{ij}^m \geq \frac{q_K(q_0)^{2/\alpha} - (q_K)^{2/\alpha} q_0}{q_K - q_0}$$

$$u_{ij}^m - \frac{(q_k)^{2/\alpha} - (q_{k-1})^{2/\alpha}}{q_k - q_{k-1}} q_{ij}^m \leq \frac{q_k(q_{k-1})^{2/\alpha} - (q_k)^{2/\alpha} q_{k-1}}{q_k - q_{k-1}} \quad (1 \leq k \leq K).$$

Similarly, we can introduce a new variable v_{ij}^m for the nonlinear discrete term $\log_2\left(1 + \frac{g_{ij}P}{\eta WQ} q_{ij}^m\right)$ and construct corresponding convex hull constraints for v_{ij}^m. The details of this part are left in a homework exercise.

Denote $\mathbf{x}$ and $\mathbf{q}$ as the vectors of variables x_{ij}^m and q_{ij}^m, respectively. We thus derive the following linear relaxation for Problem z:

Minimize $\quad \displaystyle\sum_{i \in \mathcal{N}} \sum_{m \in \mathcal{M}_i} \sum_{j \in \mathcal{T}_i^m} \frac{1}{Q^{2/\alpha}} u_{ij}^m$

subject to $\quad$ Convex hull constraints for $u_{ij}^m \quad (i \in \mathcal{N}, m \in \mathcal{M}_i, j \in \mathcal{T}_i^m)$

$$\sum_{j \in \mathcal{T}_i^m} x_{ij}^m \leq 1 \qquad\qquad (i \in \mathcal{N}, m \in \mathcal{M}_i)$$

$$q_{ij}^m - \left(\frac{d_{ij}}{R_T^{max}}\right)^\alpha Q x_{ij}^m \geq 0 \qquad (i \in \mathcal{N}, m \in \mathcal{M}_i, j \in \mathcal{T}_i^m)$$

$$q_{ij}^m - Q x_{ij}^m \leq 0 \qquad\qquad (i \in \mathcal{N}, m \in \mathcal{M}_i, j \in \mathcal{T}_i^m)$$

$$\sum_{h\in\mathcal{T}_k^m} q_{kh}^m + \left[1 - \left(\frac{d_{kj}}{R_I^{max}}\right)^{\alpha}\right] Q x_{ij}^m \le Q \quad (i\in\mathcal{N}, m\in\mathcal{M}_i, j\in\mathcal{T}_i^m, k\in\mathcal{I}_j^m, k\neq i)$$

$$\sum_{l\in\mathcal{L}}^{s(l)\neq j, d(l)\neq i} f_{ij}(l) - \sum_{m\in\mathcal{M}_{ij}} W v_{ij}^m \le 0 \qquad (i\in\mathcal{N}, j\in\mathcal{T}_i)$$

Convex hull constraints for v_{ij}^m $(i\in\mathcal{N}, m\in\mathcal{M}_i, j\in\mathcal{T}_i^m)$

$$\sum_{j\in\mathcal{T}_i} f_{ij}(l) = r(l) \qquad (l\in\mathcal{L}, i = s(l))$$

$$\sum_{j\in\mathcal{T}_i}^{j\neq s(l)} f_{ij}(l) - \sum_{k\in\mathcal{T}_i}^{k\neq d(l)} f_{ki}(l) = 0 \quad (l\in\mathcal{L}, i\in\mathcal{N}, i\neq s(l), d(l))$$

$$u_{ij}^m, v_{ij}^m \ge 0 \qquad (i\in\mathcal{N}, m\in\mathcal{M}_i, j\in\mathcal{T}_i^m)$$

$$f_{ij}(l) \ge 0 \qquad (l\in\mathcal{L}, i\in\mathcal{N}, i\neq d(l), j\in\mathcal{T}_i, j\neq s(l))$$

$$(\mathbf{x},\mathbf{q}) \in \Omega_z,$$

where Ω_z is the set of all possible values of $(\mathbf{x},\mathbf{q})$ in Problem z. For example, Ω_1 for the original problem (Problem 1) is $\{(\mathbf{x},\mathbf{q}) : 0 \le x_{ij}^m \le 1, 0 \le q_{ij}^m \le Q\}$. The above linear relaxation formulation is an LP problem, which can be easily solved and its objective value serves as a lower bound for Problem z.

5.5.2 Local search algorithm

Denote the relaxation solution as $\hat{\psi}_z$. Note that $\hat{\psi}_z$ may not be feasible because the $\mathbf{x}$- and $\mathbf{q}$-values in $\hat{\psi}_z$ may not be integers after relaxation. We now show how to obtain a feasible solution ψ_z based on $\hat{\psi}_z$. The objective of this feasible solution ψ_z will serve as an upper bound for $\hat{\psi}_z$.

To construct a feasible solution ψ_z, we use the same routing solution as that in $\hat{\psi}_z$, i.e., we let $\mathbf{f} = \hat{\mathbf{f}}$. We then need to determine integer values for $\mathbf{x}$ and $\mathbf{q}$ in ψ_z such that constraints (5.12)–(5.15) hold and that the routing solution $\mathbf{f}$ is feasible, i.e., (5.16) holds for each link $i \to j$.

Initially, we set each $q_{ij}^m = (q_{ij}^m)_L$ and each $x_{ij}^m = (x_{ij}^m)_L$, where $[(q_{ij}^m)_L, (q_{ij}^m)_U]$ and $[(x_{ij}^m)_L, (x_{ij}^m)_U]$ are the respective value-sets of q_{ij}^m and x_{ij}^m in Problem z. Based on discussion in Section 5.5.3, the initial $\mathbf{x}$- and $\mathbf{q}$-values satisfy constraints (5.12)–(5.15). Next, we consider (5.16). Based on these q_{ij}^m-values, we can compute the capacity $\sum_{m\in\mathcal{M}_i} W \log_2\left(1 + \frac{g_{ij}P}{\eta W Q} q_{ij}^m\right)$ of each link $i \to j$. The requirement on a link $i \to j$ is $\sum_{l\in\mathcal{L}}^{s(l)\neq j, d(l)\neq i} f_{ij}(l)$. If a link's requirement is larger than its current capacity, we attempt to satisfy (5.16) by increasing q_{ij}^m under its value-set limitation (with necessary adjustments on other variables' value-sets to satisfy constraint (5.15)). In particular, we need to keep track of the maximum allowed transmission power. This

---

Algorithm 5.2 An algorithm to obtain **x** and **q**.

A local search algorithm
Initialization

1. Set $q_{ij}^m = (q_{ij}^m)_L$ and $x_{ij}^m = (x_{ij}^m)_L$.

2. Compute the capacity $\sum_{m \in \mathcal{M}_i} W \log_2 \left(1 + \frac{g_{ij}P}{\eta WQ} q_{ij}^m \right)$ and the flow rate requirement $\sum_{l \in \mathcal{L}}^{s(l) \neq j, d(l) \neq i} f_{ij}(l)$ on each link $i \to j$.

Main iteration

3. If flow rate requirement on each link is satisfied, then a feasible solution is found.

4. Otherwise, among all the links with flow rate requirements larger than link capacities, identify a link $i \to j$ that has the largest requirement. //Try to increase the capacity for link $i \to j$ as follows.

5. Increase q_{ij}^m among the currently used bands in nonincreasing order of $\hat{q}_{ij}^m$ and under the limitation that $q_{ij}^m \leq \lceil \hat{q}_{ij}^m \rceil$.

6. If the achieved capacity on link $i \to j$ is sufficient, then go to Step 3.

7. Otherwise, use an available but currently unused band in nonincreasing order of $\hat{q}_{ij}^m$.★

8. For the selected band m, increase q_{ij}^m under the constraint $\left(\frac{d_{ij}}{R_T^{max}} \right)^\alpha Q \leq q_{ij}^m \leq \lceil \hat{q}_{ij}^m \rceil$ to satisfy (5.13) and set $x_{ij}^m = 1$ to satisfy (5.14).

9. Set $x_{ih}^m = 0$ for $h \in \mathcal{T}_i, h \neq j$ to satisfy (5.12), and let $(q_{kh}^m)_U \leq \left\lfloor \left(\frac{d_{kj}}{R_I^{max}} \right)^\alpha Q \right\rfloor$ for $k \in \mathcal{I}_j^m, k \neq i, h \in \mathcal{T}_k^m$ to satisfy (5.15).

10. If the achieved capacity on link $i \to j$ is sufficient, then go to Step 3.

11. Otherwise, increase q_{ij}^m among the currently used bands in nonincreasing order of $\hat{q}_{ij}^m$ and under the constraint $q_{ij}^m \leq (q_{ij}^m)_U$.

12. If the achieved capacity on link $i \to j$ is sufficient, then go to Step 3.

13. Otherwise, link $i \to j$ cannot be satisfied and thus a feasible solution cannot be found.

Note: ★ A band m is available on link $i \to j$ if for any transmitting node k using this band, node j is not in its interference range.

---

is done in Step 9 of Algorithm 5.2, where if a new band m is used for link $i \to j$, then it is necessary to pose a limit on each neighboring transmitter k such that the interference from k is negligible. As a result, the $(q_{kh}^m)_U$-value for $k \in \mathcal{I}_j^m, k \neq i, h \in \mathcal{T}_k^m$ may be decreased to satisfy (5.15). Suppose that each updated $(q_{kh}^m)_U$ is no less than $(q_{kh}^m)_L$. Then we have a new solution and we can calculate the objective value of this solution. If we can satisfy (5.16) for all links, then we have a feasible solution. Otherwise (i.e., if (5.16) on any link cannot be satisfied), we declare that we cannot find a feasible solution and thus we set the objective value to ∞. The details of this local search algorithm are presented in Algorithm 5.2.

5.5.3 Selection of partitioning variables

If the relaxation error for Problem z is not small, the gap between its lower and upper bounds may be large, i.e., $LB < (1 - \varepsilon)UB$. To narrow this gap, we create two new subproblems z_1 and z_2 from Problem z, in the hope that these two new problems will have smaller relaxation errors, and thus will yield tighter bounds for the objective function. To generate Problems z_1 and z_2, we identify a partitioning variable based on its ascribed relaxation error (Step 11 in Algorithm 5.1). The partitioning process terminates whenever we have $LB \geq (1 - \varepsilon)UB$.

The partitioning variables include all the **x**- and **q**-variables. For our problem, we find that the **x**-variables are more important than the **q**-variables, in terms of their impact on the objective value. Thus, instead of identifying a partitioning variable based on its relaxation error only, it makes sense to consider the **x**-variables before the **q**-variables. That is, we first select one of the **x**-variables for partitioning, as applicable.

In particular, for the relaxation solution $\hat{\psi}_z$, we choose an x_{ij}^m-variable having the largest relaxation error $\min\{\hat{x}_{ij}^m, 1 - \hat{x}_{ij}^m\}$ among all the **x**-variables and let its value-set in Problems z_1 and z_2 be $\{0\}$ and $\{1\}$, respectively. Since the resulting value-set for this x_{ij}^m only has one element in each new problem, this variable can be replaced by a constant. As a result, some constraints may also be removed. It should be noted that the new value-set of x_{ij}^m may narrow other variables' value-sets, based on constraints on (5.12)–(5.15). That is, if the new value-set of x_{ij}^m is $\{0\}$, then we have $q_{ij}^m = 0$ to satisfy (5.14). Thus, we have $(q_{ij}^m)_L = (q_{ij}^m)_U = 0$. If the new value-set of x_{ij}^m is $\{1\}$, then we have $x_{ih}^m = 0$ for $h \in \mathcal{T}_i, h \neq j$ to satisfy (5.12); $q_{ij}^m \geq \left(\frac{d_{ij}}{R_T^{max}}\right)^{\alpha} Q$ to satisfy (5.13); and $q_{kh}^m \leq \left(\frac{d_{kj}}{R_T^{max}}\right)^{\alpha} Q$ for $k \in \mathcal{I}_j^m, k \neq i, h \in \mathcal{T}_k$ to satisfy (5.15). Thus, we have $(x_{ih}^m)_L = (x_{ih}^m)_U = 0$; $(q_{ij}^m)_L$ is updated by $\max\left\{(q_{ij}^m)_L, \left(\frac{d_{ij}}{R_T^{max}}\right)^{\alpha} Q\right\}$; and $(q_{kh}^m)_U$ is updated by $\min\left\{(q_{kh}^m)_U, \left(\frac{d_{kj}}{R_T^{max}}\right)^{\alpha} Q\right\}$. If these updates make the new value-set of a variable empty, then the corresponding new subproblem is clearly infeasible. One sample scenario is $\left(\frac{d_{ij}}{R_T^{max}}\right)^{\alpha} Q > (q_{ij}^m)_U$. Then the updated $(q_{ij}^m)_L > (q_{ij}^m)_U$, which yields an empty value-set for q_{ij}^m. As a result, we can fathom an infeasible subproblem (i.e., remove it from the problem list).

After we are done with examining all the **x**-variables, i.e., all the **x**-variables are at a value 0 or 1 in the relaxation solution, then we select one of the **q**-variables for partitioning. In particular, for the relaxation solution $\hat{\psi}_z$, the relaxation error of a discrete term q_{ij}^m is

$$\min\{\hat{q}_{ij}^m - \lfloor \hat{q}_{ij}^m \rfloor, \lfloor \hat{q}_{ij}^m \rfloor + 1 - \hat{q}_{ij}^m\};$$

the relaxation error of a nonlinear discrete term $u_{ij}^m = (q_{ij}^m)^{2/\alpha}$ is

$$|\hat{u}_{ij}^m - (\hat{q}_{ij}^m)^{2/\alpha}|;$$

and the relaxation error of a nonlinear discrete term $v_{ij}^m = \log_2\left(1 + \frac{g_{ij}P}{\eta W Q} q_{ij}^m\right)$ is

$$\left| \hat{v}_{ij}^m - \log_2\left(1 + \frac{g_{ij}P}{\eta W Q}\hat{q}_{ij}^m\right)\right|.$$

Among these three types of relaxation errors, we identify the largest one and choose the corresponding q_{ij}^m as the partitioning variable. Assuming the value-set of q_{ij}^m in Problem z is $\{q_0, q_1, \ldots, q_K\}$, its value-set in Problems z_1 and z_2 will be $\{q_0, q_1, \ldots, \lfloor \hat{q}_{ij}^m \rfloor\}$ and $\{\lfloor \hat{q}_{ij}^m \rfloor + 1, \lfloor \hat{q}_{ij}^m \rfloor + 2, \ldots, q_K\}$, respectively. The new value-set of q_{ij}^m may narrow other variables' value-sets due to constraint (5.15). That is, if we increase the lower bound for one q-variable in the summation in constraint (5.15), the upper bounds of other q-variables in this summation may be decreased. If these updates make the new value-set of a variable empty, then the corresponding new subproblem is clearly infeasible. Again, we only keep feasible subproblems in the problem list.

A feasible subproblem has at least one solution that satisfies constraints (5.12)–(5.15). We can further verify that if we set each $q_{ij}^m = (q_{ij}^m)_L$ and each $x_{ij}^m = (x_{ij}^m)_L$, then constraints (5.12)–(5.15) are satisfied. In particular, constraints (5.12)–(5.14) hold due to updates in the partitioning process. Since constraint (5.15) holds by a solution with some q_{ij}^m values no less than $(q_{ij}^m)_L$ and some x_{ij}^m values no less than $(x_{ij}^m)_L$, decreasing these values will not violate (5.15).

5.6 Numerical examples

5.6.1 Simulation setting

In this section, we consider a randomly generated 20-node ad hoc network with each node located in a 50 x 50 area. For ease of exposition, we normalize all units for distance, bandwidth, rate, and power based on (5.1) and (5.8) with appropriate dimensions. An instance of network topology is given in Fig. 5.4 with each node's location listed in Table 5.2. We assume that there are $|\mathcal{M}| = 10$ frequency bands in the network and each band has a bandwidth of $W = 50$. At each node, only a subset of these frequency bands is available. In the simulation, this is done by randomly selecting a subset of bands for each node. Table 5.2 shows the available bands at each node. Within this network, we assume that there are $|\mathcal{L}| = 5$ user sessions, with source and destination nodes chosen randomly. The rate of each session is randomly generated within $[10, 100]$. Table 5.3 specifies an instance of each session's source node, destination node, and rate requirement.

We assume that $R_T^{max} = 20$, $R_I^{max} = 40$, and that the path-loss index $\alpha = 4$. The threshold P_T is assumed to be $P_T = \eta W = 50\eta$. Thus, we

Figure 5.4 A 20-node ad hoc
network.

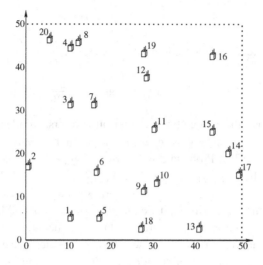

Table 5.2 Each node's location and available frequency bands for the 20-node network.

Node index	Location	Available bands
1	(10.5, 4.3)	I, II, III, IV, V, VI, VII, VIII, IX, X
2	(1.7, 17.3)	II, III, IV, V, VI, VII, X
3	(10.7, 30.8)	I, III, IV, V, VI, VII, VIII, IX, X
4	(10.2, 45.3)	I, III, IV, V, VI, VII, VIII, IX, X
5	(17.8, 4)	I, II, V, VI, VII, VIII, IX
6	(17.2, 15.2)	I, II, IV, VIII
7	(16.9, 30.8)	I, II, III, IV, V, VI, VII, VIII, IX, X
8	(12.3, 47.3)	I, III, IV, V, VII, VIII, IX
9	(28.2, 11.5)	I, III, V, VII
10	(32.1, 13.8)	I, II, III, IV, VI, VII, VIII, IX, X
11	(30.4, 25.6)	I, II, III, V, VI, VIII, IX, X
12	(29.7, 36)	I, II, III, IV, VI, VI
13	(41.7, 3.1)	I, II, III, V, VI, VIII, IX, X
14	(47.5, 20)	I, IV, V, VIII, IX, X
15	(43.3, 25.3)	II, III, IV, V, VI, VII, VIII, IX, X
16	(44.1, 42.7)	I, II, IV, VI, VII, VIII, IX, X
17	(49.6, 15.8)	I, II, III, IV, V, VI, VII, VIII
18	(28.7, 2.5)	I, II, III, VI, VII, VIII, IX, X
19	(28, 43.5)	II, IV, V, VI, VIII
20	(5, 46.9)	II, IV, V, VI, VII

have $P_I = \left(\frac{R_T^{max}}{R_I^{max}}\right)^\alpha P_T W = \frac{50}{16}\eta$ and the maximum transmission power $P = (R_T^{max})^\alpha P_T W = 8 \cdot 10^6 \eta$. We set $\varepsilon = 0.05$, which guarantees that the obtained solution is within 5% of optimality.

Table 5.3 Source node, destination node, and rate requirement of the five sessions.

Session	Source node	Destination node	Rate requirement
1	7	16	28
2	8	5	12
3	15	13	56
4	2	18	75
5	9	11	29

Figure 5.5 Objective value as a function of Q.

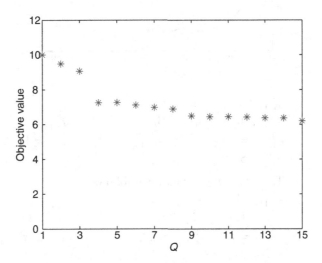

5.6.2 Results

In this set of results, we apply the proposed solution procedure to the 20-node network described above for different levels of power control granularity (Q). Note that $Q = 1$ corresponds to the case where there is no power control, i.e., a node always uses its peak power P for transmission. When Q is sufficiently large, power control approaches a continuum. Fig. 5.5 shows the results. First, we note that the granularity of power control has a significant impact on the optimal objective value. Comparing the case when there is no power control ($Q = 1$) and the case of $Q = 15$, we find that there is nearly a 40% reduction in the optimal objective value. Second, although the optimal objective value is a nonincreasing function of Q, when Q becomes sufficiently large (e.g., 10 in this network setting), further increase in Q does not have much reduction in the optimal objective value. This suggests that, for practical purposes, the number of required power control levels does not need to be a large number.

The rest of our results in this section are for $Q = 10$. For transmission power, we obtain the following results:

Figure 5.6 Flow routing for the five sessions in the 20-node network.

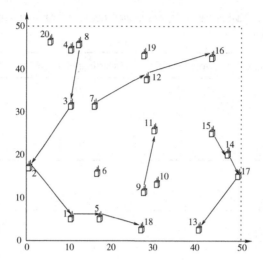

$$q_{9,11}^1 = 3, \qquad q_{12,16}^2 = 4, \qquad q_{7,12}^3 = 3,$$
$$q_{2,1}^4 = 4, \qquad q_{2,1}^5 = 4, \qquad q_{5,18}^6 = 2,$$
$$q_{17,13}^7 = 4, \qquad q_{8,3}^8 = 5, q_{14,17}^8 = 1, \qquad q_{1,5}^9 = 1, q_{15,14}^9 = 1,$$
$$q_{3,2}^{10} = 5.$$

The scheduling results are as follows:

$$x_{9,11}^1 = 1, \qquad x_{12,16}^2 = 1, \qquad x_{7,12}^3 = 1,$$
$$x_{2,1}^4 = 1, \qquad x_{2,1}^5 = 1, \qquad x_{5,18}^6 = 1,$$
$$x_{17,13}^7 = 1, \qquad x_{8,3}^8 = 1, x_{14,17}^8 = 1, \qquad x_{1,5}^9 = 1, x_{15,14}^9 = 1,$$
$$x_{3,2}^{10} = 1.$$

The flow routing topology is shown in Fig. 5.6. The corresponding flow rates are as follows:

$$f_{7,12}(1) = 28, \qquad f_{12,16}(1) = 28,$$
$$f_{8,3}(2) = 12, \qquad f_{3,2}(2) = 12, \qquad f_{2,1}(2) = 12, \qquad f_{1,5}(2) = 12,$$
$$f_{15,14}(3) = 56, \qquad f_{14,17}(3) = 56, \qquad f_{17,13}(3) = 56,$$
$$f_{2,1}(4) = 75, \qquad f_{1,5}(4) = 75, \qquad f_{5,18}(4) = 75,$$
$$f_{9,11}(5) = 29.$$

Note that a link may be used by multiple sessions. For example, link $2 \to 1$ is used by Sessions 2 and 4. As a result, the total data rate on link $2 \to 1$ is $f_{2,1}(2) + f_{2,1}(4) = 12 + 75 = 87$.

The following observations on the numerical results show the close coupling relationships between per-node power control and the upper layers. In one observation (scheduling), we can see that links $8 \to 3$ and $14 \to 17$ are active on the same band 8. This is feasible because the interference range at node 14 is 22.49 under $q_{14,17}^8 = 1$, which is smaller than 38.35 (the distance between nodes 3 and 14). We note that if there is no power control, i.e., node 14 uses the peak transmission power, then node 3 will be in the interference range of node 14, which is $R_I^{max} = 40$, and will lead to a scheduling conflict. In another observation (routing), we see that, for Session 2 (node 8 to node 5), the routing path is $8 \rightarrow 3 \rightarrow 2 \rightarrow 1 \rightarrow 5$. Here, a shorter path $8 \rightarrow 3 \rightarrow 6 \rightarrow 5$ is not used. This is because path $8 \rightarrow 3 \rightarrow 6 \rightarrow 5$ is interfered by transmissions on other paths. The optimal solution tends to choose paths that are not close to each other. Finally, for Session 3 (node 15 to node 13), the routing path is $15 \rightarrow 14 \rightarrow 17 \rightarrow 13$, while a shorter path $15 \rightarrow 17 \rightarrow 13$ is not used. This is because node 15 can use a smaller transmission power to transmit to a closer neighboring node 14 with a smaller interference footprint. This allows link $1 \rightarrow 5$ to be active on the same band 9 with link $15 \rightarrow 14$.

5.7 Chapter summary

This is the first chapter of the second part of this book, which presents some methods to develop $(1 - \varepsilon)$-optimal solutions. In this chapter, we presented the branch-and-bound framework and showed how it could be applied to solve discrete and combinatorial optimization problems. Such problems are typically considered as the most difficult problems in nonconvex optimization, and the branch-and-bound framework offers a general purpose and effective approach. The effectiveness of branch-and-bound resides in the careful design of each component in its framework, such as computation of a lower bound, local search of an upper bound, and selection of partitioning variables (in the case of a minimization problem). It should be noted that the worst-case complexity of a branch-and-bound-based method remains exponential, although a judicious design of each component could achieve reasonable computational time in practice.

In the case study, we considered a per-node power control problem for a multi-hop CRN. This problem has a large design space that involves a tight coupling relationship among power control, scheduling, and flow routing, which is typical for a cross-layer optimization problem. We developed a mathematical model and a problem formulation, which is a mixed-integer nonlinear programming (MINLP) problem. We showed how to apply the branch-and-bound framework to design a solution procedure. Under the branch-and-bound framework, we showed how to derive a linear relaxation to compute a lower bound, how to perform a local search to find a feasible solution (upper bound), and how to select suitable partitioning variables. Despite its worst-case exponential complexity, the solution procedure that we developed here is a viable approach to solve the per-node power control problem for a multi-hop CRN (as demonstrated in the numerical examples).

5.8 Problems

5.1 Describe the branch-and-bound framework for an optimization problem with a maximization objective (instead of minimization in this chapter).

5.2 In the case study, we employed the so-called "protocol model." Read [152] and discuss the pros and cons of this model.

5.3 Show that for a session, if flow balance holds at the source node [Eq. (5.5)] and at the relay nodes [Eq. (5.6)], then flow balance also holds at this session's destination node [Eq. (5.7)].

5.4 Referring to the RHS of (5.8), explain why there is no interference term in the denominator inside the log function to calculate SINR.

5.5 In this chapter, we used the so-called network-wide bandwidth-footprint-product (BFP) as the objective function. However, it is possible that two or more neighboring nodes transmitting on the same frequency band may have a partial overlap of their footprints. As a result, the overlapped area will be counted multiple times in the objective function. Give a justification on why this still makes sense.

5.6 In the formulation of the original problem, which constraints are associated with a single layer? Which constraints couple multiple layers together?

5.7 In the formulation of the original problem, we discretize transmission power from a continuous variable to a discrete variable. Then in the relaxation step of the branch-and-bound solution procedure, we change the discrete transmission power variable to a continuous variable. Explain the purpose of this back and forth change between continuous and discrete variables for the transmission power.

5.8 We showed how to construct a linear relaxation for the discrete term $(q_{ij}^m)^{2/\alpha}$. Provide the details for deriving a linear relaxation for the discrete term $\log_2\left(1 + \frac{g_{ij}P}{\eta W Q}q_{ij}^m\right)$ in the problem formulation.

5.9 There is another approach to obtain a linear relaxation for the discrete term $(q_{ij}^m)^{2/\alpha}$, where $q_{ij}^m \in \{q_0, q_1, \ldots, q_K\}$. Introduce binary variables y_{ij}^{mk}, where $y_{ij}^{mk} = 1$ if $q_{ij}^m = q_k$ and $y_{ij}^{mk} = 0$ otherwise. Thus, $\sum_{k=0}^{K} y_{ij}^{mk} = 1$ and $q_{ij}^m = \sum_{k=0}^{K} y_{ij}^{mk} q_k$. We also introduce a new variable u_{ij}^m for $(q_{ij}^m)^{2/\alpha}$ and have $u_{ij}^m = \sum_{k=0}^{K} y_{ij}^{mk}(q_k)^{2/\alpha}$. A linear relaxation can be obtained by relaxing y_{ij}^{mk} to a continuous variable in [0, 1]. Compare this approach and the linear relaxation approach in Section 5.5.1 and analyze which provides a tighter relaxation.

5.10 In local search, why do we initialize q_{ij}^m at $(q_{ij}^m)_L$ and increase it upward? Comment on the strategy of starting q_{ij}^m at $(q_{ij}^m)_U$ and decreasing it downward.

5.11 In local search, we focus on increasing the power values (see Steps 5, 8, and 11 in Algorithm. 5.2) to ensure that constraint (5.16) holds for each link $i \to j$. Explain how other constraints in the original problem hold.

5.12   In the selection of partitioning variables, why are the **x**-variables more important than the **q**-variables? After branching a problem into two new subproblems, how can we revise the value-sets of the other **x**- and **q**-variables by using constraints (5.12)–(5.15)?

5.13   When we examine the relaxation errors for the **q**-variables, why can $|\hat{u}_{ij}^m - (\hat{q}_{ij}^m)^{2/\alpha}|$ be greater than zero?

6 Reformulation-Linearization Technique and applications

Destiny is not a matter of chance, it is a matter of choice. It is not a thing to be waited for, it is a thing to be achieved.

William Jennings Bryan

6.1 An introduction of Reformulation-Linearization Technique (RLT)

In Chapter 5, we presented a framework for branch-and-bound algorithms for solving nonconvex programming problems. There are several key components in this framework that are problem specific and need to be carefully custom-designed. In particular, we need to develop a tight relaxation solution, i.e., one that yields a tight upper bound for a maximization problem (or a tight lower bound for a minimization problem). For the case study in Chapter 5, we presented two different linear relaxations for two particular types of nonlinear terms (see Section 5.5.1). However, such relaxations may not be applicable to some other nonlinear terms. Hence, we address this important question in this chapter. That is, is there a relaxation approach that can be applied to a general class of nonlinear terms?

In this chapter, we show that the Reformulation-Linearization Technique (RLT) [140–142; 146], developed by a co-author of this book (Prof. Hanif Sherali) is such an approach. RLT is a systematic approach for deriving tight linear relaxations for any monomial (i.e., a polynomial term of the form $\prod_{i=1}^{n}(x_i)^{c_i}$ in variables x_i, where the c_i-exponents are constant integers; extensions to rational exponents are presented in [144]).

The idea of RLT is best explained with a simple example. Suppose that we have a nonlinear (bilinear) term $x_1 x_2$ in a nonconvex program involving variables x_1 and x_2. To derive a linear relaxation via RLT, we introduce a new variable $X_{\{1,2\}}$ to represent $x_1 x_2$ and add suitable linear constraints for this new variable. Suppose that we have derived (or are given) the lower and upper bounds for variables x_1 and x_2, i.e., $(x_1)_L \leq x_1 \leq (x_1)_U$ and $(x_2)_L \leq x_2 \leq (x_2)_U$, respectively. Then the following so-called *bound-factor constraints* must hold:

$$[x_1 - (x_1)_L] \cdot [x_2 - (x_2)_L] \geq 0,$$
$$[x_1 - (x_1)_L] \cdot [(x_2)_U - x_2] \geq 0,$$
$$[(x_1)_U - x_1] \cdot [x_2 - (x_2)_L] \geq 0,$$
$$[(x_1)_U - x_1] \cdot [(x_2)_U - x_2] \geq 0.$$

Substituting $X_{\{1,2\}}$ for $x_1 x_2$ in the above constraints, we obtain

$$(x_1)_L \cdot x_2 + (x_2)_L \cdot x_1 - X_{\{1,2\}} \leq (x_1)_L \cdot (x_2)_L, \tag{6.1}$$

$$(x_1)_U \cdot x_2 + (x_2)_L \cdot x_1 - X_{\{1,2\}} \geq (x_1)_U \cdot (x_2)_L, \tag{6.2}$$

$$(x_1)_L \cdot x_2 + (x_2)_U \cdot x_1 - X_{\{1,2\}} \geq (x_1)_L \cdot (x_2)_U, \tag{6.3}$$

$$(x_1)_U \cdot x_2 + (x_2)_U \cdot x_1 - X_{\{1,2\}} \leq (x_1)_U \cdot (x_2)_U. \tag{6.4}$$

Therefore, a linear relaxation for the original problem can be obtained by replacing the bilinear term $x_1 x_2$ by $X_{\{1,2\}}$ throughout the problem and adding the above four linear constraints. Note that (6.1)–(6.4) are valid restrictions that relate the original variables x_1 and x_2 and the new *RLT-variable* $X_{\{1,2\}}$. In particular, $X_{\{1,2\}} = x_1 x_2$ holds true when either x_1 or x_2 equals its corresponding lower or upper bound value. This feature, which holds in a more general form (see [141]), plays a decisive role in the tightness of RLT.

Next, we show how to obtain a linear relaxation for x_1^2 in a nonconvex program. Viewing x_1^2 as $x_1 x_1$, we derive a linear relaxation using RLT, by introducing a new variable $X_{\{1,1\}}$ to represent x_1^2, where the subscript of X repeats the index 1 twice corresponding to the exponent in x_1^2, and by adding suitable linear constraints for relating the new variable $X_{\{1,1\}}$ to the original variable x_1. This is done similar to the above, by regarding x_1^2 as a special case of $x_1 x_2$ when $x_2 \equiv x_1$. Therefore, a linear relaxation for the original problem can be obtained by replacing all occurrences of x_1^2 within the polynomial program by $X_{\{1,1\}}$ and by adding the following three bound-factor linear constraints. Note that for this special case, both (6.2) and (6.3) become (6.6):

$$2(x_1)_L \cdot x_1 - X_{\{1,1\}} \leq [(x_1)_L]^2, \tag{6.5}$$

$$[(x_1)_U + (x_1)_L] \cdot x_1 - X_{\{1,1\}} \geq (x_1)_U \cdot (x_1)_L, \tag{6.6}$$

$$2(x_1)_U \cdot x_1 - X_{\{1,1\}} \leq [(x_1)_U]^2. \tag{6.7}$$

This strategy can be used in a likewise fashion for constructing a linear relaxation via RLT for a general monomial $\prod_{i=1}^{n} (x_i)^{c_i}$ in variables x_i, where the c_i-exponents are integer constants for $1 \leq i \leq n$. Although the general RLT process is described in [141], we present a *sequential quadrification* process [143] here for the sake of simplicity. This strategy adopts a two-step approach for the linearization process. In the first step, we introduce new variables and identities to represent $\prod_{i=1}^{n}(x_i)^{c_i}$ using quadratic terms. In the second step, we introduce RLT-variables as above to represent the quadratic terms and generate the related RLT bound-factor constraints in order to derive a linear relaxation. The following example illustrates the general idea:

Example 6.1

Suppose that we have a nonlinear term $x_1 x_2^5 x_3$ in a nonconvex program with variables x_1, x_2, and x_3. We begin by factoring $x_1 x_2^5 x_3$ into quadratic relationships as follows by defining new variables (note that this is not a unique representation [143]):

$$X_{\{2,2\}} = x_2^2, \tag{6.8}$$

$$X_{\{2,2,2,2\}} = X_{\{2,2\}}^2, \tag{6.9}$$

$$X_{\{1,2\}} = x_1 x_2, \tag{6.10}$$

$$X_{\{1,2,2,2,2\}} = X_{\{1,2\}} X_{\{2,2,2,2\}}, \tag{6.11}$$

$$X_{\{1,2,2,2,2,2,3\}} = X_{\{1,2,2,2,2,2\}} x_3. \tag{6.12}$$

Note that in terms of the original variables x_1, x_2, and x_3, we have that $X_{\{1,2,2,2,2,2,3\}} = x_1 x_2^5 x_3$. Hence, we replace $x_1 x_2^5 x_3$ in the given polynomial program with the new variable $X_{\{1,2,2,2,2,2,3\}}$. Furthermore, as before, we include suitable linearized bound-factor constraints to represent the quadratic relationships in the above equalities. For example, for (6.8), we include the following three linear constraints, as in (6.5)–(6.7) above:

$$2(x_2)_L \cdot x_2 - X_{\{2,2\}} \le [(x_2)_L]^2,$$

$$[(x_2)_U + (x_2)_L] \cdot x_2 - X_{\{2,2\}} \ge (x_2)_U \cdot (x_2)_L,$$

$$(x_2)_U \cdot x_2 - X_{\{2,2\}} \le [(x_2)_U]^2.$$

The similar three linear constraints for representing (6.9) are left as a homework exercise (note that implied lower and upper bounds would need to be derived for the variables $X_{\{1,2\}}$, $X_{\{2,2,2,2\}}$, and $X_{\{1,2,2,2,2\}}$). For (6.10), we generate the following four bound-factor linear constraints as in (6.1)–(6.4):

$$(x_1)_L \cdot x_2 + (x_2)_L \cdot x_1 - X_{\{1,2\}} \le (x_1)_L \cdot (x_2)_L,$$

$$(x_1)_U \cdot x_2 + (x_2)_L \cdot x_1 - X_{\{1,2\}} \ge (x_1)_U \cdot (x_2)_L,$$

$$(x_1)_L \cdot x_2 + (x_2)_U \cdot x_1 - X_{\{1,2\}} \ge (x_1)_L \cdot (x_2)_U,$$

$$(x_1)_U \cdot x_2 + (x_2)_U \cdot x_1 - X_{\{1,2\}} \le (x_1)_U \cdot (x_2)_U.$$

The similar four linear constraints for representing each of the relationships (6.11) and (6.12) are relegated to the exercises. This produces a linear relaxation for the original problem with five additional variables and 24 additional linear constraints.

For a more comprehensive discussion and surveys of RLT and related theoretical results, we refer the readers to [146; 148; 149]. In the rest of this chapter, we will show how RLT can be applied to solve problems in a wireless network.

6.2 Case study: Capacity maximization for multi-hop cognitive radio networks under the physical model

We study a capacity maximization problem under the physical model, which is also called the signal-to-interference-and-noise ratio (SINR) model. In this model, concurrent transmissions are allowed and interference (due to transmissions by nonintended transmitter) is treated as noise. SINR at a receiver not only depends on the transmission power at the corresponding transmitter, but also depends on the transmission power at other transmitters. A transmission is successful if and only if SINR at the receiver is greater than or equal to a threshold. The achieved transmission capacity is a function of SINR (via Shannon capacity formula).

Consider a multi-hop CRN where each node has access to a set of available bands (likely heterogeneous). We are interested in how to maximize the rates of a set of user communication sessions, with joint consideration at the physical layer (via power control), the link layer (via frequency band scheduling), and the network layer (via flow routing). We give a mathematical characterization of these layers and formulate a mixed-integer nonlinear programming (MINLP) problem. For this optimization problem, we first identify core optimization variables and the core optimization space based on the physical significance of the variables. We devise an algorithm using RLT in concert with the branch-and-bound framework to obtain a $(1 - \varepsilon)$-optimal solution.

As we learned in Chapter 5, although the branch-and-bound framework is standard, many components within this framework need to be custom-designed for the specific problem. For our problem, we develop the following components. (1) By applying RLT, we develop a tight linear relaxation so as to obtain a tight upper bound for our objective. (2) To compute the lower bounds, we design a local search algorithm by analyzing and removing infeasibility in the resulting linear relaxation solution. (3) For problem partitioning, we develop a branching strategy based on the physical significance of the partitioning variables. With these carefully designed components, the overall branch-and-bound solution procedure is able to compute a $(1 - \varepsilon)$-optimal solution much faster than using a brute-force exhaustive search.

The remainder of this chapter is organized as follows. Section 6.3 gives a mathematical characterization of power control, scheduling, and routing in the SINR model for a multi-hop CRN. In Section 6.4, we perform problem reformulation and obtain a cleaner and more compact formulation. Section 6.5 analyzes the core optimization space and describes the algorithm to obtain a $(1 - \varepsilon)$-optimal solution. Section 6.6 presents some numerical results. Section 6.7 summarizes this chapter.

6.3 Mathematical models

Denote $\mathcal{N}$ as the set of nodes of a multi-hop CRN. Each node $i \in \mathcal{N}$ senses its environment and finds a set of available frequency bands $\mathcal{M}_i$ that it can use, which may not be identical to those at other nodes. We assume that the bandwidth of each frequency band (channel) is W. Denote $\mathcal{M}$ the union of all frequency bands among all the nodes in the network, i.e., $\mathcal{M} = \bigcup_{i \in \mathcal{N}} \mathcal{M}_i$. Denote $\mathcal{M}_{ij} = \mathcal{M}_i \bigcap \mathcal{M}_j$, which is the set of frequency bands that is common between nodes i and j and thus can be used for communication between the two nodes. In the rest of this section, we present mathematical modeling for each layer and formulate a throughput maximization problem. Table 6.1 lists the notation.

6.3.1 Power control, scheduling, and their relationship in the SINR model

Power control on each transmitting node at the physical layer affects SINR at a receiving node. These SINR values in turn will affect scheduling decisions at the link layer. That is, if a node is scheduled to receive, then its SINR must be at least $S_{\min}$ (minimum threshold requirement). Therefore, power control and scheduling are tightly coupled via SINR and cannot be modeled separately.

Scheduling at a node can be done either in the frequency domain or time domain. In this chapter, we consider scheduling in the frequency domain in the form of assigning frequency bands (channels). Note that a time domain-based formulation can be done in a similar fashion.

In the SINR model, there may still be concurrent transmissions within the same channel (and thus interference). Denote scheduling variables x_{ij}^m as follows:

$$x_{ij}^m = \begin{cases} 1 & \text{if node } i \text{ transmits data to node } j \text{ on band } m, \\ 0 & \text{otherwise.} \end{cases}$$

We assume that a node can use a band for transmission (or reception) to (or from) only one other node. That is,

$$\sum_{k \in \mathcal{N}, k \neq i}^{k:m \in \mathcal{M}_k} x_{ki}^m + \sum_{j \in \mathcal{N}, j \neq i}^{j:m \in \mathcal{M}_j} x_{ij}^m \leq 1 \quad (i \in \mathcal{N}, m \in \mathcal{M}_i). \tag{6.13}$$

For power control, we assume that the transmission power at a node can only be tuned to a finite number of levels between 0 and $P_{\max}$. To model this *discrete* power control, we introduce an integer parameter Q that represents the total number of power levels to which a transmitter can be adjusted,

Table 6.1 Notation.

Symbol	Definition
c_{ij}^m	Capacity of link $i \to j$ under p_{ij}^m
$d(l)$	Destination node of session $l \in \mathcal{L}$
$f_{ij}(l)$	Data rate for session l on link $i \to j$
g_{ij}	Propagation loss from node i to node j
K	Rate-scaling factor for all sessions
LB_z	Lower bound for Problem z
LB	The maximum lower bound among all problems
$\mathcal{L}$	A set of user communication sessions in the network
$\mathcal{M}_i$	A set of available bands at node $i \in \mathcal{N}$
$\mathcal{M}$	$= \bigcup_{i \in \mathcal{N}} \mathcal{M}_i$, the set of frequency bands in the network
$\mathcal{M}_{ij}$	$= \mathcal{M}_i \bigcap \mathcal{M}_j$, the set of frequency bands on link $i \to j$
$\mathcal{N}$	A set of nodes in the network
p_{ij}^m	The transmission power from node i to node j on band m
$P_{\max}$	The maximum transmission power at a transmitter
q_{ij}^m	Discrete transmission power level from node i to node j on band m
Q	Number of discrete transmission power levels at a transmitter
$r(l)$	Minimum rate requirement of session l
$s(l)$	Source node of session $l \in \mathcal{L}$
s_{ij}^m	SINR from node i to node j on band m
$S_{\min}$	The minimum required SINR threshold
t_i^m	$= \sum_{j \in \mathcal{N}, j \neq i}^{j: m \in \mathcal{M}_j} p_{ij}^m$, sum of transmission power at node i on band m
UB_z	An upper bound for Problem z
UB	The maximum upper bound among all problems
W	Bandwidth of a frequency band
x_{ij}^m	Binary variable to indicate whether or not band m is used on link $i \to j$
ε	A small positive constant reflecting desired accuracy
η	Ambient Gaussian noise density
ψ_z	A local search solution for Problem z
ψ_ε	A $(1 - \varepsilon)$-optimal solution
Ω_z	The core optimization space of the relaxed Problem z

i.e., the transmission power can be $0, \frac{1}{Q} P_{\max}, \frac{2}{Q} P_{\max}, \ldots, P_{\max}$. Denote $q_{ij}^m \in \{0, 1, 2, \ldots, Q\}$ as the integer levels corresponding to their respective transmission powers. Clearly, when node i does not transmit data to node j on band m, $q_{ij}^m = 0$. Thus, power control and scheduling are coupled with each other via the following relationship:

$$q_{ij}^m \begin{cases} \in [1, Q] & \text{if } x_{ij}^m = 1, \\ = 0 & \text{otherwise,} \end{cases} \quad (i, j \in \mathcal{N}, i \neq j, m \in \mathcal{M}_{ij}). \quad (6.14)$$

Consider a transmission from node i to node j on band m. When there is interference from concurrent transmissions on the same band, the SINR at node j, denoted as s_{ij}^m, is

$$s_{ij}^m = \frac{g_{ij} \frac{q_{ij}^m}{Q} P_{\max}}{\eta W + \sum_{k \in \mathcal{N}, k \neq i}^{k:m \in \mathcal{M}_k} \sum_{h \in \mathcal{N}, h \neq k}^{h:m \in \mathcal{M}_h} g_{kj} \frac{q_{kh}^m}{Q} P_{\max}} \qquad (i, j \in \mathcal{N}, i \neq j, m \in \mathcal{M}_{ij}), \quad (6.15)$$

where η is the ambient Gaussian noise density and g_{ij} is the propagation loss from node i to node j.

Note that, in theory, for any small SINR, the corresponding capacity is still positive (by Shannon's capacity formula). But in practice, if SINR is too small, then the achieved capacity will also be very small. In this case, such a weak link will not be very useful to carry traffic flow. Thus, we may use a threshold to remove such weak links from consideration. In this regard, we introduce a threshold for SINR, i.e., a transmission from node i to node j on band m is considered *successful* if and only if $s_{ij}^m \geq S_{\min}$. We thus have the following coupling relationship for scheduling (x_{ij}^m) and SINR (s_{ij}^m):

$$x_{ij}^m = 1 \iff s_{ij}^m \geq S_{\min} \qquad (i, j \in \mathcal{N}, i \neq j, m \in \mathcal{M}_{ij}). \tag{6.16}$$

6.3.2 Routing and link capacity

We assume that there is a set $\mathcal{L}$ of active user communication (unicast) sessions in the network. Denote $s(l)$ and $d(l)$ as the source and destination nodes of session $l \in \mathcal{L}$, and $r(l)$ as the minimum rate requirement (in b/s) for session l. In our study, we aim to maximize a common scaling factor K for all session rates. That is, we aim to determine the maximum K such that a rate of $K \cdot r(l)$ can be transported from $s(l)$ to $d(l)$ for *each* session $l \in \mathcal{L}$. Again, for optimality and flexibility, we allow flow splitting and multi-path routing inside the network.

Mathematically, this can be modeled as follows. Denote $f_{ij}(l)$ as the data rate on link $i \to j$ that is attributed to session l. If node i is the source node of session l, i.e., $i = s(l)$, then

$$\sum_{\substack{j \in \mathcal{N}, j \neq i}}^{j:\mathcal{M}_{ij} \neq \emptyset} f_{ij}(l) = K \cdot r(l) \qquad (l \in \mathcal{L}, i = s(l)). \tag{6.17}$$

If node i is an intermediate relay node for flow attributed to session l, i.e., $i \neq s(l)$ and $i \neq d(l)$, then

$$\sum_{\substack{j \in \mathcal{N}, j \neq i, s(l)}}^{j:\mathcal{M}_{ij} \neq \emptyset} f_{ij}(l) = \sum_{\substack{k \in \mathcal{N}, k \neq i, d(l)}}^{k:\mathcal{M}_{ki} \neq \emptyset} f_{ki}(l) \qquad (l \in \mathcal{L}, i \in \mathcal{N}, i \neq s(l), d(l)).$$

$$\tag{6.18}$$

If node i is the destination node of session l, i.e., $i = d(l)$, then

$$\sum_{\substack{k \in \mathcal{N}, k \neq i}}^{k:\mathcal{M}_{ki} \neq \emptyset} f_{ki}(l) = K \cdot r(l) \qquad (l \in \mathcal{L}, i = d(l)). \tag{6.19}$$

In addition to the above flow balance equations at each node $i \in \mathcal{N}$ for session $l \in \mathcal{L}$, we impose a constraint to assure that the aggregated flow rates on each radio link do not exceed this link's capacity, i.e., for a link $i \to j$, we have

$$\sum_{\substack{l \in \mathcal{L}}}^{s(l) \neq j, d(l) \neq i} f_{ij}(l) \leq \sum_{m \in \mathcal{M}_{ij}} W \log_2(1 + s_{ij}^m) \quad (i, j \in \mathcal{N}, i \neq j, \mathcal{M}_{ij} \neq \emptyset).$$

(6.20)

This constraint shows the coupling relationship between flow routing and SINR.

6.3.3 A throughput maximization problem

For throughput maximization, suppose we are interested in maximizing a common scaling factor K for all sessions under some given minimum rate requirements $r(l)$. That is, we want to determine the maximum factor K such that a rate of $K \cdot r(l)$ can be transmitted from $s(l)$ to $d(l)$ for *each* session $l \in \mathcal{L}$ in the network. Putting together the objective and all the constraints for power control, scheduling, and flow routing, we have the following formulation:

Maximize $\quad K$

subject to $\quad$ Constraints (6.13)–(6.20)

$x_{ij}^m \in \{0, 1\}, q_{ij}^m \in \{0, 1, 2, \ldots, Q\}, t_i^m, s_{ij}^m \geq 0 \quad (i, j \in \mathcal{N}, i \neq j, m \in \mathcal{M}_{ij})$

$K, f_{ij}(l) \geq 0 \quad (l \in \mathcal{L}, i, j \in \mathcal{N}, i \neq j, i \neq d(l), j \neq s(l), \mathcal{M}_{ij} \neq \emptyset).$

6.4 Reformulation

A formulation like the one in Section 6.3.3 is the first step in formulating our cross-layer optimization problem. But it is in a rather "raw" form and more work needs to be done to reformulate it into a more compact form that is amenable to solution development. In this section, we analyze each constraint in detail and perform some necessary and important reformulations.

- The constraint described in (6.14) is not suitable for mathematical programming. We reformulate it with the following linear constraint:

$$x_{ij}^m \leq q_{ij}^m \leq Q x_{ij}^m \quad (i, j \in \mathcal{N}, i \neq j, m \in \mathcal{M}_{ij}). \quad (6.21)$$

It is easy to verify that this constraint is equivalent to (6.14).

- Constraint (6).15 is in the form of a fraction. In a mathematical program, a product form is more convenient to handle. We can rewrite (6).15 as follows:

$$
s_{ij}^m = \frac{g_{ij}\frac{q_{ij}^m}{Q}P_{\max}}{\eta W + \sum_{k \in \mathcal{N}, k \neq i}^{k:m \in \mathcal{M}_k} \sum_{h \in \mathcal{N}, h \neq k}^{h:m \in \mathcal{M}_h} g_{kj}\frac{q_{kh}^m}{Q}P_{\max}}
$$

$$
= \frac{g_{ij}q_{ij}^m}{\frac{\eta W Q}{P_{\max}} + \sum_{k \in \mathcal{N}, k \neq i}^{k:m \in \mathcal{M}_k} \sum_{h \in \mathcal{N}, h \neq k}^{h:m \in \mathcal{M}_h} g_{kj}q_{kh}^m} \quad (i, j \in \mathcal{N}, i \neq j, m \in \mathcal{M}_{ij}).
$$

This is equivalent to

$$
\frac{\eta W Q}{P_{\max}}s_{ij}^m + \sum_{k \in \mathcal{N}, k \neq i}^{k:m \in \mathcal{M}_k} \sum_{h \in \mathcal{N}, h \neq k}^{h:m \in \mathcal{M}_h} g_{kj}q_{kh}^m s_{ij}^m - g_{ij}q_{ij}^m = 0
$$
$$
(i, j \in \mathcal{N}, i \neq j, m \in \mathcal{M}_{ij}). \quad (6.22)
$$

Note that in (6.22), q_{kh}^m and s_{ij}^m are variables, while all other symbols are constants. Thus, we have a double sum of nonlinear terms $q_{kh}^m s_{ij}^m$ in (6.22). To reduce the number of nonlinear terms, denote

$$
t_k^m = \sum_{h \in \mathcal{N}, h \neq k}^{h:m \in \mathcal{M}_h} q_{kh}^m \quad (k \in \mathcal{N}, m \in \mathcal{M}_k). \quad (6.23)
$$

Then (6.22) can be rewritten as

$$
\frac{\eta W Q}{P_{\max}}s_{ij}^m + \sum_{k \in \mathcal{N}, k \neq i}^{k:m \in \mathcal{M}_k} g_{kj}t_k^m s_{ij}^m - g_{ij}q_{ij}^m = 0 \quad (i, j \in \mathcal{N}, i \neq j, m \in \mathcal{M}_{ij}), \quad (6.24)
$$

which now only involves a single sum of nonlinear terms $t_k^m s_{ij}^m$.
- Similar to (6.14), the constraint described in (6.16) is not suitable for mathematical programming. It can be shown that (6.16) can be eliminated if we have (6.21), (6).24, and the following new constraint:

$$
s_{ij}^m \geq S_{\min}x_{ij}^m \quad (i, j \in \mathcal{N}, i \neq j, m \in \mathcal{M}_{ij}). \quad (6.25)
$$

The verification of this fact is left as a homework problem.
- Finally, we can easily prove that (6.17) and (6.18) imply (6.19). Thus, we can remove constraint (6.19) and only retain (6.17) and (6.18) in the formulation.

With these careful reformulations, we now have a more compact problem formulation, which is shown in Table 6.2.

Table 6.2 Problem formulation.

Maximize $\quad K$

subject to $\quad \sum_{k \in \mathcal{N}, k \neq i}^{k:m \in \mathcal{M}_k} x_{ki}^m + \sum_{j \in \mathcal{N}, j \neq i}^{j:m \in \mathcal{M}_j} x_{ij}^m \leq 1 \qquad\qquad (i \in \mathcal{N}, m \in \mathcal{M}_i)$

$\quad q_{ij}^m - x_{ij}^m \geq 0 \qquad\qquad\qquad\qquad\qquad (i, j \in \mathcal{N}, i \neq j, m \in \mathcal{M}_{ij})$

$\quad q_{ij}^m - Q x_{ij}^m \leq 0 \qquad\qquad\qquad\qquad\quad (i, j \in \mathcal{N}, i \neq j, m \in \mathcal{M}_{ij})$

$\quad \sum_{j \in \mathcal{N}, j \neq i}^{j:m \in \mathcal{M}_j} q_{ij}^m - t_i^m = 0 \qquad\qquad\qquad (i \in \mathcal{N}, m \in \mathcal{M}_i)$

$\quad \frac{\eta W Q}{P_{\max}} s_{ij}^m + \sum_{k \in \mathcal{N}, k \neq i}^{k:m \in \mathcal{M}_k} g_{kj} t_k^m s_{ij}^m - g_{ij} q_{ij}^m = 0 \qquad (i, j \in \mathcal{N}, i \neq j, m \in \mathcal{M}_{ij})$

$\quad S_{\min} x_{ij}^m - s_{ij}^m \leq 0 \qquad\qquad\qquad\qquad\quad (i, j \in \mathcal{N}, i \neq j, m \in \mathcal{M}_{ij})$

$\quad \sum_{j \in \mathcal{N}, j \neq i}^{j:\mathcal{M}_{ij} \neq \emptyset} f_{ij}(l) - r(l) K = 0 \qquad\qquad\quad (l \in \mathcal{L}, i = s(l))$

$\quad \sum_{j \in \mathcal{N}, j \neq i, s(l)}^{j:\mathcal{M}_{ij} \neq \emptyset} f_{ij}(l) - \sum_{k \in \mathcal{N}, k \neq i, d(l)}^{k:\mathcal{M}_{ki} \neq \emptyset} f_{ki}(l) = 0 \quad (l \in \mathcal{L}, i \in \mathcal{N}, i \neq s(l), d(l))$

$\quad \sum_{l \in \mathcal{L}}^{s(l) \neq j, d(l) \neq i} f_{ij}(l) - \sum_{m \in \mathcal{M}_{ij}} W \log_2(1 + s_{ij}^m) \leq 0 \quad (i, j \in \mathcal{N}, i \neq j, \mathcal{M}_{ij} \neq \emptyset)$

$\quad x_{ij}^m \in \{0, 1\}, q_{ij}^m \in \{0, 1, 2, \ldots, Q\}, t_i^m, s_{ij}^m \geq 0 \quad (i, j \in \mathcal{N}, i \neq j, m \in \mathcal{M}_{ij})$

$\quad K, f_{ij}(l) \geq 0 \qquad (l \in \mathcal{L}, i, j \in \mathcal{N}, i \neq j, i \neq d(l), j \neq s(l), \mathcal{M}_{ij} \neq \emptyset)$

6.5 A solution procedure

For the optimization problem in Table. 6.2, $K, x_{ij}^m, q_{ij}^m, t_i^m, s_{ij}^m$, and $f_{ij}(l)$ are optimization variables and $Q, \eta, W, S_{\min}, P_{\max}, g_{ij}$, and $r(l)$ are constants. This formulation is a mixed-integer nonlinear programming (MINLP), which is NP-hard in general [46]. In Section 6.5.1, we first analyze the intricate relationship among the variables and identify the core variables among all the variables. We show that the dependent variables can be derived once these core variables are fixed. We call the optimization space for the core variables the core optimization space. In Section 6.5.2, we present the main algorithm on how to determine an optimal solution in the core optimization space. Several key components in the main algorithm are described in Sections 6.5.3, 6.5.4, and 6.5.5.

6.5.1 Core variables

For the complex MINLP problem, its variables include $x_{ij}^m, q_{ij}^m, t_i^m, s_{ij}^m, f_{ij}(l)$, and K. However, a close investigation of these variables show that they are inter-dependent. In particular, we find that the x_{ij}^m- and q_{ij}^m-variables are "core" variables and the other variables $\{t_i^m, s_{ij}^m, f_{ij}(l), K\}$ can all be derived based on these core variables. How to derive these dependent variables based on the core variables is left as a homework problem. As a result, we can focus our study on an optimization space defined by the core variables, x_{ij}^m and q_{ij}^m, which is a much smaller space.

6.5.2 A solution

In this section, we describe a solution procedure based on the branch-and-bound framework. Recall that under branch-and-bound, we aim to provide a $(1 - \varepsilon)$-optimal solution, where ε is a small positive constant that reflects our desired minimal accuracy $(1 - \varepsilon)$ in computing the final solution. A branch-and-bound framework for a minimization problem was presented in Section 5.1. In this chapter, we need to solve a maximization problem, which follows a similar scheme and is described in Algorithm 6.1.

Algorithm 6.1 A solution

Initialization

1. Let the initial best solution $\psi_\varepsilon \equiv \emptyset$ and the initial lower bound $LB \equiv -\infty$.
2. Determine initial value-set for each core variable.
3. Initialize the problem list to include the original problem, and denote this problem as Problem 1.
4. Obtain an upper bound UB_1 for Problem 1.

Main iteration

5. Select Problem z that has the maximum UB_z-value among all problems in the problem list.
6. Update upper bound $UB = UB_z$.
7. Find a feasible solution ψ_z along with a lower bound LB_z.
8. If $(LB_z > LB)$ {
9.   Update $\psi_\varepsilon = \psi_z$ and $LB = LB_z$.
10.   If $LB \geq (1 - \varepsilon)UB$, stop with a $(1 - \varepsilon)$-optimal solution ψ_ε.
11.   Otherwise, remove all problems z' with $LB \geq (1 - \varepsilon)UB_{z'}$ from the problem list. }
12. Build two new problems z_1 and z_2 from Problem z.
13. Remove Problem z from the problem list.
14. Obtain UB_{z1} and UB_{z2} for Problems z_1 and z_2.
15. If $LB < (1 - \varepsilon)UB_{z1}$, add Problem z_1 to the problem list.
    If $LB < (1 - \varepsilon)UB_{z2}$, add Problem z_2 to the problem list.
16. Go to the next main iteration.

Since our core optimization space is finite (with a finite number of core variables x_{ij}^m and q_{ij}^m, where each core variable has a finite integer value-set), the branch-and-bound algorithm is guaranteed to converge in a finite number of iterations. Several components (i.e., determining upper and lower bounds, and partitioning subproblems) in the main algorithm are yet to be developed. These components should exploit problem-specific structures to optimize performance. In the rest of this section, we show how these components can be designed.

6.5.3 Determining upper bounds

To find an upper bound for a subproblem in the proposed branch-and-bound algorithm (see Steps 4 and 14 in Algorithm 6.1), we can construct a linear relaxation by linearizing all the constraints in the model of Table 6.2. Hence, the relaxed problem can be solved as an LP problem, where the optimal solution value provides an upper bound.

Note that in Table 6.2, $t_k^m s_{ij}^m$ and $\log_2(1 + s_{ij}^m)$ are nonlinear terms. For the monomial $t_k^m s_{ij}^m$, we can apply the RLT that we introduced in Section 6.1. We define a new variable u_{ijk}^m to represent $t_k^m s_{ij}^m$ and introduce four additional bound-factor linear constraints to relate this new variable to the original variables. In particular, suppose that t_k^m and s_{ij}^m are bounded by $(t_k^m)_L \le t_k^m \le (t_k^m)_U$ and $(s_{ij}^m)_L \le s_{ij}^m \le (s_{ij}^m)_U$, respectively. Then, this process yields the following linear constraints for u_{ijk}^m:

$$(t_k^m)_L \cdot s_{ij}^m + (s_{ij}^m)_L \cdot t_k^m - u_{ijk}^m \le (t_k^m)_L \cdot (s_{ij}^m)_L,$$

$$(t_k^m)_U \cdot s_{ij}^m + (s_{ij}^m)_L \cdot t_k^m - u_{ijk}^m \ge (t_k^m)_U \cdot (s_{ij}^m)_L,$$

$$(t_k^m)_L \cdot s_{ij}^m + (s_{ij}^m)_U \cdot t_k^m - u_{ijk}^m \ge (t_k^m)_L \cdot (s_{ij}^m)_U,$$

$$(t_k^m)_U \cdot s_{ij}^m + (s_{ij}^m)_U \cdot t_k^m - u_{ijk}^m \le (t_k^m)_U \cdot (s_{ij}^m)_U.$$

For the nonlinear term $\log_2(1 + s_{ij}^m) = \frac{1}{\ln 2}\ln(1 + s_{ij}^m)$, we propose to employ three tangential supports for $\ln(1 + s_{ij}^m)$, which yields a relaxation for the underlying convex hull linear representation (see Fig. 6.1). Suppose that s_{ij}^m is bounded by $(s_{ij}^m)_L \le s_{ij}^m \le (s_{ij}^m)_U$. We introduce a variable $c_{ij}^m = \ln(1 + s_{ij}^m)$, and bound the convex hull of the region defined by the curve c_{ij}^m over $(s_{ij}^m)_L \le s_{ij}^m \le (s_{ij}^m)_U$ by using four segments as displayed in Fig. 6.1, where the segments I, II, and III are tangential supports and the segment IV is the chord. In particular, the three segments I, II, and III are tangential at points $(1 + (s_{ij}^m)_L, \ln(1 + (s_{ij}^m)_L))$, $(1 + \beta, \ln(1 + \beta))$, and $(1 + (s_{ij}^m)_U, \ln(1 + (s_{ij}^m)_U))$, where

$$\beta = \frac{[1 + (s_{ij}^m)_L] \cdot [1 + (s_{ij}^m)_U] \cdot [\ln(1 + (s_{ij}^m)_U) - \ln(1 + (s_{ij}^m)_L)]}{(s_{ij}^m)_U - (s_{ij}^m)_L} - 1$$

Figure 6.1 A convex hull for $c_{ij}^m = \ln(1 + s_{ij}^m)$.

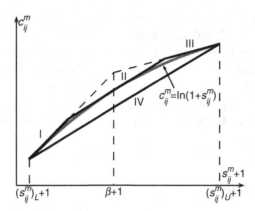

is the s-value corresponding to the intersection point of the segments I and III (see Fig. 6.1); and segment IV is the chord that joins points $(1 + (s_{ij}^m)_L, \ln(1 + (s_{ij}^m)_L))$ and $(1 + (s_{ij}^m)_U, \ln(1 + (s_{ij}^m)_U))$. The convex region defined by the four segments can be described by the following four *linear* constraints:

$$[1 + (s_{ij}^m)_L] \cdot c_{ij}^m - s_{ij}^m \le [1 + (s_{ij}^m)_L] \cdot [\ln(1 + (s_{ij}^m)_L) - 1] + 1,$$

$$(1 + \beta) \cdot c_{ij}^m - s_{ij}^m \le (1 + \beta) \cdot [\ln(1 + \beta) - 1] + 1,$$

$$[1 + (s_{ij}^m)_U] \cdot c_{ij}^m - s_{ij}^m \le [1 + (s_{ij}^m)_U] \cdot [\ln(1 + (s_{ij}^m)_U) - 1] + 1,$$

$$[(s_{ij}^m)_U - (s_{ij}^m)_L] \cdot c_{ij}^m + [\ln(1 + (s_{ij}^m)_L) - \ln(1 + (s_{ij}^m)_U)] \cdot s_{ij}^m$$
$$\ge (s_{ij}^m)_U \cdot \ln(1 + (s_{ij}^m)_L) - (s_{ij}^m)_L \cdot \ln(1 + (s_{ij}^m)_U).$$

As a result, the nonlinear logarithmic term is relaxed using linear constraints.

After relaxing all such nonlinear terms for a subproblem in the branch-and-bound process, say Problem z, we obtain a relaxed Problem $\hat{z}$ as formulated in Table 6.3, which is an LP problem. In Problem $\hat{z}$, $\mathbf{x}$ and $\mathbf{q}$ are vectors that represent all the x_{ij}^m- and q_{ij}^m-variables, respectively; $(x_{ij}^m)_L$, $(x_{ij}^m)_U$, $(q_{ij}^m)_L$, and $(q_{ij}^m)_U$ are constant bounds; $\Omega_z \equiv \{(\mathbf{x}, \mathbf{q}) : (\mathbf{x})_L \le \mathbf{x} \le (\mathbf{x})_U, (\mathbf{q})_L \le \mathbf{q} \le (\mathbf{q})_U\}$ is the core optimization space of $(\mathbf{x}, \mathbf{q})$; and the bounds on the s_{ij}^m-variables for constructing the linearization are derived from the constraints of the problem contingent on $(\mathbf{x}, \mathbf{q}) \in \Omega_z$.

Any such relaxed problem $\hat{z}$ can be solved in polynomial time (since it is an LP problem), where the optimal objective value provides an upper bound for Problem z. We denote by $LP(\hat{z})$ the corresponding optimal solution obtained for the relaxed problem $\hat{z}$.

6.5.4 Determining lower bounds

To compute a lower bound for Problem z, it is sufficient to find a feasible solution to this problem. By a feasible solution, we mean one that simply satisfies all the constraints for Problem z in Table 6.2. Although any feasible solution

Table 6.3 Linear relaxation.

Maximize $\quad\quad K$

subject to $\quad \sum_{k\in\mathcal{N},k\neq i}^{k:m\in\mathcal{M}_k} x_{ki}^m + \sum_{j\in\mathcal{N},j\neq i}^{j:m\in\mathcal{M}_j} x_{ij}^m \leq 1$ $\hfill (i\in\mathcal{N}, m\in\mathcal{M}_i)$

$q_{ij}^m - x_{ij}^m \geq 0$ $\hfill (i,j\in\mathcal{N}, i\neq j, m\in\mathcal{M}_{ij})$

$q_{ij}^m - Q x_{ij}^m \leq 0$ $\hfill (i,j\in\mathcal{N}, i\neq j, m\in\mathcal{M}_{ij})$

$\sum_{j\in\mathcal{N},j\neq i}^{j:m\in\mathcal{M}_j} q_{ij}^m - t_i^m = 0$ $\hfill (i\in\mathcal{N}, m\in\mathcal{M}_i)$

$\frac{\eta W Q}{P_{\max}} s_{ij}^m + \sum_{k\in\mathcal{N},k\neq i}^{k:m\in\mathcal{M}_k} g_{kj} u_{ijk}^m - g_{ij} q_{ij}^m = 0$ $\hfill (i,j\in\mathcal{N}, i\neq j, m\in\mathcal{M}_{ij})$

Linear constraints for u_{ijk}^m $\hfill (i,j,k\in\mathcal{N}, i\neq j, m\in\mathcal{M}_{ij}, m\in\mathcal{M}_k)$

$S_{\min} x_{ij}^m - s_{ij}^m \leq 0$ $\hfill (i,j\in\mathcal{N}, i\neq j, m\in\mathcal{M}_{ij})$

$\sum_{j\in\mathcal{N},j\neq i}^{j:\mathcal{M}_{ij}\neq\emptyset} f_{ij}(l) - r(l)K = 0$ $\hfill (l\in\mathcal{L}, i=s(l))$

$\sum_{j\in\mathcal{N},j\neq i,s(l)}^{j:\mathcal{M}_{ij}\neq\emptyset} f_{ij}(l) - \sum_{k\in\mathcal{N},k\neq i,d(l)}^{k:\mathcal{M}_{ki}\neq\emptyset} f_{ki}(l) = 0$ $\hfill (l\in\mathcal{L}, i\in\mathcal{N}, i\neq s(l), d(l))$

$\sum_{l\in\mathcal{L}}^{s(l)\neq j, d(l)\neq i} f_{ij}(l) - \sum_{m\in\mathcal{M}_{ij}} \frac{W}{\ln 2} c_{ij}^m \leq 0$ $\hfill (i,j\in\mathcal{N}, i\neq j, \mathcal{M}_{ij}\neq\emptyset)$

Linear constraints for c_{ij}^m $\hfill (i,j\in\mathcal{N}, i\neq j, m\in\mathcal{M}_{ij})$

$t_i^m, s_{ij}^m, c_{ij}^m, u_{ijk}^m \geq 0$ $\hfill (i,j,k\in\mathcal{N}, i\neq j, m\in\mathcal{M}_{ij}, m\in\mathcal{M}_k)$

$K, f_{ij}(l) \geq 0$ $\hfill (l\in\mathcal{L}, i,j\in\mathcal{N}, i\neq j, i\neq d(l), j\neq s(l), \mathcal{M}_{ij}\neq\emptyset)$

$(\mathbf{x}, \mathbf{q}) \in \Omega_z$

---

to Problem z can serve to provide a lower bound, it is desirable to find one that offers a tight lower bound, i.e., with an objective value close to the optimal value. Such a feasible solution (denoted by ψ_z) can be found by searching the neighborhood of $LP(\hat{z})$, a process that is called *local search*.

A local search algorithm begins with an initial feasible solution. Such a solution may be far away from the optimum and may not provide a tight lower bound. But we can iteratively improve the current solution to achieve a better lower bound until we can no longer improve (increase) the lower bound, whence we terminate this search process.

To obtain an initial feasible solution, we set $x_{ij}^m = (x_{ij}^m)_L$ for scheduling and $q_{ij}^m = (q_{ij}^m)_L$ for power control. Then we can compute SINR value s_{ij}^m by (6).15. When an SINR value is larger than or equal to $S_{\min}$, the achieved capacity is $W\log_2(1+s_{ij}^m)$. Otherwise (i.e., SINR $< S_{\min}$), the transmission is

---

Algorithm 6.2 Pseudocode for a local search algorithm

Initialization

1. Set $x_{ij}^m = (x_{ij}^m)_L$ and $q_{ij}^m = (q_{ij}^m)_L$.

2. Compute the ratio λ_{ij} based on (6.26) for each link $i \to j$, and denote $\lambda_{\min} = \min\{\lambda_{ij} : i, j \in \mathcal{N}, i \neq j, \mathcal{M}_{ij} \neq \emptyset\}$.

Main iteration

3. Select a link $i \to j$ such that $\lambda_{ij} = \lambda_{\min}$.

4. If we can increase q_{ij}^m on a used band {

5.   Suppose that band m has the largest q_{ij}^m-value in the solution $LP(\hat{z})$ among these bands.

6.   Increase q_{ij}^m such that $q_{ij}^m \leq (q_{ij}^m)_U$ and the newly updated $\lambda_{kh} > \lambda_{\min}$ for all other links $k \to h$. }

7. else, if we can increase q_{ij}^m on an available and unused band {

8.   Suppose that band m has the largest q_{ij}^m-value in the solution $LP(\hat{z})$ among these bands.

9.   Increase q_{ij}^m such that $q_{ij}^m \leq (q_{ij}^m)_U$ and the newly updated $\lambda_{kh} > \lambda_{\min}$ for all other links $k \to h$.

10.   If q_{ij}^m increases, then set $x_{ij}^m = 1$. }

11. else the iteration terminates.

---

considered unsuccessful. Note that although the flow rates $f_{ij}(l)$ in the relaxed solution $LP(\hat{z})$ guarantee flow balance at each node, such flow rates may exceed the capacities on some links under the initial x_{ij}^m- and q_{ij}^m-values. To find feasible flow rates for the current x_{ij}^m- and q_{ij}^m-values, we compare the achievable link capacity (under this current solution) to the aggregated flow rates $f_{ij}(l)$ on each link $i \to j$ by computing the ratio between the two (denoted as λ_{ij}) as follows:

$$\lambda_{ij} = \frac{\sum_{m \in \mathcal{M}_{ij}} W \log_2(1 + s_{ij}^m)}{\sum_{l \in \mathcal{L}}^{s(l) \neq j, d(l) \neq i} f_{ij}(l)}. \tag{6.26}$$

If $\lambda_{ij} < 1$ for some link $i \to j$, then the aggregated flow rates exceed the link capacity and the link capacity constraint on $i \to j$ is violated. In this case, we need to scale down the flow rates on link $i \to j$ (to satisfy the link capacity constraint) and the flow rates on all other links (to maintain flow balance in the network) by a value $\lambda \leq \lambda_{ij}$. On the other hand, we want to have a λ as large as possible so as to maximize the scaling factor (our objective). Such a value is the bottleneck value λ_{ij} among all links (denoted as $\lambda_{\min} = \min\{\lambda_{ij} : i, j \in \mathcal{N}, i \neq j, \mathcal{M}_{ij} \neq \emptyset\}$). We now have a complete solution $\lambda_{\min} \cdot f_{ij}(l), (x_{ij}^m)_L, (q_{ij}^m)_L$ for routing, scheduling, and power control, respectively. The achieved objective is $\lambda_{\min} \cdot K$, where K is the objective value in the relaxed solution $LP(\hat{z})$.

In the next iteration, we aim to improve the current solution. Note that if we can increase $\lambda_{\min}$, then the current solution is improved. Suppose that link $i \to j$ is the link with $\lambda_{ij} = \lambda_{\min}$. To increase λ_{ij}, we try to increase the transmission power q_{ij}^m on some band m under the constraint $q_{ij}^m \leq (q_{ij}^m)_U$. Based on the constraints in Table 6.2, we may update the values of other variables to maintain feasibility. For example, by the first constraint in Table 6.2, we need to increase x_{ij}^m from 0 to 1 if q_{ij}^m is increased to a positive value. As a consequence of increasing q_{ij}^m, the interference with other transmissions on band m is increased and thus the achieved capacities on other links are decreased. Thus, q_{ij}^m can be successfully increased only if for any other link $k \to h$, its updated λ_{kh} does not fall below the current $\lambda_{\min}$. If the current solution can be improved (with a larger $\lambda_{\min}$), then we continue to the next iteration and try for further improvement. Otherwise, the local search algorithm terminates. The pseudocode for this local search algorithm is given in Algorithm 6.2.

6.5.5 Partitioning approach

In a standard branch-and-bound procedure, partitioning (see Step 12 in Algorithm 6.1) is done by choosing a variable having the largest relaxation error and its value in the relaxed solution $LP(\hat{z})$ is used to split its value-set into two smaller sets. The reason for selecting a variable having the largest relaxation error is that such a variable is most likely to influence the gap between the upper and lower bounds. Thus, by partitioning its value-set, the relaxation error might become smaller. This branching process also partitions the optimization space for Problem z into two subspaces, thus resulting in two new problems z_1 and z_2, respectively.

Such a partitioning technique is of general purpose and does not exploit any problem-specific property that could be used to prioritize the partitioning variables. We find that if we weigh the significance of each variable when choosing a partitioning variable, the complexity of the overall algorithm can be significantly decreased. In particular, for our problem, we find that the x-variables are more important than the q-variables. So we should first decide whether or not a band is used for transmission. Only if a band is used can we further explore the transmission power on this band. Thus, we first choose some suitable x-variable for partitioning, and when none of them remains (i.e., all the x-variables are either 0 or 1 in the relaxation solution), we then consider the q-variables for partitioning. Incidentally, we note that commercial software (e.g., CPLEX) also provide the opportunity for users to specify priorities among potential partitioning variables.

When choosing a specific x_{ij}^m-variable for partitioning, we select one having the maximum relaxation error, which is defined as $\min\{\hat{x}_{ij}^m, 1 - \hat{x}_{ij}^m\}$, where $\hat{x}_{ij}^m$ is the relaxed solution value for x_{ij}^m in $LP(\hat{z})$. Note that this error is positive if and only if the variable is not presently binary-valued. Once partitioned, the value-set for this variable is set to either 0 or 1 in the two respective subproblems.

We further observe that by fixing the value of x_{ij}^m, some other variables may also be fixed. Note that by (6.13), a node can only receive from or transmit to one node on the same band. Based on this observation, if the value of x_{ij}^m is set to 1, then we have $x_{ki}^m = 0$ for $k \in \mathcal{N}, k \neq i$, if $m \in \mathcal{M}_k$; $x_{ip}^m = 0$ for $p \in \mathcal{N}, p \neq j$, if $m \in \mathcal{M}_p$; $x_{jh}^m = 0$ for $h \in \mathcal{N}, h \neq j$, if $m \in \mathcal{M}_h$; and $x_{qj}^m = 0$ for $q \in \mathcal{N}, q \neq i$, if $m \in \mathcal{M}_q$. On the other hand, if the value of x_{ij}^m is set to 0, then we have $q_{ij}^m = 0$ based on (6.14) (or (6.21)).

When we are done with all the x-variables, we then consider partitioning on the q-variables. Similar to what we did on the x-variables, we select a q-variable that has the maximum relaxation error for partitioning. The relaxation error of a variable q_{ij}^m is defined as $\min\{\hat{q}_{ij}^m - \lfloor\hat{q}_{ij}^m\rfloor, \lfloor\hat{q}_{ij}^m\rfloor + 1 - \hat{q}_{ij}^m\}$, where $\hat{q}_{ij}^m$ is the value of q_{ij}^m in the relaxed solution $LP(\hat{z})$. Again, note that this relaxation error is positive if and only if $\hat{q}_{ij}^m$ is not binary-valued. The value-set of q_{ij}^m in Problem z is $\{(q_{ij}^m)_L, (q_{ij}^m)_L + 1, \ldots, (q_{ij}^m)_U\}$. Hence, its new value-set in the two subproblems will be $\{(q_{ij}^m)_L, (q_{ij}^m)_L + 1, \ldots, \lfloor q_{ij}^m\rfloor\}$ and $\{\lfloor q_{ij}^m\rfloor + 1, \lfloor q_{ij}^m\rfloor + 2, \ldots, (q_{ij}^m)_U\}$, respectively.

6.6 Numerical results

In this section, we present some numerical results for the solution procedure.

6.6.1 Simulation setting

We consider 20-, 30-, and 50-node CRNs with each node randomly located in a 50×50 area (see Tables 6.4, 6.6, and 6.11). For ease of exposition, we normalize all units for distance, bandwidth, rate, and power based on (6.20) with appropriate dimensions. We assume that for the 20-, 30-, and 50-node CRNs, there are $|\mathcal{M}| = 10$, 20, and 30 frequency bands in the network and that each band has a bandwidth of $W = 50$. At each node, only a subset of these bands is available. For the 20- and 30-node CRNs, we assume that there are five user communication sessions (see Tables 6.5 and 6.7) and for the 50-node CRN, we assume that the number of user communication sessions is 10 (see Table 6.12). The source node and destination node for each session are randomly selected, and the minimum rate requirement of each session is randomly generated within [1, 10].

We assume that $g_{ij} = d_{ij}^{-4}$ and that the SINR threshold $S_{\min} = 3$ [54]. We assume that under the maximum transmission power, a node at a distance of 20 can receive data when there is no interference. Thus, we have $\frac{(20)^{-4}P_{\max}}{\eta W} = S_{\min}$, i.e., the maximum transmission power is $P_{\max} = S_{\min} \cdot (20)^4 \cdot \eta W = 4.8 \cdot 10^5 \eta W$. We assume that power control can be done at $Q = 10$ levels.

Table 6.4 Location and available frequency bands at each node for a 20-node network.

Node	Location	Available bands
1	(0.1, 9.9)	1, 2, 3, 4, 7, 8, 9, 10
2	(29.2, 31.7)	1, 2, 3, 4, 5, 7, 8, 10
3	(3, 31.1)	1, 4, 5, 6
4	(11.8, 40.1)	1, 2, 3, 4, 6, 9, 10
5	(15.8, 9.7)	1, 2, 3, 5, 6, 8, 9
6	(16.3, 19.5)	3, 5, 6, 8, 9
7	(0.6, 27.4)	1, 4, 8, 9, 10
8	(22.6, 40.9)	1, 2, 3, 5, 7, 9, 10
9	(35.3, 10.3)	2, 9
10	(31.9, 19.6)	1, 2, 3, 4, 5, 6, 7, 8, 9, 10
11	(28.1, 25.6)	1, 2, 3, 4, 5, 6, 7, 8, 9, 10
12	(32.3, 38)	1, 8, 9, 10
13	(47.2, 2.6)	3, 5, 10
14	(44.7, 15)	2, 3, 6, 7, 8
15	(44.7, 24)	1, 2, 3, 4, 5, 6, 7, 8, 9, 10
16	(47.9, 43.8)	1, 3
17	(46.4, 16.8)	1, 7, 9
18	(11.5, 12.2)	2, 5, 6, 10
19	(28.2, 14.8)	4, 5, 6, 7, 8, 9, 10
20	(2.5, 14.5)	1, 7, 10

For our proposed algorithm, we set $\varepsilon \equiv 0.1$, which guarantees that the solution is at least 90% optimal.

We note that the brute-force approach cannot solve the problem even for the 20-node CRN. The solution space of the capacity problem in Section 6.3.3 includes all possible sets of values for $(x_{ij}^m, q_{ij}^m, K, f_{ij}(l))$. Thus, the number of solutions examined in the brute-force approach is clearly more than the number of all possible sets of values for the q_{ij}^m-variables. For the 20-node CRN, the number of q_{ij}^m-variables is about $20 \cdot (20 - 1) \cdot 5 = 1900$, where 5 is an approximation of the average number of available bands on a link. Each q_{ij}^m-variable has $(Q + 1)$ or 11 possible values. Thus, the number of all possible sets of values for the q_{ij}^m-variables is about 11^{1900}. Therefore, for the 20-node CRN, the number of solutions examined in the brute-force approach is at least 11^{1900}. Even if each solution can be examined in 10^{-6} second, the run time is $11^{1900} \cdot 10^{-6} > 11^{1894}$ seconds, which is $11^{1894}/(365 \cdot 24 \cdot 60 \cdot 60) > 11^{1894}/10^8 > 11^{1886}$ years! Therefore, a brute-force approach cannot be used to solve our problem except for toy-sized instances.

6.6.2 Results

For the 20-node network with five sessions, our solution achieves a scaling factor $K = 13.24$. Based on the minimum rate requirement $r(l)$ in

Table 6.5 Source node, destination node, and minimum rate requirement for each session in the 20-node network.

Session l	Source node $s(l)$	Dest. node $d(l)$	Min. rate req. $r(l)$
1	16	10	9
2	18	3	1
3	12	11	4
4	13	17	3
5	15	6	2

Figure 6.2 The routing topology for the 20-node five-session network.

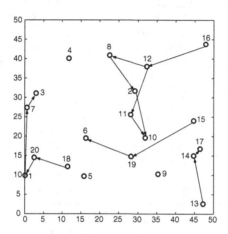

Table 6.5, the corresponding flow rates $K \cdot r(l)$ for the five sessions are 119.16, 13.24, 52.96, 39.72, 26.48, respectively. Fig. 6.2 shows the routing topology for the final solution. The flow rates for each session on the links along its path are as follows:

Session $l = 1$: $f_{16,12}(1) = 119.16$, $f_{12,8}(1) = 103.30$, $f_{12,11}(1) = 15.86$, $f_{8,2}(1) = 103.30$, $f_{11,10}(1) = 15.86$, $f_{2,10}(1) = 103.30$;

Session $l = 2$: $f_{18,20}(2) = 13.24$, $f_{20,1}(2) = 13.24$, $f_{1,7}(2) = 13.24$, $f_{7,3}(2) = 13.24$;

Session $l = 3$: $f_{12,11}(3) = 52.96$;

Session $l = 4$: $f_{13,14}(4) = 39.72$, $f_{14,17}(4) = 39.72$;

Session $l = 5$: $f_{15,19}(5) = 26.48$, $f_{19,6}(5) = 26.48$.

It is easy to verify that flow balance holds at all nodes. Note that flow splitting and multi-path routing are used for Session 1, which has the largest rate requirement.

Our solution also prescribes the scheduling variables x_{ij}^m as follows, where we indicate only the nonzero x_{ij}^m-variables:

Band $m = 1$: $x_{7,3}^1 = 1$, $x_{16,12}^1 = 1$;

Band $m = 2$: $x_{8,2}^2 = 1$;

Band $m = 3$: $x_{13,14}^3 = 1$; Band $m = 4$: $x_{1,7}^4 = 1$, $x_{2,10}^4 = 1$;

Band $m = 5$: $x_{11,10}^5 = 1$;
Band $m = 6$: $x_{15,19}^6 = 1$;
Band $m = 7$: $x_{14,17}^7 = 1$, $x_{20,1}^7 = 1$;
Band $m = 8$: $x_{12,11}^8 = 1$;
Band $m = 9$: $x_{12,8}^9 = 1$, $x_{19,6}^9 = 1$;
Band $m = 10$: $x_{18,20}^{10} = 1$.

The transmission power levels on the respective frequency bands are as follows:
Band $m = 1$: $q_{7,3}^1 = 1$, $q_{16,12}^1 = 7$;
Band $m = 2$: $q_{8,2}^2 = 2$;
Band $m = 3$: $q_{13,14}^3 = 2$;
Band $m = 4$: $q_{1,7}^4 = 7$, $q_{2,10}^4 = 2$;
Band $m = 5$: $q_{11,10}^5 = 1$;
Band $m = 6$: $q_{15,19}^6 = 9$;
Band $m = 7$: $q_{14,17}^7 = 1$, $q_{20,1}^7 = 1$;
Band $m = 8$: $q_{12,11}^8 = 3$;
Band $m = 9$: $q_{12,8}^9 = 1$, $q_{19,6}^9 = 3$;
Band $m = 10$: $q_{18,20}^{10} = 1$.

Note that the same frequency band may be used by concurrent transmissions. For example, since $x_{7,3}^1 = 1$ and $x_{16,12}^1 = 1$, we have that both nodes 7 and 16 are transmitting on band 1. Such concurrent transmissions are allowed as long as the SINR at each receiving node is no less than $S_{\min}$. For example, the SINR at the receiving node 12 on band 1 is $s_{16,12}^1 = \frac{g_{16,12} \cdot q_{16,12}^1}{\frac{\eta W Q}{P_{\max}} + g_{7,12} \cdot q_{7,3}^1} = \frac{(1.303 \cdot 10^{-5}) \cdot 7}{2.083 \cdot 10^{-5} + (8.011 \cdot 10^{-7}) \cdot 1} = 4.22$, which is larger than $S_{\min} = 3$. Thus, the transmission $16 \rightarrow 12$ on band 1 is successful. Following the same token, we can compute all SINR s_{ij}^m-values as follows:
Band $m = 1$: $s_{7,3}^1 = 118.47$, $s_{16,12}^1 = 4.22$;
Band $m = 2$: $s_{8,2}^2 = 5.84$;
Band $m = 3$: $s_{13,14}^3 = 3.75$;
Band $m = 4$: $s_{1,7}^4 = 3.33$, $s_{2,10}^4 = 3.14$;
Band $m = 5$: $s_{11,10}^5 = 18.87$;
Band $m = 6$: $s_{15,19}^6 = 3.39$;
Band $m = 7$: $s_{14,17}^7 = 1261.14$, $s_{20,1}^7 = 65.46$;
Band $m = 8$: $s_{12,11}^8 = 4.90$;
Band $m = 9$: $s_{12,8}^9 = 3.56$, $q_{19,6}^9 = 4.74$;
Band $m = 10$: $s_{18,20}^{10} = 6.45$.

As expected, we see that the achieved SINR at each receiving node on each band is larger than $S_{\min} = 3$ in the derived solution.

For each link, we can further verify that the flow rates on this link do not exceed its capacity. For example, for link $16 \to 12$, there is a flow rate $f_{16,12}(1) = 119.16$ on this link. The achieved capacity is $W \log_2(1 + s_{16,12}^1) = W \log_2(1 + 4.22) = 119.21$.

The above results are for the 20-node network. The results for the 30- and 50-node networks are similar and we abbreviate our discussion. For the 30-node five-session network, the available bands at each node and the location of each node are displayed in Table 6.6. The source, destination, and minimum rate requirement for each session are shown in Table 6.7. The routing topology for the derived solution is depicted in Fig. 6.3. The achieved scaling factor is 31.18. Accordingly, based on the minimum rate requirement $r(l)$ in Table 6.7, the flow rates $K \cdot r(l)$ for the five sessions

Table 6.6 Location and available frequency bands at each node for a 30-node network.

Node	Location	Available bands
1	(7, 0.7)	1, 2, 6, 7, 16, 17, 19, 20
2	(5, 4)	3, 5, 9, 12, 14, 15
3	(6.8, 14)	1, 2, 6, 7, 8, 11, 16, 17, 19, 20
4	(15.7, 3.3)	1, 2, 7, 16, 20
5	(9.5, 17)	3, 4, 5, 9, 12
6	(19.4, 17.1)	1, 2, 6, 7, 8, 16, 19, 20
7	(34.7, 14.6)	3, 4, 5, 9, 12, 14
8	(4.9, 25.9)	3, 4, 12
9	(46.6, 42.1)	10, 18
10	(8.3, 38.3)	3, 4, 5, 9, 14
11	(26.7, 11.1)	1, 6, 7, 8, 11, 16, 17, 19, 20
12	(36.4, 47.3)	10, 13, 18
13	(24.3, 21.2)	1, 2, 6, 8, 11, 19
14	(23.1, 0.8)	3, 5, 9, 14
15	(21.4, 19.2)	4, 9, 12, 14
16	(30.3, 28.1)	7, 8, 11, 16, 17, 19, 20
17	(32, 41.1)	7, 11, 16, 17, 19, 20
18	(14.1, 33.7)	3, 4, 5
19	(23, 46.4)	3, 12, 15
20	(30.3, 9.3)	5, 9
21	(17.6, 29.2)	1, 2, 6, 7, 8, 11, 16, 17, 19, 20
22	(27.1, 27.8)	9, 12, 14, 15
23	(26.9, 45.9)	3, 4, 5, 9, 10, 12, 13, 14, 15, 17
24	(43.3, 32.4)	1, 2, 11, 16, 17, 20
25	(45.4, 8.2)	3, 4, 5, 9, 12, 14
26	(43.4, 35)	3, 5, 9, 15
27	(41.3, 45.1)	1, 16, 20
28	(14.4, 30.3)	1, 2, 6, 7, 8, 11, 16, 17, 20
29	(41.6, 41.7)	3, 4, 5, 9, 10, 12, 14, 15, 18
30	(25.9, 12)	1, 2, 6, 7, 8, 11, 16, 17, 19, 20

Table 6.7 Source node, destination node, and minimum rate requirement for each session in the 30-node network.

Session l	Source node $s(l)$	Dest. node $d(l)$	Min. rate req. $r(l)$
1	16	28	4
2	24	11	7
3	13	1	1
4	19	29	8
5	26	15	1

Table 6.8 Flow rates for each session in the 30-node five-session network.

Session l	Session rate $K \cdot r(l)$	Flow rates of session l on each link $f_{ij}(l)$
1	124.72	$f_{16,21}(1) = 124.72$, $f_{21,28}(1) = 124.72$;
2	218.26	$f_{24,16}(2)=113.91$, $f_{24,17}(2)=104.36$, $f_{17,16}(2)=104.36$, $f_{16,21}(2) = 24.22$, $f_{16,13}(2) = 194.05$, $f_{21,28}(2) = 24.22$, $f_{28,13}(2) = 24.22$, $f_{13,11}(2) = 160.59$, $f_{13,30}(2) = 57.68$, $f_{30,11}(2) = 57.68$;
3	31.18	$f_{13,30}(3) = 31.18$, $f_{30,4}(3) = 31.18$, $f_{4,1}(3) = 31.18$;
4	249.44	$f_{19,23}(4) = 167.49$, $f_{19,29}(4) = 81.95$, $f_{23,26}(4) = 81.26$, $f_{23,29}(4) = 86.23$, $f_{26,29}(4) = 81.26$;
5	31.18	$f_{26,22}(5) = 31.18$, $f_{22,15}(5) = 31.18$.

Figure 6.3 The routing topology for the 30-node five-session network.

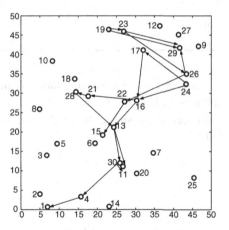

are 124.72, 218.26, 31.18, 249.44, 31.18, respectively. The flow rates for each session on links along its path are shown in Table 6.8. The nonzero scheduling variables x_{ij}^m and the nonzero power control variables q_{ij}^m are specified in Tables 6.9 and 6.10, respectively.

For the 50-node ten-session network, the available bands at each node and the location of each node are displayed in Table 6.11. The source, destination, and minimum rate requirement for each session are specified in Table 6.12.

Table 6.9 Scheduling for the 30-node five-session network.

Band	Scheduling	Band	Scheduling	Band	Scheduling
1	$x^1_{4,1}=1, x^1_{21,28}=1;$	8	$x^8_{16,13}=1;$	15	$x^{15}_{23,26}=1;$
2	$x^2_{28,13}=1;$	9	$x^9_{26,22}=1;$	16	$x^{16}_{24,17}=1, x^{16}_{30,11}=1;$
3	$x^3_{19,29}=1;$	10	$x^{10}_{ij}=0;$	17	$x^{17}_{24,16}=1;$
4	$x^4_{23,29}=1;$	11	$x^{11}_{17,16}=1;$	18	$x^{18}_{ij}=0;$
5	$x^5_{26,29}=1;$	12	$x^{12}_{19,23}=1;$	19	$x^{19}_{13,30}=1;$
6	$x^6_{13,11}=1;$	13	$x^{13}_{ij}=0;$	20	$x^{20}_{30,4}=1.$
7	$x^7_{16,21}=1;$	14	$x^{14}_{22,15}=1;$		

Table 6.10 Transmission power levels for the 30-node five-session network.

Band	Transmission power	Band	Transmission power
1	$q^1_{4,1}=1, q^1_{21,28}=1;$	11	$q^{11}_{17,16}=2;$
2	$q^2_{28,13}=3;$	12	$q^{12}_{19,23}=1;$
3	$q^3_{19,29}=9;$	13	$q^{13}_{ij}=0;$
4	$q^4_{23,29}=4;$	14	$q^{14}_{22,15}=1;$
5	$q^5_{26,29}=1;$	15	$q^{15}_{23,26}=10;$
6	$q^6_{13,11}=2;$	16	$q^{16}_{24,17}=3, q^{16}_{30,11}=1;$
7	$q^7_{16,21}=4;$	17	$q^{17}_{24,16}=7;$
8	$q^8_{16,13}=2;$	18	$q^{18}_{ij}=0;$
9	$q^9_{26,22}=7;$	19	$q^{19}_{13,30}=1;$
10	$q^{10}_{ij}=0;$	20	$q^{20}_{30,4}=4.$

The routing topology for the derived solution is depicted in Fig. 6.4. The achieved scaling factor is 13.36. The detailed solution regarding power control, scheduling, and routing are similar to the 20- and 30-node networks, and are hence omitted.

6.7 Chapter summary

When applying a branch-and-bound framework to solve nonconvex programming problems, it is necessary to develop a tight relaxation, i.e., one that yields a tight upper bound for a maximization problem (or a tight lower bound for a minimization problem). In the last chapter (see Section 5.5.1), we showed two different linear relaxations for two particular types of nonlinear terms. However, such relaxations may not be applicable to some other nonlinear terms.

In this chapter, we described the Reformulation-Linearization Technique (RLT) to derive tight linear relaxations for any monomial. Simply put, RLT

Table 6.11 Location and available frequency bands at each node for a 50-node network.

Node	Location	Available bands
1	(11.1, 21.7)	2, 3, 4, 8, 25
2	(0.1, 4)	6, 7, 10, 13, 14, 20, 23, 24, 26, 28
3	(7.2, 16.6)	6, 10, 14, 20, 23, 24, 26
4	(11, 32.2)	6, 7, 10, 13, 14, 20, 23, 24, 26, 28
5	(16.3, 3.6)	10, 13, 14, 20, 23
6	(14.5, 24.7)	8, 11, 25
7	(14.9, 13.7)	5, 9, 12, 16, 17, 18, 22, 27, 29, 30
8	(19.5, 14.9)	7, 24, 28
9	(26.6, 13.4)	1, 19, 21, 25
10	(22.5, 29.3)	1, 3, 4, 8, 11, 15, 19
11	(24.6, 40.5)	3, 8, 25
12	(38.4, 13.1)	2, 8, 11, 15
13	(4, 3.9)	9, 12, 16, 22, 27, 29, 30
14	(6.1, 18.6)	9, 12, 16, 17, 18, 22, 27, 30
15	(38.5, 22.6)	2, 4, 11, 15, 19, 21, 25
16	(1.2, 24.3)	5, 9, 12, 17, 22, 29, 30
17	(4.9, 42.3)	5, 27
18	(18.5, 1.4)	5, 9, 12, 17, 18, 27, 30
19	(16.9, 29.1)	3, 4, 10, 11, 12, 15
20	(33.5, 10.4)	7, 13, 14, 20, 23, 24, 26, 28
21	(25.6, 12.8)	6, 7, 20, 23, 24, 28
22	(45.2, 45.5)	2, 8, 15, 19
23	(43.6, 22.7)	1, 2, 3, 4, 11, 15, 19, 21
24	(10.6, 40.5)	4, 15, 19, 21, 25
25	(18.2, 32.7)	9, 12, 18, 22, 27
26	(25.2, 27.2)	10, 14, 20, 24, 26
27	(22.5, 42.2)	5, 9, 12, 16, 18, 27, 29, 30
28	(30, 31.5)	6, 13, 24, 26, 28
29	(35, 22.1)	6, 10
30	(25.7, 6.2)	5, 9, 12, 17, 18, 22, 27, 29, 30
31	(34.1, 12.4)	9, 12, 16, 17, 30
32	(26.4, 30)	5, 9, 12, 16, 17, 18, 22, 27, 29, 30
33	(14.1, 40.7)	1, 2, 25
34	(34.4, 46.5)	9, 17, 18, 30
35	(19, 22.5)	1, 6, 7, 10, 13, 14, 20, 23, 24, 28
36	(39.9, 25.1)	6, 13, 14, 20, 23, 24, 26, 28
37	(20.3, 18.2)	1, 2, 3, 4, 8, 11, 15, 19, 21, 27
38	(10, 20.5)	6, 7, 10, 13, 14, 20, 23, 24, 26, 28
39	(20.5, 21.4)	1, 2, 3, 4, 8, 11, 15, 19, 21, 25
40	(37.1, 28.6)	7, 10, 13, 14, 20, 23, 24, 26
41	(44.1, 16.1)	1, 15, 21
42	(41.1, 6)	9, 29
43	(43, 18.8)	5, 9, 12, 16, 18, 22
44	(45.4, 24.2)	9, 12, 16, 17, 18, 30
45	(36.2, 41.2)	5, 9, 17, 27, 29, 30
46	(27.5, 32.3)	12, 16, 17, 18, 29, 30
47	(47.8, 13.8)	22, 27, 29, 30
48	(8.9, 14.8)	5, 30
49	(6.8, 6.2)	5, 9, 12, 16, 17, 27, 30
50	(11.7, 35.8)	1, 2, 3, 4, 8, 11, 15, 19, 21, 25

Table 6.12 Source node, destination node, and minimum rate requirement of each session in the 50-node network.

Session l	Source node $s(l)$	Dest. node $d(l)$	Min. rate req. $r(l)$
1	21	4	4
2	5	26	7
3	19	20	6
4	33	6	10
5	37	10	9
6	23	11	2
7	25	46	3
8	42	43	9
9	44	27	8
10	47	30	1

Figure 6.4 The routing topology for the 50-node ten-session network.

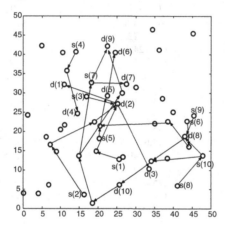

can be applied to any polynomial term of the form $\prod_{i=1}^{n}(x_i)^{c_i}$ in variables x_i, where the c_i-exponents are constant integers. Given such generality, RLT is a powerful tool in deriving tight linear relaxations.

As a case study, we considered a throughput maximization problem in a multi-hop CRN under the SINR model. We developed a mathematical formulation for joint optimization of power control, scheduling, and flow routing. We presented a solution procedure based on the branch-and-bound framework and applied RLT in deriving tight linear relaxations for a product of variables. In this case study, we also learned how to identify the core optimization space for the underlying problem and how to exploit different physical interpretations of the core variables in developing a solution.

6.8 Problems

6.1   For the first simple example of RLT in Section 6.1, we showed that a nonlinear (bilinear) term $x_1 x_2$ can be represented by a single variable $X_{\{1,2\}}$ and four linear constraints (6.1)–(6.4). Show that $X_{\{1,2\}} = x_1 x_2$ holds true when either x_1 or x_2 equals to its corresponding lower or upper bound value.

6.2 For Example 6.1, please show the three linear constraints for representing the relationship $X_{\{2,2,2,2\}} = X_{\{2,2\}}^2$, and the four linear constraints for representing $X_{\{1,2,2,2,2,2\}} = X_{\{1,2\}} X_{\{2,2,2,2\}}$ and $X_{\{1,2,2,2,2,2,3\}} = X_{\{1,2,2,2,2,2\}} x_3$, respectively. Derive appropriate lower and upper bounds on variables as necessary for this representation.

6.3 In the formulation of the multi-hop CRN problem, describe the set of constraints associated with each layer. Which set of constraints couples multiple layers together?

6.4 By introducing new variables t_k^m, we can replace (6.22) by (6).24 and decrease the number of nonlinear terms. Analyze the number of nonlinear terms before and after this substitution. What is the effect of this reformulation?

6.5 Verify that (6.21), (6).24, and (6.25) imply (6.16).

6.6 For all the variables in the MINLP formulation for the multi-hop CRN problem, what do we mean by "core" variables? What is the benefit of identifying core variables in developing a solution?

6.7 Show how the values for the dependent variables t_i^m, s_{ij}^m, $f_{ij}(l)$, and K can be derived from the values for the core variables x_{ij}^m and q_{ij}^m.

6.8 When determining upper bounds for a subproblem, we need to perform linear relaxation for two different nonlinear terms. Describe the technique we used in linear relaxation of each nonlinear term. Can each technique in linear relaxation be applied for both nonlinear terms?

6.9 In the local search, why do we initialize q_{ij}^m at $(q_{ij}^m)_L$ and increase it upward? Discuss the case of initializing q_{ij}^m at $(q_{ij}^m)_U$ and decreasing it downward.

6.10 When choosing a partitioning variable, why do we consider that the **x**-variables are more important than the **q**-variables?

6.11 After partitioning a problem into two new subproblems on a x_{ij}^m-variable, how can the lower and upper bounds for the other core variables be adjusted by using the first five constraints in Table 6.2?

7 Linear approximation

The great pleasure in life is doing what people say you cannot do.

Walter Bagehot

7.1 Review of linear approximation for nonlinear terms

In contrast to LP, nonlinear optimization problems are usually hard and tedious to solve. Therefore, one strategy for nonlinear programs is to replace all non-linear terms by using some linear approximation approach [11]. Depending on the nature of the nonlinear program, the obtained linear approximation problem may or may not provide a feasible solution to the original nonlinear problem. In the case that the LP solution is feasible, we may prove that it is a near-optimal solution to the original problem. Even if it is infeasible, we may construct a feasible solution via local search and further prove the constructed solution is near-optimal.

In this chapter, we will explore such an approach to some nonlinear programs with special structures. This approach is also known as *grid linearization* [88] or *separable programming* [11]. We consider the following problem:

$$\text{Minimize } f(\mathbf{x})$$
$$\text{subject to } g_i(\mathbf{x}) \le p_i \quad (1 \le i \le I)$$
$$h_j(\mathbf{x}) = q_j \quad (1 \le j \le J)$$
$$\mathbf{x} \ge 0,$$

where the objective function $f(\mathbf{x})$ and constraint functions $g_i(\mathbf{x})$ are additively separable, the constraint functions $h_j(\mathbf{x})$ are linear, and p_i and q_j are constants. In other words, this implies that each $f(\mathbf{x})$ and $g_i(\mathbf{x})$ can be expressed as a sum of functions as follows:

Figure 7.1 Piecewise linear approximation.

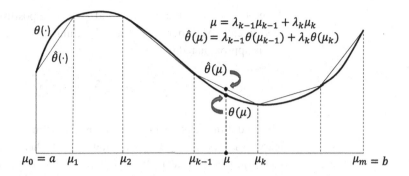

$$f(\mathbf{x}) = \sum_{n=1}^{N} f_n(x_n)$$

$$g_i(\mathbf{x}) = \sum_{n=1}^{N} g_{in}(x_n) \leq p_i \qquad (1 \leq i \leq I).$$

A general approach to approximate a nonlinear function is to use a *piecewise linear* function [11]. Let us consider a continuous function $\theta(\mu)$, where $\mu \in [a, b]$ (see Fig. 7.1). In order to define an approximating function $\hat{\theta}(\mu)$, we divide the interval $[a, b]$ into smaller intervals $[\mu_{k-1}, \mu_k]$, $1 \leq k \leq m$, through the grid points $\mu_0, \mu_1, \mu_2, \ldots, \mu_m$, where $\mu_0 = a$ and $\mu_m = b$. Note that any $\mu \in [\mu_{k-1}, \mu_k]$ can be represented by

$$\mu = \lambda_{k-1}\mu_{k-1} + \lambda_k\mu_k$$
$$\lambda_{k-1} + \lambda_k = 1$$
$$\lambda_{k-1}, \lambda_k \geq 0 \,.$$

The original function $\theta(\mu)$ in the interval $[\mu_{k-1}, \mu_k]$ can be approximated by

$$\hat{\theta}(\mu) = \lambda_{k-1}\theta(\mu_{k-1}) + \lambda_k\theta(\mu_k).$$

It is quite apparent that as the number of grid points increases, the accuracy of the approximation improves.

In general, for any $\mu \in [a, b]$, we can represent it as

$$\mu = \sum_{k=0}^{m} \lambda_k\mu_k \qquad (7.1)$$

$$\sum_{k=0}^{m} \lambda_k = 1, \qquad (7.2)$$

where only two continuous λ_{k-1} and λ_k can be positive (when μ falls in the kth segment). All other λ_l, $l \neq k-1, k$, must be zero. The original function $\theta(\mu)$ is approximated by

$$\hat{\theta}(\mu) = \sum_{k=0}^{m} \lambda_k \theta(\mu_k) \, . \tag{7.3}$$

To model the relationship that when μ falls in the kth segment, we define binary variables z_k, $1 \leq k \leq m$, to indicate whether μ falls within the kth segment $[\mu_{k-1}, \mu_k]$. That is, $z_k = 1$ if $\mu_{k-1} \leq \mu < \mu_k$ and $z_k = 0$ otherwise. Since μ can only fall in one of the m segments, we have

$$\sum_{k=1}^{m} z_k = 1 \, . \tag{7.4}$$

To model the relationship between λ_k and z_k, we note that when μ falls in the kth segment (i.e., $z_k = 1$), λ_{k-1} and λ_k may be positive, while all other λ_l ($l \neq k-1, k$) must be zero. That is, $\lambda_0 > 0$ only if $z_1 = 1$; $\lambda_k > 0$ only if $z_k = 1$ or $z_{k+1} = 1$, $k = 1, \ldots, m-1$; and $\lambda_m > 0$ only if $z_m = 1$. These relationships can be written as follows:

$$\lambda_0 \leq z_1 \tag{7.5}$$

$$\lambda_k \leq z_k + z_{k+1} \quad (1 \leq k < m) \tag{7.6}$$

$$\lambda_m \leq z_m. \tag{7.7}$$

Therefore, piecewise linear approximation can be formulated by constraints (7.1)–(7.7). We note here that there are other improved strategies to model such piecewise linear functions, which afford partial convex hull representations [147].

Using such piecewise linear approximations, we can derive a linear approximating problem for the original problem by replacing each nonlinear function $f_n(x_n)$ and $g_{in}(x_n)$ with its piecewise linear approximation. The function $h_j(\mathbf{x})$ is already linear and no effort is needed to approximate it. Therefore, the new problem is a linear mixed-integer program.

Suppose we solve the linear approximating problem and obtain its solution $\hat{\psi}$. It has been shown that if suitable convexity requirements hold (i.e., both the objective functions $f_n(x_n)$ and the constraint functions $g_{in}(x_n)$ are convex), then even without the z-variables and the constraints (7.4)–(7.7), the solution $\hat{\psi}$ is a feasible solution to the original problem [11]. Note that each function $h_j(\mathbf{x})$ is linear. Thus, the above requirements hold if and only if the original problem is convex. In this case, we can further analyze the performance of $\hat{\psi}$. Since the accuracy of piecewise linear approximation is closely related with the number of grid points used in approximating each function, we can expect that the performance of $\hat{\psi}$ improves as the number of grid points increases. However, as the number of grid points increases, the number of variables in the linear approximating problem also increases, leading to a higher complexity. Thus, by adjusting the number of grid points in the linear approximation, we can make a tradeoff between accuracy in near-optimality and complexity.

For the case that the original problem is not convex, $\hat{\psi}$ is likely to be infeasible to the original problem. But since the new problem is an approximation of the original problem, $\hat{\psi}$ is near-feasible, i.e., the violation of each constraint is very small. Then we may construct a feasible solution ψ based on $\hat{\psi}$ by a local search, which is a problem-specific procedure. In the following sections, we will demonstrate the use of piecewise linear approximations and local search to solve a nonconvex program for a wireless sensor network (WSN).

7.2 Case study: Renewable sensor networks with wireless energy transfer

Wireless sensor networks (WSNs) today are mainly powered by batteries. Due to limited energy storage capacity in a battery at each node, a WSN can only remain operational for a limited amount of time. To prolong its lifetime, there have been a flourish of research efforts in the last decade (see, e.g., [24], [163], [50], [71], [151]). Despite these intensive efforts, lifetime remains a performance bottleneck of a WSN and is perhaps one of the key factors that hinder its wide-scale deployment.

Although energy-harvesting (or energy scavenging) techniques (see, e.g., [76], [77], [21, Chapter 9], [121]) have been proposed to extract energy from the environment, their success remains limited in practice. This is because the proper operations of any energy-harvesting technique are highly dependent on the environment. Further, the size of an energy-harvesting device may pose concern in deployment, particularly when the size of such a device is of much larger physical scale than the sensor node it is attempting to power.

Quite unexpectedly, the recent breakthrough in the area of *wireless energy transfer* technology, developed by Kurs *et al.* [85], has opened up a revolutionary paradigm for prolonging sensor network lifetime. Basically, Kurs *et al.*'s work showed that by exploiting a novel technique called *magnetic resonant coupling,* wireless power transfer (i.e., the ability to transfer electric power from one storage device to another *without any plugs or wires*) is both feasible and practical. In addition to wireless power transfer, they experimentally showed that the source energy storage device does not need to be in contact with the energy receiving device (e.g., a distance of 2 meters) for efficient energy transfer. Moreover, wireless power transfer is immune to the neighboring environment and does not require a line of sight between the power charging and receiving nodes. Recent advances in this technology further show that it can be made portable, with applications to palm-size devices such as cell phones [49].

The impact of wireless energy transfer on WSNs or other energy-constrained wireless networks is immense. Instead of generating energy locally at a node (as in the case of energy-harvesting), we can bring clean electric energy efficiently generated elsewhere to a sensor node periodically, and charge its battery without the constraint of wires and plugs. As we can imagine, the applications of wireless energy transfer are numerous. For example, wireless energy transfer has already been applied to replenish battery energy in medical sensors and implantable devices [178] in the healthcare industry.

Inspired by this new breakthrough in energy transfer technology, this chapter re-examines the network lifetime paradigm for a WSN. We envision employing a mobile vehicle carrying a power charging station to periodically visit each sensor node and charge it wirelessly. This mobile wireless charging vehicle (WCV) can either be manned by a human or be entirely autonomous. In this chapter, we investigate the fundamental question of whether such a new technology can be applied to remove the lifetime performance bottleneck of a WSN. That is, through periodic wireless re-charge, we show that each sensor node will always have an energy level above a minimum threshold so that the WSN remains operational forever. Some of the highlights of this chapter are as follows:

- We introduce the concept of a *renewable energy cycle* where the remaining energy level in a sensor node's battery exhibits some periodicity over a time cycle. We offer both necessary and sufficient conditions for a renewable energy cycle and show that feasible solutions satisfying these conditions can offer renewable energy cycles and, thus, unlimited sensor network lifetime.
- We investigate an optimization problem, with the objective of maximizing the ratio of the WCV's vacation time (time spent at its home station) over the cycle time. In terms of achieving the maximum ratio, we prove that the optimal traveling path for the WCV in each renewable cycle is the shortest Hamiltonian cycle. We also derive several interesting properties associated with an optimal solution, such as the optimal objective being independent of traveling direction on the shortest Hamiltonian cycle and the existence of an energy bottleneck node in the network.
- Under the optimal traveling path, our optimization problem now only needs to consider flow routing and the charging time for each sensor node. We formulate an optimization problem for joint flow routing and charging schedule for each sensor node. The problem is shown to be a nonlinear optimization problem and is NP-hard in general. We show how to apply a piecewise linear approximation technique for each nonlinear term and obtain a tight linear relaxation. Based on this linear relaxation, we obtain a feasible solution and prove that it can achieve near-optimality for any desired level of accuracy.

The remainder of this chapter is organized as follows. In Section 7.3, we review recent advances in wireless energy transfer technology. In Section 7.4, we describe the scope of our problem for a renewable sensor network. Section 7.5 introduces the concept of a renewable energy cycle and presents some interesting properties. Section 7.6 shows that an optimal traveling path should be along the shortest Hamiltonian path. In Section 7.7, we present our problem formulation and a near-optimal solution. Section 7.8 shows how to construct the initial transient cycle preceding the first renewable cycle. In Section 7.9, we present numerical examples to demonstrate the properties of a renewable wireless sensor network under our solution. Section 7.10 summarizes this chapter.

7.3 Wireless energy transfer: a primer

Efforts at transferring power wirelessly can be dated back to the early 1900s (long before wired electric power grid) when Nikola Tesla experimented with large-scale wireless power distribution [156]. Due to its large electric fields, which is undesirable for efficient energy transfer, Tesla's invention was never put into practice.

Since then, there was hardly any progress in wireless energy transfer for many decades. In the early 1990s, the need for wireless power transfer re-emerged when portable electronic devices became widely spread (see, e.g., [159]). The most well-known example is the electric toothbrush. However, due to stringent requirements such as close contact, accurate alignment in charging direction, and uninterrupted line of sight, most of the wireless power transfer technologies at the time (based on inductive coupling) only found limited applications.

Recently, wireless power transfer based on radio frequency (RF) between 850 MHz – 950 MHz (with a center frequency of 915 MHz) has been explored [125]. Under such radiative energy transfer technology, an RF transmitter broadcasts radio waves in the 915 MHz ISM band and an RF receiver tunes to the same frequency band to harvest radio power. However, it was found in [93], [124], and [157] that a receiver operating under such radiative energy transfer technology can only obtain about 45 mW power when it is 10 cm away from the RF transmitter, with about 1% power transfer efficiency. A similar experimental finding was reported in [66]. The technology is also sensitive to obstructions between sources and devices, requires complicated tracking mechanisms if relative positions change, and poses more stringent safety concerns. Due to these issues, the potential of RF-based power transfer technology is limited.

The foundation of this chapter is based on a recent breakthrough technology by Kurs *et al.* [85], which was published in *Science* in 2007 and has since caught worldwide attention. In [85], Kurs *et al.* experimentally demonstrated that efficient nonradiative energy transfer was not only possible, but was

Figure 7.2 An example sensor network with a mobile WCV.

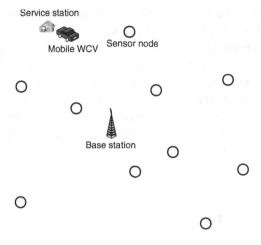

also practical. They used two magnetic resonant objects having the same resonant frequency to exchange energy efficiently, while dissipating relatively little energy in extraneous off-resonant objects. They showed that efficient power transfer implemented in this way can be nearly omnidirectional, irrespective of the environment and even none-line-of-sight (NLOS). The power transfer efficiency, however, decreases with distance. A highlight of their experiment was to fully power a 60-W light bulb from a distance of two meters away, with about 40% power transfer efficiency.

Since the first demo by Kurs *et al.* in 2007, there has been some rapid advance on wireless energy transfer, particularly in the area of making it portable. In particular, Kurs *et al.* launched a start-up company [171] and in 2009 they developed and demonstrated wireless energy transfer for portable devices such as cell phones [49]. Note that the source coil remains sizable, but the device coil is already portable (corresponding to our WCV and sensor node, respectively).

With the recent establishment of Wireless Power Consortium [172] to set the international standard for interoperable wireless charging, it is expected that wireless power transfer will revolutionize how energy is replenished in the near future.

7.4 Problem description

We consider a set of sensor nodes $\mathcal{N}$ distributed over a two-dimensional area (see Fig. 7.2). Each sensor node has a battery capacity of $E_{\max}$ and is fully charged initially. Also, denote $E_{\min}$ as the minimum energy at a sensor node battery (for it to be operational). For simplicity, we define network lifetime as the time until the energy level of any sensor node in the network falls below $E_{\min}$ [24; 133; 164]. Although a more general definition of network lifetime (e.g., [34]) is available, we decided to choose a simple network lifetime definition in this chapter, which is sufficient to show the potential of wireless energy transfer in a sensor network.

Each sensor node i generates sensing data with a rate of R_i (in b/s), $i \in \mathcal{N}$. Within the sensor network, there is a fixed base station (B), which is the sink node for all data generated by the sensor nodes. Multi-hop data routing can be employed for forwarding data by the sensor nodes. Denote f_{ij} as the flow rate from sensor node i to sensor node j and f_{iB} as the flow rate from sensor node i to the base station B, respectively. Then we have the following flow balance constraint at each sensor node i:

$$\sum_{\substack{k \in \mathcal{N} \\ k \neq i}} f_{ki} + R_i = \sum_{\substack{j \in \mathcal{N} \\ j \neq i}} f_{ij} + f_{iB} \qquad (i \in \mathcal{N}). \tag{7.8}$$

Each sensor node consumes energy for data transmission and reception. Denote p_i the energy consumption rate at sensor node $i \in \mathcal{N}$. In this chapter, we use the energy consumption model in Section 2.3.1. We have

$$p_i = \rho \sum_{\substack{k \in \mathcal{N} \\ k \neq i}} f_{ki} + \sum_{\substack{j \in \mathcal{N} \\ j \neq i}} C_{ij} f_{ij} + C_{iB} f_{iB} \qquad (i \in \mathcal{N}), \tag{7.9}$$

where ρ is the energy consumption for receiving a unit of data, C_{ij} (or C_{iB}) is the energy consumption for transmitting a unit of data from node i to node j (or the base station B). Further, $C_{ij} = \beta_1 + \beta_2 D_{ij}^{\alpha}$, where D_{ij} is the distance between nodes i and j, β_1 is a distance-independent constant term, β_2 is a coefficient of the distance-dependent term, and α is the path-loss index. In the model, $\rho \sum_{k \in \mathcal{N}}^{k \neq i} f_{ki}$ is the energy consumption rate for reception, and $\sum_{j \in \mathcal{N}}^{j \neq i} C_{ij} f_{ij} + C_{iB} f_{iB}$ is the energy consumption rate for transmission.

To re-charge the battery at each sensor node, a mobile wireless charging vehicle (WCV) is employed in the network. The WCV starts from a service station (S), and the traveling speed of the WCV is V (in m/s). When it arrives at a sensor node, say i, it will spend an amount of time τ_i to charge the sensor node's battery wirelessly via wireless power transfer [85]. Denote U as the energy transfer rate of the WCV. After τ_i, the WCV leaves node i and travels to the next node on its path. We assume that the WCV has sufficient energy to charge all sensor nodes in the network.

After the WCV visits all the sensor nodes in the network, it will return to its service station to be serviced (e.g., replacing or re-charging its battery) and get ready for the next trip. We call this resting period *vacation time*, denoted as τ_{vac}. After this vacation, the WCV will go out for its next trip.

Denote τ as the time for a trip cycle of the WCV. A number of important questions need to be addressed for such a network. First and foremost, we would ask whether it is possible for each sensor node never to run out of its energy. If this is possible, then the sensor network will have unlimited lifetime and will never cease to be operational. Second, if the answer to the first question is positive, then is there an optimal plan (including traveling path, stop schedule) such that some useful objective can be maximized or minimized? For example, in this chapter, we would like to maximize the percentage of time in a cycle that the WCV spends on its vacation (i.e., $\frac{\tau_{\mathrm{vac}}}{\tau}$), or equivalently, to minimize the percentage of time that the WCV is out in the field.

Table 7.1 lists the notation used in this chapter.

7.5 Renewable cycle construction

In this section, we focus on the renewable cycle construction. We assume the WCV starts from the service station, visits each sensor node once in a cycle, and ends at the service station (see Fig. 7.3). Further, we assume that the data flow routing in the network is invariant with time, with both routing and flow rates being part of our optimization problem.

The middle sawtooth graph (in dashed line) in Fig. 7.4 shows the energy level of a sensor node i during the first two renewable cycles. Note that there is an initialization cycle (marked in the grey area) before the first renewable cycle. That initialization cycle will be constructed in Section 7.8 once we have an optimal solution to the renewable cycles.

Denote $P = (\pi_0, \pi_1, \ldots, \pi_N, \pi_0)$ as the physical path traversed by the WCV over a trip cycle, which starts from and ends at the service station (i.e., $\pi_0 = S$) and the ith node traversed by the WCV along path P is π_i, $i \in \mathcal{N}$. Denote $D_{\pi_0 \pi_1}$ as the distance between the service station and the first sensor node visited along P and $D_{\pi_k \pi_{k+1}}$ as the distance between the kth and $(k+1)$th sensor nodes, respectively. Denote a_i as the arrival time of the WCV at node i in the first renewable energy cycle. We have

$$a_{\pi_i} = \tau + \sum_{k=0}^{i-1} \frac{D_{\pi_k \pi_{k+1}}}{V} + \sum_{k=1}^{i-1} \tau_{\pi_k} \, . \tag{7.10}$$

Denote D_P as the physical distance of path P and $\tau_P = D_P/V$ as the time used for traveling over distance D_P. Recall that τ_{vac} is the vacation time the WCV spends at its service station. Then the cycle time τ can be written as

$$\tau = \tau_P + \tau_{\mathrm{vac}} + \sum_{i \in \mathcal{N}} \tau_i \, , \tag{7.11}$$

where $\sum_{i \in \mathcal{N}} \tau_i$ is the total amount of time the WCV spends near all the sensor nodes in the network for wireless energy transfer.

Table 7.1 Notation.

a_i	Arrival time of the WCV at node i during the first renewable cycle
B	Denotes the base station
C_{ij} (or C_{iB})	Energy consumption for transmitting a unit of data from node i to node j (or the base station B)
D_{ij} (or D_{iS})	Distance from node i to node j (or the service station S)
D_P	The traveling distance during a renewable cycle under a traveling path P
D_{TSP}	The minimal traveling distance during a renewable cycle
$e_i(t)$	Energy level of sensor node i at time t with fully re-charged battery
E_i	Starting energy of sensor node i in a renewable cycle with fully re-charged battery
$E_{\max}$	Full battery capacity of sensor nodes
$E_{\min}$	Minimal energy reserve to keep a sensor node operational
f_{ij} (or f_{iB})	Flow rate from sensor node i to node j (or the base station B)
h	The reciprocal of τ
m	The number of piecewise linear segments used to approximate
$\mathcal{N}$	The set of sensor nodes in the network
p_i	Energy consumption rate at sensor node i
P	The traveling path of the WCV
R_i	Data rate generated at sensor node i
S	Denotes the service station
u_i	Charge rate at sensor node i during the initial transient cycle
U	Full charge rate of the WCV
V	Moving speed of the WCV
$z_{i,k}$	Binary variable indicating whether η_i falls within the kth segment
α	Path-loss index
β_1	A distance-independent constant term in energy consumption for data transmission
β_2	A coefficient of the distance-dependent term in energy consumption for data transmission
ρ	The power consumption coefficient for receiving data
ϵ	Targeted approximation error, $0 < \epsilon \ll 1$
ζ_i	An approximation of η_i^2
η_i	The ratio of the charging time at node i to the entire cycle time.
η_{vac}	The ratio of the vacation time to the entire cycle time.
$\lambda_{i,k}$	The weight of the grid point $\frac{k}{m}$ for η_i
π_i	The ith node traversed by the WCV along path P
τ	Overall time spent during a renewable cycle
τ_i	Time span for the WCV to be present at the sensor node i for re-charging battery
τ_P	Traveling time of the WCV during a renewable cycle under a traveling path P
τ_{TSP}	Minimal traveling time of the WCV during a renewable cycle
τ_{vac}	Vacation time during a renewable cycle

Figure 7.3 A WCV periodically visits each sensor node and charges its battery via wireless energy transfer.

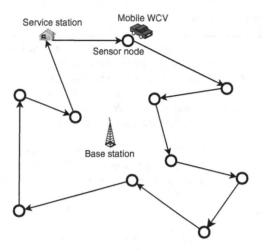

Figure 7.4 The energy level of a sensor node i during the first two renewable cycles (partially re-charged vs. fully re-charged).

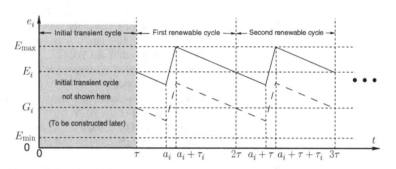

We formally define a renewable energy cycle as follows:

Definition 7.1

The energy level of a sensor node $i \in \mathcal{N}$ exhibits a renewable energy cycle if it meets the following two requirements: (i) it starts and ends with the same energy level over a period of τ; and (ii) it never falls below $E_{\min}$.

During a renewable cycle, the amount of charged energy at a sensor node i during τ_i must be equal to the amount of energy consumed in the cycle (so as to ensure the first requirement in Definition 7.1). That is

$$\tau \cdot p_i = \tau_i \cdot U \qquad (i \in \mathcal{N}) . \tag{7.12}$$

Note that when the WCV visits a node i at time a_i during a renewable energy cycle, it does not have to re-charge the sensor node's battery to $E_{\max}$. This is illustrated in Fig. 7.4, where G_i denotes the starting energy of sensor node i in a renewable cycle and $g_i(t)$ denotes the energy level at time t (dashed sawtooth graph). During a cycle $[\tau, 2\tau]$, we see that the energy level has only two slopes: (i) a slope of $-p_i$ when the WCV is not at this node (i.e., noncharging period), and (ii) a slope of $(U - p_i)$ when the WCV is charging this node at a rate of U (i.e., charging period). It is clear that $g_i(a_i) \le g_i(t) \le g_i(a_i + \tau_i)$, i.e., node i's energy level is lowest at time a_i and is highest at time $a_i + \tau_i$.

Also shown in Fig. 7.4 is another renewable energy cycle (in solid sawtooth graph) where the battery energy is charged to $E_{\max}$ during a WCV's visit. For this energy curve, denote E_i as the starting energy of node i in a renewable cycle and $e_i(t)$ as the energy level at time t, respectively. Let φ^*_{Full} be an optimal solution with fully re-charged battery in each cycle that maximizes the ratio of the WCV's vacation time over the cycle time. Let φ^* be an optimal solution, where there is no requirement on whether or not a node's battery is fully re-charged. Naturally, the optimal objective obtained by φ^*_{Full} is no more than the optimal objective obtained by φ^* due to the additional requirement (battery is fully re-charged) in φ^*_{Full}. Surprisingly, the following lemma shows that φ^*_{Full} is equally good as φ^* in terms of maximizing the ratio of the WCV's vacation time over the cycle time. Thus, for our optimization problem, it is sufficient to consider a solution with fully re-charged battery.

Lemma 7.1

Solution φ^*_{Full} can achieve the same maximum ratio of vacation time to cycle time as that for solution φ^*.

Proof. Our proof has two parts. (i) First, we show that the maximal ratio of vacation time to cycle time achieved by solution φ^* is greater than or equal to that achieved by solution φ^*_{Full}. (ii) Second, we show the converse is also true, i.e., the maximal ratio of vacation time to cycle time achieved by solution φ^*_{Full} is also greater than or equal to that achieved by solution φ^*. If both (i) and (ii) hold, then the lemma is proved.

Since φ^*_{Full} is an optimal solution with the additional requirement (fully re-charged battery in each cycle), the maximal ratio of vacation time to the cycle time obtained by φ^*_{Full} is no more than that obtained by φ^*. Thus, (i) holds.

We now prove (ii). Instead of considering optimal solution φ^*, we will prove that any ratio achieved by a feasible solution φ can also be achieved by a feasible fully re-charged solution $\hat{\varphi}$. If this is true, in the special case that $\varphi = \varphi^*$ is an optimal solution, we have that the maximal ratio achieved by φ^* can also be achieved by a feasible fully re-charged solution. Therefore, (ii) will hold.

The proof is based on construction. Suppose $\varphi = (P, a_i, G_i, f_{ij}, f_{iB}, \tau, \tau_i, \tau_p, \tau_{\text{vac}}, p_i)$ is a feasible solution to our problem. We construct $\hat{\varphi} = (\hat{P}, \hat{a}_i, E_i, \hat{f}_{ij}, \hat{f}_{iB}, \hat{\tau}, \hat{\tau}_i, \hat{\tau}_{\hat{p}}, \hat{\tau}_{\text{vac}}, \hat{p}_i)$ by letting $\hat{P} = P$, $\hat{a}_i = a_i$, $E_i = E_{\max} + p_i$

$(a_i + \tau_i - \tau) - U\tau_i$, $\hat{f}_{ij} = f_{ij}$, $\hat{f}_{iB} = f_{iB}$, $\hat{\tau} = \tau$, $\hat{\tau}_i = \tau_i$, $\hat{\tau}_{\hat{p}} = \tau_p$, $\hat{\tau}_{vac} = \tau_{vac}$, and $\hat{p}_i = p_i$. Note that under $\hat{\varphi}$, the maximal energy level of node i occurs at time $(\hat{a}_i + \hat{\tau}_i)$, which is $e_i(\hat{a}_i + \hat{\tau}_i) = E_i + U\hat{\tau}_i - p_i(\hat{a}_i + \hat{\tau}_i - \hat{\tau}) = [E_{max} + p_i(a_i + \tau_i - \tau) - U\tau_i] + U\tau_i - p_i(a_i + \tau_i - \tau) = E_{max}$. Thus, $\hat{\varphi}$ is a fully re-charged solution. Moreover, it is clear that $\frac{\hat{\tau}_{vac}}{\hat{\tau}} = \frac{\tau_{vac}}{\tau}$ since $\hat{\tau} = \tau$ and $\hat{\tau}_{vac} = \tau_{vac}$. Now, all we need to do is to verify that $\hat{\varphi}$ is a feasible renewable cycle.

To show that $\hat{\varphi}$ is feasible, we need to verify that $\hat{\varphi}$ meets constraints (7.8), (7.9), and (7.11), as well as $e_i(t) \geq E_{min}$ for $i \in \mathcal{N}$, $t \geq \tau$. Further, to show that $\hat{\varphi}$ is a renewable cycle, we need to verify that $e_i(k\tau) = E_i$ for $k \in \mathbb{N}$. We now verify each of these requirements. Since φ is a feasible solution, it satisfies (7.8), (7.9), and (7.11). In $\hat{\varphi}$, we have $\hat{\tau} = \tau$, $\hat{\tau}_i = \tau_i$, $\hat{\tau}_{\hat{p}} = \tau_p$, $\hat{\tau}_{vac} = \tau_{vac}$, $\hat{f}_{ij} = f_{ij}$, $\hat{f}_{iB} = f_{iB}$, and $\hat{p}_i = p_i$. Then $\hat{\varphi}$ also satisfies (7.8), (7.9), and (7.11).

We now show $e_i(t) \geq E_{min}$ for $i \in \mathcal{N}$, $t \geq \tau$. Since φ is a feasible solution, $E_{max} \geq g_i(a_i + \tau_i) = G_i + U\tau_i - p_i(a_i + \tau_i - \tau)$. Thus, we have $E_{max} + p_i(a_i + \tau_i - \tau) - U\tau_i \geq G_i$. Since E_i is set as $E_i = E_{max} + p_i(a_i + \tau_i - \tau) - U\tau_i$, we have $E_i \geq G_i$. Moreover, since $\hat{p}_i = p_i$, the energy at a node i in $\hat{\varphi}$, $e_i(t)$, is parallel to $g_i(t)$ in φ. Because of the parallelism and $E_i \geq G_i$, we have $e_i(t) \geq g_i(t)$ for $i \in \mathcal{N}$. Since φ is a feasible solution, we have $g_i(t) \geq E_{min}$ for $i \in \mathcal{N}$. Thus, $e_i(t) \geq g_i(t) \geq E_{min}$ for $i \in \mathcal{N}$. Therefore, $\hat{\varphi}$ is feasible.

Because of the parallelism and $g_i(k\tau) = G_i$ for $k = 1, 2, \ldots$, we have $e_i(k\tau) = E_i$ for $k = 1, 2, \ldots$. Thus, $\hat{\varphi}$ is a feasible renewable cycle, and (ii) holds. $\qquad\square$

Based on Lemma 7.1, we will only consider a renewable cycle where each node is fully re-charged when it is visited by the WCV. Since the energy level at node i is at its lowest at time a_i, to ensure the second requirement in Definition 7.1 we must have $e_i(a_i) = E_i - (a_i - \tau)p_i \geq E_{min}$. Since for a renewable cycle,

$$E_i = e_i(2\tau) = e_i(a_i + \tau_i) - (2\tau - a_i - \tau_i)p_i = E_{max} - (2\tau - a_i - \tau_i)p_i, \tag{7.13}$$

we have $e_i(a_i) = E_{max} - (2\tau - a_i - \tau_i)p_i - (a_i - \tau)p_i = E_{max} - (\tau - \tau_i)p_i$. Therefore,

$$E_{max} - (\tau - \tau_i) \cdot p_i \geq E_{min} \quad (i \in \mathcal{N}). \tag{7.14}$$

To construct a renewable energy cycle, we need to consider the traveling path P, the arrival time a_i, the starting energy E_i, the flow rates f_{ij} and f_{iB}, time intervals τ, τ_i, τ_p, and τ_{vac}, and power consumption p_i. By (7.10) and (7.13), a_i and E_i are variables that depend on P, τ, and τ_i. Thus, a_i and E_i can be excluded from a solution φ, with the result that

$\varphi = (P, f_{ij}, f_{iB}, \tau, \tau_i, \tau_p, \tau_{\text{vac}}, p_i)$. Although more variables can be removed from φ, we keep this representation for the sake of future discussion.

For a renewable energy cycle, we have the following lemma:

Lemma 7.2

A cycle is a renewable energy cycle if and only if constraints (7.11), (7.12), and (7.14) are met for each sensor node $i \in \mathcal{N}$.

Proof. The "only if" part of the lemma can be proved by showing that a renewable cycle meets (7.11), (7.12), and (7.14). This has already been shown in the description of the renewable cycle.

We now prove the "if" part of the lemma, i.e., if (7.11), (7.12), and (7.14) hold, then (i) and (ii) in Definition 7.1 will also hold, thus the cycle is a renewable energy cycle. Since (7.11) holds, the given cycle satisfies the time constraint. Constraint (7.12) ensures that the amount of energy charged to each sensor node i during τ_i is equal to the amount of energy consumed by sensor node i in the cycle. So the energy level of each sensor node i at the end of the cycle is the same as that at the beginning of the cycle. Therefore, requirement (i) in Definition 7.1 is satisfied.

During the first renewable cycle, the lowest energy level at node i occurs at time a_i, which is

$$
\begin{aligned}
e_i(a_i) &= e_i(\tau) - (a_i - \tau)p_i \\
&= e_i(2\tau) - (a_i - \tau)p_i \\
&= E_{\max} - (2\tau - a_i - \tau_i)p_i - (a_i - \tau)p_i \\
&= E_{\max} - (\tau - \tau_i)p_i \\
&\geq E_{\min},
\end{aligned}
$$

where the second equality holds by requirement (i), which we just proved, the third equality holds by (7.13), the fourth equality holds by $e_i(a_i + \tau_i) = E_{\max}$ in a fully re-charged solution, and the last inequality holds by (7.14). Since the lowest energy level of node i occurs at time a_i and is still no less than $E_{\min}$, requirement (ii) in Definition 7.1 is met. The proof of the "if" part of the lemma is complete. $\qquad\square$

The following property shows that in an optimal solution, there exists at least one "bottleneck" node in the network, where the energy level at this node drops exactly to $E_{\min}$ upon the WCV's arrival.

Property 7.1

In an optimal solution, there exists at least one node in the network with its battery energy dropping to $E_{\min}$ when the WCV arrives at this node.

Proof. The proof is based on contradiction (i.e., if this is not true, then we can further increase the objective value, thus leading to contradiction).

Suppose there exists an optimal solution $\varphi^* = (P^*, f_{ij}^*, f_{iB}^*, \tau^*, \tau_i^*, \tau_p^*, \tau_{vac}^*, p_i^*)$, where none of the nodes in the network has its energy ever drop to E_{min}, i.e., $e_i^*(t) > E_{min}$ for all $i \in \mathcal{N}$, $t \geq \tau$. Then we can construct a new solution $\hat{\varphi} = (\hat{P}, \hat{f}_{ij}, \hat{f}_{iB}, \hat{\tau}, \hat{\tau}_i, \hat{\tau}_{\hat{p}}, \hat{\tau}_{vac}, \hat{p}_i)$ by choosing $\gamma = \min_{i \in \mathcal{N}} \{\frac{E_{max} - E_{min}}{(\tau^* - \tau_i^*) p_i^*}\} - 1$ and letting $\hat{P} = P^*$, $\hat{f}_{ij} = f_{ij}^*$, $\hat{f}_{iB} = f_{iB}^*$, $\hat{\tau} = (1 + \gamma)\tau^*$, $\hat{\tau}_i = (1 + \gamma)\tau_i^*$, $\hat{\tau}_{\hat{p}} = \tau_p^*$, $\hat{\tau}_{vac} = \tau_{vac}^* + \gamma(\tau^* - \sum_{i \in \mathcal{N}} \tau_i^*)$, and $\hat{p}_i = p_i^*$.

Now we show $\gamma > 0$. Since $e_i^*(t) > E_{min}$ for all $i \in \mathcal{N}$, $t \geq \tau$, we have $e_i^*(a_i) = E_{max} - (\tau^* - \tau_i^*) p_i^* > E_{min}$ for all $i \in \mathcal{N}$, i.e., $\min_{i \in \mathcal{N}} \{E_{max} - (\tau^* - \tau_i^*) p_i^*\} > E_{min}$. It follows that $E_{max} - \max_{i \in \mathcal{N}} \{(\tau^* - \tau_i^*) p_i^*\} > E_{min}$, or $\frac{E_{max} - E_{min}}{\max_{i \in \mathcal{N}} \{(\tau^* - \tau_i^*) p_i^*\}} > 1$. Thus, $\gamma = \min_{i \in \mathcal{N}} \{\frac{E_{max} - E_{min}}{(\tau^* - \tau_i^*) p_i^*}\} - 1 = \frac{E_{max} - E_{min}}{\max_{i \in \mathcal{N}} \{(\tau^* - \tau_i^*) p_i^*\}} - 1 > 0$.

The feasibility of $\hat{\varphi}$ can be verified similarly to that in the proof of Lemma 7.1.

We now show that this new feasible solution $\hat{\varphi}$ can offer a better (increased) objective value. By (7.11), we have $\frac{\hat{\tau}_{vac}}{\hat{\tau}} = 1 - \frac{\hat{\tau}_{\hat{p}}}{\hat{\tau}} - \frac{\sum_{i \in \mathcal{N}} \hat{\tau}_i}{\hat{\tau}}$. Since $\hat{\tau} = (1 + \gamma)\tau^*$, $\hat{\tau}_i = (1 + \gamma)\tau_i^*$, $\hat{\tau}_{\hat{p}} = \tau_p^*$, it follows that $\frac{\hat{\tau}_{vac}}{\hat{\tau}} = 1 - \frac{\tau_p^*}{(1 + \gamma)\tau^*} - \frac{\sum_{i \in \mathcal{N}} (1 + \gamma)\tau_i^*}{(1 + \gamma)\tau^*} > 1 - \frac{\tau_p^*}{\tau^*} - \frac{\sum_{i \in \mathcal{N}} \tau_i^*}{\tau^*} = \frac{\tau_{vac}^*}{\tau^*}$, i.e., $\frac{\hat{\tau}_{vac}}{\hat{\tau}} > \frac{\tau_{vac}^*}{\tau^*}$. This contradicts the assumption that φ^* is an optimal solution. $\qquad\square$

7.6 Optimal traveling path

In this section, we show that the WCV must move along the shortest Hamiltonian cycle in an optimal solution. This is formally stated in the following theorem:

Theorem 7.1

In an optimal solution with the maximal $\frac{\tau_{vac}}{\tau}$, the WCV must move along the shortest Hamiltonian cycle that crosses all the sensor nodes and the service station.

Proof. Theorem 7.1 can be proved by contradiction. That is, if there is an optimal solution $\varphi^* = (P^*, f_{ij}^*, f_{iB}^*, \tau^*, \tau_i^*, \tau_{p*}, \tau_{vac}^*, p_i^*)$, where the WCV does not move along the shortest Hamiltonian cycle, then we can construct a new solution $\hat{\varphi} = (\hat{P}, \hat{f}_{ij}, \hat{f}_{iB}, \hat{\tau}, \hat{\tau}_i, \hat{\tau}_{\hat{p}}, \hat{\tau}_{vac}, \hat{p}_i)$, with the WCV moving along the shortest Hamiltonian cycle and with an improved objective.

By assumption, P^* in φ^* does not follow the shortest Hamiltonian cycle. The new solution is constructed as follows. Let $\hat{P}$ follow the shortest hamiltonian cycle (by either direction), $\hat{f}_{ij} = f_{ij}^*$, $\hat{f}_{iB} = f_{iB}^*$, $\hat{\tau} = \tau^*$, $\hat{\tau}_i = \tau_i^*$, $\hat{p}_i = p_i^*$, $\hat{\tau}_{\hat{p}}$ is the traveling time spent on path $\hat{P}$, and

$$\hat{\tau}_{\text{vac}} = \tau_{\text{vac}}^* + \tau_{p*} - \hat{\tau}_{\hat{p}} . \qquad (7.15)$$

Now we show the constructed solution $\hat{\varphi}$ is feasible. To verify the feasibility, we need to show that $\hat{\varphi}$ satisfies the flow conservation constraint (7.8), the time constraint (7.11), and the energy constraints (7.12), and (7.14). Since φ^* is a feasible solution, it satisfies (7.8), (7.11), (7.12), and (7.14). Since we have $\hat{f}_{ij} = f_{ij}^*$, $\hat{f}_{iB} = f_{iB}^*$, $\hat{\tau} = \tau^*$, $\hat{\tau}_i = \tau_i^*$, and $\hat{p}_i = p_i^*$ in $\hat{\varphi}$, the constraints (7.8), (7.12), and (7.14) also hold by $\hat{\varphi}$. To show that $\hat{\varphi}$ also satisfies (7.11), we have $\hat{\tau}_{\hat{p}} + \sum_{i \in \mathcal{N}} \hat{\tau}_i + \hat{\tau}_{\text{vac}} = \hat{\tau}_{\hat{p}} + \sum_{i \in \mathcal{N}} \tau_i^* + \left(\tau_{\text{vac}}^* + \tau_{p*} - \hat{\tau}_{\hat{p}} \right) = \sum_{i \in \mathcal{N}} \tau_i^* + \tau_{\text{vac}}^* + \tau_{p*} = \tau^* = \hat{\tau}$, where the first equality follows from (7.15), the second equality follows by the feasibility of φ^* and (7.11), and the third equality follows by $\hat{\tau} = \tau^*$ during construction.

To show $\hat{\tau}_{\text{vac}}/\hat{\tau} > \tau_{\text{vac}}^*/\tau^*$, recall that $\hat{P}$ follows the shortest Hamiltonian cycle while P^* does not, i.e., the traveling distance $D_{p*} > \hat{D}_{\hat{p}}$. Therefore, the traveling time $\tau_{p*} > \hat{\tau}_{\hat{p}}$. Then by (7.15), $\hat{\tau}_{\text{vac}} = \tau_{\text{vac}}^* + \tau_{p*} - \hat{\tau}_{\hat{p}} > \tau_{\text{vac}}^*$, or $\hat{\tau}_{\text{vac}}/\hat{\tau} > \tau_{\text{vac}}^*/\hat{\tau} = \tau_{\text{vac}}^*/\tau^*$. But $\hat{\tau}_{\text{vac}}/\hat{\tau} > \tau_{\text{vac}}^*/\tau^*$ contradicts the assumption that φ^* is optimal. This completes the proof. □

Theorem 7.1 says that the WCV should move along the shortest Hamiltonian cycle, which can be obtained by solving the well-known Traveling Salesman Problem (TSP) (see, e.g., [28], [119]). Denote D_{TSP} as the traveling distance in the shortest Hamiltonian cycle and let $\tau_{\text{TSP}} = D_{\text{TSP}}/V$. Then with the optimal traveling path, (7.11) becomes

$$\tau_{\text{TSP}} + \tau_{\text{vac}} + \sum_{i \in \mathcal{N}} \tau_i = \tau, \qquad (7.16)$$

and the solution for a renewable cycle becomes $\varphi = (P_{\text{TSP}}, f_{ij}, f_{iB}, \tau, \tau_i, \tau_{\text{TSP}}, \tau_{\text{vac}}, p_i)$. Since the optimal traveling path is determined, the solution can be simplified as $\varphi = (f_{ij}, f_{iB}, \tau, \tau_i, \tau_{\text{vac}}, p_i)$.

We note that the shortest Hamiltonian cycle may not be unique. Since any shortest Hamiltonian cycle has the same total path distance and traveling time τ_{TSP}, the selection of a particular shortest Hamiltonian cycle does not affect constraint (7.16), and yields the same optimal objective. This insight is formally stated in the following corollary:

Corollary 7.1
Any shortest Hamiltonian cycle can achieve the same optimal objective.

We also note that to travel the shortest Hamiltonian cycle, there are two (opposite) outgoing directions for the WCV to start from its home service station. Since the proof of Theorem 7.1 is independent of the starting direction for the WCV, either direction will yield an optimal solution with the same objective value, although some variables in each optimal solution will have different values. We have the following corollary:

Corollary 7.2

The WCV can follow either direction to traverse the shortest Hamiltonian cycle, both of which will achieve the same optimal objective. There exist two optimal solutions corresponding to the two opposite directions, with identical values of f_{ij}, f_{iB}, τ, τ_i, τ_{TSP}, τ_{vac}, p_i, but different values of a_i (by (7.10)) and E_i (by (7.13)) due to difference in their respective renewable cycles, where $i, j \in \mathcal{N}, i \neq j$.

7.7 Problem formulation and solution

7.7.1 Mathematical formulation

Summarizing the objective and all the constraints in Sections 7.4, 7.5, and 7.6, our problem can be formulated as follows:

OPT

$$\text{Maximize } \frac{\tau_{\text{vac}}}{\tau}$$

$$\text{subject to } \sum_{\substack{j \in \mathcal{N} \\ j \neq i}} f_{ij} + f_{iB} - \sum_{\substack{k \in \mathcal{N} \\ k \neq i}} f_{ki} = R_i \quad (i \in \mathcal{N})$$

$$\rho \cdot \sum_{\substack{k \in \mathcal{N} \\ k \neq i}} f_{ki} + \sum_{\substack{j \in \mathcal{N} \\ j \neq i}} C_{ij} \cdot f_{ij}$$

$$+ C_{iB} \cdot f_{iB} - p_i = 0 \quad (i \in \mathcal{N}) \tag{7.17}$$

$$\tau - \sum_{i \in \mathcal{N}} \tau_i - \tau_{\text{vac}} = \tau_{\text{TSP}} \tag{7.18}$$

$$\tau \cdot p_i - U \cdot \tau_i = 0 \quad (i \in \mathcal{N}) \tag{7.19}$$

$$(\tau - \tau_i) \cdot p_i \leq E_{\max} - E_{\min} \quad (i \in \mathcal{N}) \tag{7.20}$$

$$f_{ij}, f_{iB}, \tau_i, \tau, \tau_{\text{vac}}, p_i \geq 0 \quad (i, j \in \mathcal{N}, i \neq j).$$

In this problem, flow rates f_{ij} and f_{iB}, time intervals τ, τ_i, and τ_{vac}, and power consumption p_i are optimization variables, and $R_i, \rho, C_{ij}, C_{iB}, U, E_{\max}, E_{\min}$, and τ_{TSP} are constants. This problem has both nonlinear objective ($\frac{\tau_{\text{vac}}}{\tau}$) and nonlinear terms ($\tau p_i$ and $\tau_i p_i$) in constraints (7.19) and (7.20).

Note that there are two possible outcomes for optimization problem OPT: either an optimal solution exists or OPT is infeasible. There are several scenarios where the latter outcome may occur, e.g., (i) the energy charging rate of

WCV is too small or the energy consumption rate of a node is too large; (ii) the time interval between WCV's visits at any node is too large. As a result, some constraints in Problem OPT will not hold. These are physical limitations for a WCV to achieve a renewable network lifetime for a WSN.

We note that in the above formulation, only constant τ_{TSP} is related to the shortest Hamiltonian cycle. Since this value does not depend on the traveling direction along the Hamiltonian cycle, an optimal solution to problem OPT will work for either direction and yields a different renewable cycle for each direction.

7.7.2 Reformulation

We first use a change-of-variable technique to simplify the formulation. For the nonlinear objective $\frac{\tau_{\mathrm{vac}}}{\tau}$, we define

$$\eta_{\mathrm{vac}} = \frac{\tau_{\mathrm{vac}}}{\tau} . \tag{7.21}$$

For (7.18), we divide both sides by τ and have $\tau_{\mathrm{TSP}} \cdot \frac{1}{\tau} + \eta_{\mathrm{vac}} + \sum_{i \in \mathcal{N}} \frac{\tau_i}{\tau} = 1$. To remove the nonlinear terms $\frac{1}{\tau}$ and $\frac{\tau_i}{\tau}$ in the above equation, we define

$$\eta_i = \frac{\tau_i}{\tau} \quad (i \in \mathcal{N}) \tag{7.22}$$

$$h = \frac{1}{\tau} . \tag{7.23}$$

Then (7.18) is reformulated as $\tau_{\mathrm{TSP}} \cdot h + \eta_{\mathrm{vac}} + \sum_{i \in \mathcal{N}} \eta_i = 1$, or equivalently,

$$h = \frac{1 - \sum_{i \in \mathcal{N}} \eta_i - \eta_{\mathrm{vac}}}{\tau_{\mathrm{TSP}}} . \tag{7.24}$$

Similarly, (7.19) and (7.20) can be reformulated (by dividing both sides by τ) as

$$p_i = U \cdot \eta_i \qquad\qquad (i \in \mathcal{N}), \tag{7.25}$$

$$(1 - \eta_i) \cdot p_i \le (E_{\max} - E_{\min}) \cdot h \quad (i \in \mathcal{N}) \tag{7.26}$$

By (7.24) and (7.25), constraint (7.26) can be rewritten as $(1 - \eta_i) \cdot U \eta_i \le (E_{\max} - E_{\min}) \frac{1 - \sum_{k \in \mathcal{N}} \eta_k - \eta_{\mathrm{vac}}}{\tau_{\mathrm{TSP}}}$, or

$$\eta_{\mathrm{vac}} \le 1 - \sum_{k \in \mathcal{N}} \eta_k - \frac{U \cdot \tau_{\mathrm{TSP}}}{E_{\max} - E_{\min}} \cdot \eta_i \cdot (1 - \eta_i) \quad (i \in \mathcal{N}) .$$

By (7.25), constraint (7.17) can be rewritten as

$$\rho \sum_{\substack{k \in \mathcal{N} \\ k \ne i}} f_{ki} + \sum_{\substack{j \in \mathcal{N} \\ j \ne i}} C_{ij} f_{ij} + C_{iB} f_{iB} - U \eta_i = 0 \quad (i \in \mathcal{N}).$$

Algorithm 7.1 Algorithm to problem OPT

Once we solve problem OPT-R, we can obtain the solution to problem OPT (i.e., calculate the values for τ, τ_i, τ_{vac}, and p_i) as follows: h by (7.24), τ by (7.23), τ_i by (7.22), τ_{vac} by (7.21), and p_i by (7.25).

Now problem OPT is reformulated as follows:

OPT-R

Maximize η_{vac}

$$\text{subject to } \sum_{\substack{j \in \mathcal{N} \\ j \neq i}} f_{ij} + f_{iB} - \sum_{\substack{k \in \mathcal{N} \\ k \neq i}} f_{ki} = R_i \qquad (i \in \mathcal{N})$$

$$\rho \sum_{\substack{k \in \mathcal{N} \\ k \neq i}} f_{ki} + \sum_{\substack{j \in \mathcal{N} \\ j \neq i}} C_{ij} f_{ij} + C_{iB} f_{iB} - U \eta_i = 0 \quad (i \in \mathcal{N}) \quad (7.27)$$

$$\eta_{\text{vac}} \leq 1 - \sum_{k \in \mathcal{N}} \eta_k - \frac{U \cdot \tau_{\text{TSP}}}{E_{\max} - E_{\min}} \cdot \eta_i \cdot (1 - \eta_i) \; (i \in \mathcal{N}) \quad (7.28)$$

$$f_{ij}, f_{iB} \geq 0, 0 \leq \eta_i, \eta_{\text{vac}} \leq 1 \qquad (i, j \in \mathcal{N}, i \neq j).$$

In this problem, f_{ij}, f_{iB}, η_i, and η_{vac} are optimization variables, and R_i, ρ, C_{ij}, C_{iB}, U, $E_{\max}$, $E_{\min}$, and τ_{TSP} are constants. Algorithm 7.1 shows how to obtain a solution to problem OPT once we obtain a solution to problem OPT-R.

After reformulation, the objective function and the constraints become linear except (7.28), where we have a second-order η_i^2 term, with $0 \leq \eta_i \leq 1$. In the next section, we present an efficient linear approximation technique to approximate this second-order nonlinear term (with performance guarantee). This is the main technique that we wish to show in this chapter. Subsequently, we develop an efficient near-optimal solution to our optimization problem.

Remark 7.1

In our optimization problem, data routing and charging time are closely coupled. We may want to de-couple routing from the charging problem and require certain energy-efficient routing, e.g., the minimum energy routing.[1] However, minimum energy routing cannot guarantee optimality. This is because, to maximize η_{vac}, by (7.28), we need to minimize $\max_{i \in \mathcal{N}} \{\sum_{k \in \mathcal{N}} \eta_k + \frac{U \cdot \tau_{\text{TSP}}}{E_{\max} - E_{\min}} \cdot \eta_i \cdot (1 - \eta_i)\}$, i.e., minimize $\sum_{k \in \mathcal{N}} \eta_k + \max_{i \in \mathcal{N}} \{\frac{U \cdot \tau_{\text{TSP}}}{E_{\max} - E_{\min}} \cdot \eta_i \cdot (1 - \eta_i)\}$. But under minimum energy routing, we can only guarantee that $\sum_{i \in \mathcal{N}} (\rho \sum_{k \neq i}^{k \in \mathcal{N}} f_{ki} + \sum_{j \neq i}^{j \in \mathcal{N}} C_{ij} f_{ij} + C_{iB} f_{iB})$ is minimized. By the relationship in (7.27), minimizing $\sum_{i \in \mathcal{N}} (\rho \sum_{k \neq i}^{k \in \mathcal{N}} f_{ki} + \sum_{j \neq i}^{j \in \mathcal{N}} C_{ij} f_{ij} + C_{iB} f_{iB})$ is equivalent to minimizing $\sum_{i \in \mathcal{N}} \eta_i$, which is only part of $\sum_{k \in \mathcal{N}} \eta_k + \max_{i \in \mathcal{N}} \{ \frac{U \cdot \tau_{\text{TSP}}}{E_{\max} - E_{\min}} \cdot \eta_i \cdot (1 - \eta_i) \}$. Therefore, minimum energy routing cannot guarantee the optimality of our problem. This insight will be confirmed by our numerical examples in Section 7.9.

7.7.3 A near-optimal solution

Roadmap   Our roadmap to solve problem OPT is as follows. First, we employ the piecewise linear approximation technique (see Section 7.1) for the quadratic terms (η_i^2) in problem OPT-R. This approximation relaxes the corresponding nonlinear constraints into linear constraints, which allows for the problem to be solved by a solver such as CPLEX [31]. Based on the solution from CPLEX, we construct a feasible solution to problem OPT. In Section 7.7.4, we prove the near-optimality of this feasible solution.

Piecewise linear approximation for η_i^2   Note that the only nonlinear terms in the formulation are η_i^2, $i \in \mathcal{N}$. Further, η_i lies in the interval [0, 1], which is small. This motivates us to employ the piecewise linear approximation technique for the quadratic terms η_i^2.

Based on the discussion in Section 7.1, for a piecewise linear approximation with m grid points (see Fig. 7.5), we have the following constraints for η_i^2:

$$\sum_{k=1}^{m} z_{ik} = 1 \tag{7.29}$$

$$\eta_i = \sum_{k=0}^{m} \lambda_{ik} \cdot \frac{k}{m} \tag{7.30}$$

$$\sum_{k=0}^{m} \lambda_{ik} = 1 \tag{7.31}$$

$$\zeta_i = \sum_{k=0}^{m} \lambda_{ik} \cdot \frac{k^2}{m^2} \tag{7.32}$$

$$\lambda_{i0} \leq z_{i1} \tag{7.33}$$

$$\lambda_{ik} \leq z_{ik} + z_{i,k+1} \quad (1 \leq k < m) \tag{7.34}$$

$$\lambda_{im} \leq z_{im} , \tag{7.35}$$

where z_{ik} is a binary variable indicating whether η_i falls within the kth segment and $\lambda_{ik} \in [0, 1]$ is a weight associated with grid point $\frac{k}{m}$. The setting of m will determine the level of accuracy and will be studied in Section 7.7.4.

Since $y = x^2$ is a convex function, the piecewise linear approximation curve (η_i, ζ_i) lies above the curve (η_i, η_i^2), $0 \leq \eta_i \leq 1$. Thus, we have $\zeta_i \geq \eta_i^2$ (see Fig. 7.5). The following lemma characterizes the approximation error $\zeta_i - \eta_i^2$ as a function of m.

Figure 7.5 An illustration of
piecewise linear approximation
(with $m = 4$) for the curve
$(\eta_i, \eta_i^2), 0 \leq \eta_i \leq 1$.

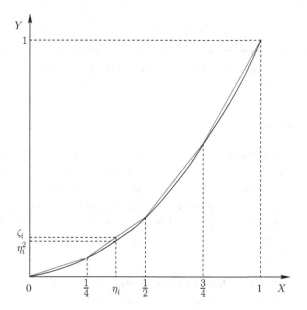

Lemma 7.3

$\zeta_i - \eta_i^2 \leq \dfrac{1}{4m^2}$ for $i \in \mathcal{N}$.

Proof. Assume η_i falls in the kth segment, $k = 1, 2, \ldots, m$. Then we have

$$
\begin{aligned}
\zeta_i - \eta_i^2 &= \left(\lambda_{i,k-1} \cdot \frac{(k-1)^2}{m^2} + \lambda_{i,k} \cdot \frac{k^2}{m^2} \right) - \left(\lambda_{i,k-1} \cdot \frac{k-1}{m} + \lambda_{i,k} \cdot \frac{k}{m} \right)^2 \\
&= (\lambda_{i,k-1} - \lambda_{i,k-1}^2) \cdot \frac{(k-1)^2}{m^2} + (\lambda_{i,k} - \lambda_{i,k}^2) \cdot \frac{k^2}{m^2} \\
&\quad - 2\lambda_{i,k-1}\lambda_{i,k} \cdot \frac{k-1}{m} \cdot \frac{k}{m} \\
&= \left((1 - \lambda_{i,k}) - (1 - \lambda_{i,k})^2 \right) \cdot \frac{(k-1)^2}{m^2} + (\lambda_{i,k} - \lambda_{i,k}^2) \cdot \frac{k^2}{m^2} \\
&\quad - 2(1 - \lambda_{i,k})\lambda_{i,k} \cdot \frac{k-1}{m} \cdot \frac{k}{m} \\
&= (\lambda_{i,k} - \lambda_{i,k}^2) \cdot \left(\frac{(k-1)^2}{m^2} + \frac{k^2}{m^2} - 2 \cdot \frac{k-1}{m} \cdot \frac{k}{m} \right) \\
&= (\lambda_{i,k} - \lambda_{i,k}^2) \cdot \left(\frac{k}{m} - \frac{k-1}{m} \right)^2 \\
&= (\lambda_{i,k} - \lambda_{i,k}^2) \cdot \frac{1}{m^2} \\
&\leq \frac{1}{4m^2},
\end{aligned}
$$

where the first equality holds by (7.30), (7.32), and the fact that $\lambda_{il} = 0$ for $l \neq k - 1, k$, the third equality holds by (7.31) and $\lambda_{il} = 0$ for $l \neq k - 1, k$, and the last inequality holds by $\lambda_{i,k} - \lambda_{i,k}^2 \leq \frac{1}{4}$ when $0 \leq \lambda_{i,k} \leq 1$. This completes the proof. $\qquad \square$

Relaxed linear formulation　By replacing η_i^2 with ζ_i in (7.28), we have

$$\eta_{\text{vac}} \leq 1 - \sum_{k \in \mathcal{N}} \eta_k - \frac{U \tau_{\text{TSP}}}{E_{\max} - E_{\min}} (\eta_i - \zeta_i) \qquad (i \in \mathcal{N}). \qquad (7.36)$$

By adding new constraints (7.29)–(7.35), we obtain the following linear relaxed formulation:

OPT-L

Maximize η_{vac}

$$\text{subject to} \quad \sum_{\substack{j \in \mathcal{N} \\ j \neq i}} f_{ij} + f_{iB} - \sum_{\substack{k \in \mathcal{N} \\ k \neq i}} f_{ki} = R_i \quad (i \in \mathcal{N})$$

$$\rho \cdot \sum_{\substack{k \in \mathcal{N} \\ k \neq i}} f_{ki} + \sum_{\substack{j \in \mathcal{N} \\ j \neq i}} C_{ij} \cdot f_{ij} + C_{iB} \cdot f_{iB}$$
$$- U \cdot \eta_i = 0 \qquad (i \in \mathcal{N})$$

$$\eta_{\text{vac}} \leq 1 - \sum_{k \in \mathcal{N}} \eta_k - \frac{U \tau_{\text{TSP}}}{E_{\max} - E_{\min}} \cdot (\eta_i - \zeta_i) \quad (i \in \mathcal{N})$$

$$\sum_{k=1}^{m} z_{ik} = 1 \qquad (i \in \mathcal{N})$$

$$\eta_i - \sum_{k=0}^{m} \frac{k}{m} \cdot \lambda_{ik} = 0 \qquad (i \in \mathcal{N})$$

$$\sum_{k=0}^{m} \lambda_{ik} = 1 \qquad (i \in \mathcal{N})$$

$$\zeta_i - \sum_{k=0}^{m} \frac{k^2}{m^2} \cdot \lambda_{ik} = 0 \qquad (i \in \mathcal{N})$$

$$\lambda_{i0} - z_{i1} \leq 0 \qquad (i \in \mathcal{N})$$

$$\lambda_{ik} - z_{ik} - z_{i,k+1} \leq 0 \qquad (i \in \mathcal{N}, \ 1 \leq k < m)$$

$$\lambda_{im} - z_{im} \leq 0 \qquad (i \in \mathcal{N})$$

$$f_{ij}, f_{iB} \geq 0, 0 \leq \eta_i, \eta_{\text{vac}}, \zeta_i \leq 1 \qquad (i, j \in \mathcal{N}, i \neq j)$$

$$z_{ik} \in \{0, 1\} \quad (i \in \mathcal{N}, 1 \leq k \leq m)$$

$$0 \leq \lambda_{ik} \leq 1 \quad (i \in \mathcal{N}, 0 \leq k \leq m),$$

where f_{ij}, f_{iB}, η_i, η_{vac}, z_{ik}, λ_{ik}, and ζ_i are variables, R_i, ρ, C_{ij}, C_{iB}, U, $E_{\max}$, $E_{\min}$, and τ_{TSP} are constants. The new formulation can be solved by a solver such as CPLEX [31].

Construction of a feasible near-optimal solution   The solution to problem OPT-L is likely to be infeasible to problem OPT-R (and problem OPT). But based on this solution, we can construct a feasible solution to problem OPT.

Suppose $\hat{\psi} = (\hat{f}_{ij}, \hat{f}_{iB}, \hat{\eta}_i, \hat{\eta}_{\text{vac}}, \hat{z}_{ik}, \hat{\lambda}_{ik}, \hat{\zeta}_i)$ is the solution to problem OPT-L. By observing $(\hat{f}_{ij}, \hat{f}_{iB}, \hat{\eta}_i, \hat{\eta}_{\text{vac}})$, we find that it satisfies all constraints to problem OPT-R except (7.28). To construct a feasible solution $\psi = (f_{ij}, f_{iB}, \eta_i, \eta_{\text{vac}})$ to problem OPT-R, we let $f_{ij} = \hat{f}_{ij}$, $f_{iB} = \hat{f}_{iB}$, $\eta_i = \hat{\eta}_i$. For η_{vac}, in order to satisfy (7.28), we define

$$\eta_{\text{vac}} = \min_{i \in \mathcal{N}} \left\{ 1 - \sum_{k \in \mathcal{N}} \hat{\eta}_k - \frac{U \cdot \tau_{\text{TSP}}}{E_{\max} - E_{\min}} \hat{\eta}_i (1 - \hat{\eta}_i) \right\} .$$

It is easy to verify that this newly constructed solution ψ satisfies all the constraints for problem OPT-R. Once we have this solution to problem OPT-R, we can easily find a solution to problem OPT via Algorithm 7.1.

7.7.4 Proof of near-optimality

In this section, we quantify the performance gap between the optimal objective (unknown, denoted as η_{vac}^*) and the objective (denoted as η_{vac}) obtained by the feasible solution ψ that we derived in the last section. Naturally, we expect such a performance gap to be a function of m, i.e., the number of segments that we use in the piecewise linear approximation. This result will be stated in Lemma 7.4. Based on this result, we can obtain an important inverse result (in Theorem 7.2), which shows how to set m such that $\eta_{\text{vac}}^* - \eta_{\text{vac}} \leq \epsilon$ for a given target performance gap ϵ ($0 < \epsilon \ll 1$).

Lemma 7.4

For the feasible solution ψ with objective value η_{vac}, we have $\eta_{\text{vac}}^* - \eta_{\text{vac}} \leq \frac{U \cdot \tau_{\text{TSP}}}{4(E_{\max} - E_{\min})} \cdot \frac{1}{m^2}$.

To prove Lemma 7.4, we need two intermediate results for $\hat{\eta}_{\text{vac}}$ and η_{vac}, which are stated in Lemmas 7.5 and 7.6, respectively.

Lemma 7.5

For the optimal solution $\hat{\psi}$ to problem OPT-L, we have

$$\hat{\eta}_{\text{vac}} = 1 - \sum_{k \in \mathcal{N}} \hat{\eta}_k - \frac{U \tau_{\text{TSP}}}{E_{\max} - E_{\min}} \cdot (\eta_{\max} - \zeta_{\max}) ,$$

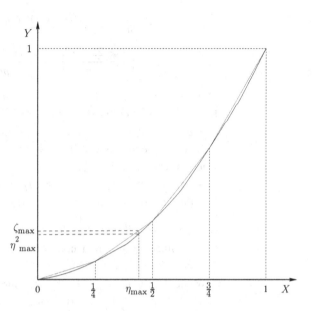

Figure 7.6 An illustration of $\eta_{\max}^2$ and its approximation $\zeta_{\max}$.ff

where

$$\eta_{\max} \equiv \max_{i \in \mathcal{N}}\{\hat{\eta}_i\}$$

and $\zeta_{\max}$ is the piecewise linear approximation of $\eta_{\max}^2$ (see Fig. 7.6).

Proof. It is easy to see that in $\hat{\psi}$, $\hat{\eta}_{\text{vac}} = \min_{i \in \mathcal{N}}\{1 - \sum_{k \in \mathcal{N}} \hat{\eta}_k - \frac{U \tau_{\text{TSP}}}{E_{\max} - E_{\min}} \cdot (\hat{\eta}_i - \hat{\zeta}_i)\}$. Thus, to prove this lemma, it is sufficient to show $\max_{i \in \mathcal{N}}\{\hat{\eta}_i - \hat{\zeta}_i\} = \eta_{\max} - \zeta_{\max}$, i.e., $\hat{\eta}_i - \hat{\zeta}_i \leq \eta_{\max} - \zeta_{\max}$ for each $i \in \mathcal{N}$.

We consider the following two cases:

Case 1: $0 < \eta_{\max} \leq \frac{1}{2}$: Based on the definition of $\eta_{\max}$, we have $0 < \hat{\eta}_i \leq \eta_{\max} \leq \frac{1}{2}$ for each $i \in \mathcal{N}$. We now show that $\hat{\eta}_i - \hat{\zeta}_i$ is a nondecreasing function when $0 < \hat{\eta}_i \leq \frac{1}{2}$. Then it follows that $\hat{\eta}_i - \hat{\zeta}_i \leq \eta_{\max} - \zeta_{\max}$.

Suppose $\hat{\eta}_i$ falls within the lth segment, i.e., $\frac{l-1}{m} \leq \hat{\eta}_i < \frac{l}{m}$ for a particular l. We have

$$\hat{\lambda}_{i,l-1} = l - m \cdot \hat{\eta}_i \text{ and } \hat{\lambda}_{i,l} = m \cdot \hat{\eta}_i - (l-1) .$$

Then we can express $\hat{\eta}_i - \hat{\zeta}_i$ as a function of $\hat{\eta}_i$.

$$\hat{\eta}_i - \hat{\zeta}_i = \hat{\eta}_i - \left(\hat{\lambda}_{i,l-1} \cdot \frac{(l-1)^2}{m^2} + \hat{\lambda}_{i,l} \cdot \frac{l^2}{m^2}\right)$$

$$= \hat{\eta}_i - \left((l - m \cdot \hat{\eta}_i) \cdot \frac{(l-1)^2}{m^2} + (m \cdot \hat{\eta}_i - (l-1)) \cdot \frac{l^2}{m^2}\right)$$

$$= \hat{\eta}_i - \frac{(l^2 - (l-1)^2) \cdot m\hat{\eta}_i - l(l-1)}{m^2}$$

$$= \left(1 - \frac{2l-1}{m}\right) \cdot \hat{\eta}_i + \frac{l(l-1)}{m^2}. \tag{7.37}$$

When $0 < \hat{\eta}_i \leq \frac{1}{2}$, we have $l \leq \frac{m+1}{2}$ if m is odd and $l \leq \frac{m}{2}$ if m is even. Then $1 - \frac{2l-1}{m} \geq 0$ for both cases. Thus, when $0 < \hat{\eta}_i \leq \frac{1}{2}$ and $\hat{\eta}_i$ falls within the lth segment, $\hat{\eta}_i - \hat{\zeta}_i$ is a nondecreasing function. Since the nondecreasing property holds for all segments within $(0, \frac{1}{2}]$, we know that $\hat{\eta}_i - \hat{\zeta}_i$ is a nondecreasing function when $0 < \hat{\eta}_i \leq \frac{1}{2}$.

Case 2: $\frac{1}{2} < \eta_{\max} \leq 1$: Suppose $\hat{\eta}_j = \eta_{\max}$. Let $\hat{\eta}_{j'} = 1 - \hat{\eta}_j$ and denote $\hat{\zeta}_{j'}$ the approximation of $\hat{\eta}_{j'}^2$. By (7.36), we have $\hat{\eta}_i \leq 1 - \hat{\eta}_j - \left[\sum_{k \in \mathcal{N}}^{k \neq i, k \neq j} \hat{\eta}_k + \hat{\eta}_{vac} + \frac{U\tau_{TSP}}{E_{\max}-E_{\min}} \cdot (\hat{\eta}_i - \hat{\zeta}_i)\right] < 1 - \hat{\eta}_j = \hat{\eta}_{j'} < \frac{1}{2}$ for each $i \in \mathcal{N}$. Since $\hat{\eta}_i - \hat{\zeta}_i$ is a nondecreasing function for $0 < \hat{\eta}_i \leq \frac{1}{2}$, we have $\hat{\eta}_i - \hat{\zeta}_i \leq \hat{\eta}_{j'} - \hat{\zeta}_{j'}$.

We now show $\hat{\eta}_{j'} - \hat{\zeta}_{j'} = \hat{\eta}_j - \hat{\zeta}_j$. Suppose $\hat{\eta}_j$ falls within the lth segment, i.e., $\frac{l-1}{m} \leq \hat{\eta}_j < \frac{l}{m}$ for a particular l. Then $\hat{\eta}_{j'}$ falls within the $(m-l+1)$th segment, i.e., $\frac{m-l}{m} \leq \hat{\eta}_{j'} < \frac{m-l+1}{m}$. Then by (7.37), we have

$$\hat{\eta}_{j'} - \hat{\zeta}_{j'}$$
$$= \left(1 - \frac{2(m-l+1)-1}{m}\right) \cdot \hat{\eta}_{j'} + \frac{(m-l+1)(m-l+1-1)}{m^2}$$
$$= \frac{-m+2l-1}{m} \cdot (1 - \hat{\eta}_j) + \frac{(m-l+1)(m-l)}{m^2}$$
$$= \left(1 - \frac{2l-1}{m}\right) \cdot \hat{\eta}_j + \frac{(-m^2+2lm-m) + (m^2-2lm+l^2+m-l)}{m^2}$$
$$= \left(1 - \frac{2l-1}{m}\right) \cdot \hat{\eta}_j + \frac{l^2-l}{m^2}$$
$$= \hat{\eta}_j - \hat{\zeta}_j.$$

Therefore, we have $\hat{\eta}_i - \hat{\zeta}_i \leq \hat{\eta}_{j'} - \hat{\zeta}_{j'} = \hat{\eta}_j - \hat{\zeta}_j = \eta_{\max} - \zeta_{\max}$.

Combining both cases, we have $\hat{\eta}_i - \hat{\zeta}_i \leq \eta_{\max} - \zeta_{\max}$ for each $i \in \mathcal{N}$. This completes the proof. $\square$

Lemma 7.6

For the constructed solution ψ to problem OPT-R, we have

$$\eta_{vac} = 1 - \sum_{k \in \mathcal{N}} \hat{\eta}_k - \frac{U\tau_{TSP}}{E_{\max} - E_{\min}} \cdot \eta_{\max} \cdot (1 - \eta_{\max}).$$

Proof. To prove the lemma, it is sufficient to show $\max_{i \in \mathcal{N}}\{\hat{\eta}_i(1 - \hat{\eta}_i)\} = \eta_{\max}(1 - \eta_{\max})$, i.e, $\hat{\eta}_i(1 - \hat{\eta}_i) \leq \eta_{\max}(1 - \eta_{\max})$ for each $i \in \mathcal{N}$.

We consider the following two cases:

Case 1: $0 < \eta_{\max} \leq \frac{1}{2}$: We have $0 < \hat{\eta}_i \leq \eta_{\max} \leq \frac{1}{2}$ for each $i \in \mathcal{N}$. Since $x(1 - x)$ is an increasing function for $0 < x \leq \frac{1}{2}$, we have $\hat{\eta}_i(1 - \hat{\eta}_i) \leq \eta_{\max}(1 - \eta_{\max})$.

Case 2: $\frac{1}{2} < \eta_{\max} < 1$: Suppose $\hat{\eta}_j = \eta_{\max}$. By (7.36), we have $\hat{\eta}_i \leq 1 - \eta_j - \left[\sum_{k \in \mathcal{N}, k \neq i, k \neq j} \hat{\eta}_k + \hat{\eta}_{\text{vac}} + \frac{U \tau_{\text{TSP}}}{E_{\max} - E_{\min}} \cdot (\hat{\eta}_i - \hat{\zeta}_i) \right] < 1 - \eta_j < \frac{1}{2}$ for each $i \in \mathcal{N}$. Since $0 < \hat{\eta}_i < 1 - \eta_j < \frac{1}{2}$ and $x(1 - x)$ is an increasing function for $0 < x < \frac{1}{2}$, we have $\hat{\eta}_i(1 - \hat{\eta}_i) \leq (1 - \eta_j)\left[1 - (1 - \eta_j)\right] = \eta_j(1 - \eta_j) = \eta_{\max}(1 - \eta_{\max})$.

Combining both cases, we have $\hat{\eta}_i(1 - \hat{\eta}_i) \leq \eta_{max}(1 - \eta_{max})$. This completes the proof. $\qquad\square$

Based on Lemmas 7.5 and 7.6, we can prove Lemma 7.4.

Proof of Lemma 7.4. Denote $\hat{\eta}_{\text{vac}}$ as the objective value obtained by solution $\hat{\psi}$ to the relaxed linear problem OPT-L. Since problem OPT-L is a relaxation of problem OPT-R, $\hat{\eta}_{\text{vac}}$ is an upper bound of η_{vac}^*, i.e., $\eta_{\text{vac}}^* \leq \hat{\eta}_{\text{vac}}$. Therefore,

$$
\eta_{\text{vac}}^* - \eta_{\text{vac}} \leq \hat{\eta}_{\text{vac}} - \eta_{\text{vac}}
$$

$$
= \left[1 - \sum_{k \in \mathcal{N}} \hat{\eta}_k - \frac{U \tau_{\text{TSP}}}{E_{\max} - E_{\min}} \cdot (\eta_{\max} - \zeta_{\max}) \right]
$$

$$
- \left[1 - \sum_{k \in \mathcal{N}} \hat{\eta}_k - \frac{U \tau_{\text{TSP}}}{E_{\max} - E_{\min}} \cdot \eta_{\max} \cdot (1 - \eta_{\max}) \right]
$$

$$
= \frac{U \tau_{\text{TSP}}}{E_{\max} - E_{\min}} (\zeta_{\max} - \eta_{\max}^2)
$$

$$
\leq \frac{U \tau_{\text{TSP}}}{4(E_{\max} - E_{\min})} \cdot \frac{1}{m^2},
$$

where the second equality holds by Lemmas 7.5 and 7.6, the fourth inequality holds by Lemma 7.3. This completes the proof. $\qquad\square$

Based on Lemma 7.4, the following theorem shows how to set m such that $\eta_{\text{vac}}^* - \eta_{\text{vac}} \leq \epsilon$ for a given target performance gap ϵ ($0 < \epsilon \ll 1$).

Theorem 7.2

For a given ϵ, $0 < \epsilon \ll 1$, if $m = \left\lceil \sqrt{\dfrac{U \tau_{\text{TSP}}}{4\epsilon(E_{\max} - E_{\min})}} \right\rceil$, then we have $\eta_{\text{vac}}^* - \eta_{\text{vac}} \leq \epsilon$.

Proof. Lemma 7.4 shows that the performance gap is $\eta_{\text{vac}}^* - \eta_{\text{vac}} \leq \frac{U \tau_{\text{TSP}}}{4(E_{\max} - E_{\min})} \cdot \frac{1}{m^2}$. Therefore, if we set $m = \left\lceil \sqrt{\dfrac{U \tau_{\text{TSP}}}{4\epsilon(E_{\max} - E_{\min})}} \right\rceil \geq \sqrt{\dfrac{U \tau_{\text{TSP}}}{4\epsilon(E_{\max} - E_{\min})}}$, then we have

---

Algorithm 7.2 Summary of the procedure of how to construct a near-optimal solution.

Construction of a near-optimal solution

1. Given a target performance gap ϵ.
2. Let $m = \left\lceil \sqrt{\dfrac{U\tau_{\mathrm{TSP}}}{4\epsilon(E_{\max}-E_{\min})}} \right\rceil$.
3. Solve problem OPT-L with m segments by CPLEX, and obtain its solution $\hat{\psi} = (\hat{f}_{ij}, \hat{f}_{iB}, \hat{\eta}_i, \hat{\eta}_{\mathrm{vac}}, \hat{z}_{ik}, \hat{\lambda}_{i,k}, \hat{\zeta}_i)$.
4. Construct a feasible solution $\psi = (f_{ij}, f_{iB}, \eta_i, \eta_{\mathrm{vac}})$ for problem OPT-R by letting $f_{ij} = \hat{f}_{ij}$, $f_{iB} = \hat{f}_{iB}$, $\eta_i = \hat{\eta}_i$ and $\eta_{\mathrm{vac}} = \min_{i \in \mathcal{N}}\{1 - \sum_{k \in \mathcal{N}} \hat{\eta}_k - \dfrac{U\tau_{\mathrm{TSP}}}{E_{\max}-E_{\min}} \cdot \hat{\eta}_i \cdot (1 - \hat{\eta}_i)\}$.
5. Obtain a near-optimal solution $(f_{ij}, f_{iB}, \tau, \tau_i, \tau_{\mathrm{vac}}, p_i)$ to problem OPT by Algorithm 7.1.

$$
\begin{aligned}
\eta_{\mathrm{vac}}^{*} - \eta_{\mathrm{vac}} &\leq \frac{U\tau_{\mathrm{TSP}}}{4(E_{\max}-E_{\min})} \cdot \frac{1}{m^2} \\
&\leq \frac{U\tau_{\mathrm{TSP}}}{4(E_{\max}-E_{\min})} \cdot \frac{4\epsilon(E_{\max}-E_{\min})}{U\tau_{\mathrm{TSP}}} \\
&= \epsilon .
\end{aligned}
$$

This completes the proof.   □

With Theorem 7.2, we display the complete solution procedure on how to obtain a near-optimal solution to problem OPT in Fig. 7.7.

7.8 Construction of initial transient cycle

In Section 7.5, we skipped discussion on how to construct the initial transient cycle before the first renewable cycle. Now with the optimal traveling path P (the shortest Hamiltonian cycle) and the feasible near-optimal solution $(f_{ij}, f_{iB}, \tau, \tau_i, \tau_{\mathrm{vac}}, p_i)$ obtained in Section 7.7, we are ready to construct the initial transient cycle.

Unlike a renewable energy cycle at node i, which starts and ends with the same energy level E_i, the initial transient starts with $E_{\max}$ and ends with E_i. Specifically, the initial transient cycle must meet the following criterion:

Criterion 7.1

At each node $i \in \mathcal{N}$, its initial transient cycle must meet the following criteria: (i) $e_i(0) = E_{\max}$ and $e_i(\tau) = E_i$; and (ii) $e_i(t) \geq E_{\min}$ for $t \in [0, \tau]$.

We now construct an initial cycle to meet the above criterion. First, we need to calculate E_i ($i \in \mathcal{N}$). From (7.13), we have $E_i = E_{\max} - (2\tau - a_i - \tau_i)p_i$, where a_i can be obtained by (7.10).

Figure 7.7 Illustration of energy behavior for the initial transient cycle and how it connects the first renewable cycle.

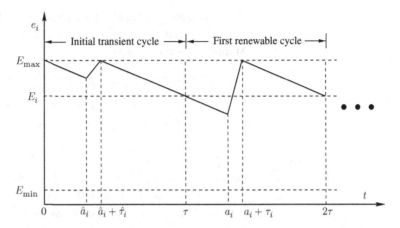

For a solution $\varphi = (P, f_{ij}, f_{iB}, \tau, \tau_i, \tau_p, \tau_{vac}, p_i, U)$ corresponding to a renewable energy cycle for $t \geq \tau$, we construct $\hat{\varphi} = (\hat{P}, \hat{f}_{ij}, \hat{f}_{iB}, \hat{\tau}, \hat{\tau}_i, \hat{\tau}_{\hat{p}}, \hat{\tau}_{vac}, \hat{p}_i, u_i)$ for the initial transient cycle for $t \in [0, \tau]$ by letting $\hat{P} = P$, $\hat{f}_{ij} = f_{ij}$, $\hat{f}_{iB} = f_{iB}$, $\hat{\tau} = \tau$, $\hat{\tau}_i = \tau_i$, $\hat{\tau}_{\hat{p}} = \tau_p$, $\hat{\tau}_{vac} = \tau_{vac}$, $\hat{p}_i = p_i$, and

$$u_i = \frac{p_i \hat{a}_i}{\tau_i} + p_i, \tag{7.38}$$

where u_i is the charging rate at node i during the initial transient cycle and $\hat{a}_i$ is the arrival time of the WCV at node i in the initial transient cycle (see Fig. 7.7).

Now we need to show $u_i \leq U$ where U is the full charging rate. First, we have

$$\hat{a}_{\pi_i} = \sum_{k=0}^{i-1} \frac{\hat{D}_{\pi_k \pi_{k+1}}}{V} + \sum_{k=1}^{i-1} \hat{\tau}_{\pi_k} = \sum_{k=0}^{i-1} \frac{D_{\pi_k \pi_{k+1}}}{V} + \sum_{k=1}^{i-1} \tau_{\pi_k} = a_{\pi_i} - \tau, \tag{7.39}$$

where the second equality holds by $\hat{P} = P$ and $\hat{\tau}_i = \tau_i$, and the last equality follows from (7.10). Further, by (7.12), we have $U \cdot \tau_i = \tau \cdot p_i = (2\tau - \tau) \cdot p_i \geq (a_i + \tau_i - \tau) \cdot p_i$. It follows that

$$(a_i - \tau) \cdot p_i \leq (U - p_i) \cdot \tau_i. \tag{7.40}$$

Then, we have

$$u_i = \frac{p_i \hat{a}_i}{\tau_i} + p_i = \frac{p_i (a_i - \tau)}{\tau_i} + p_i \leq \frac{(U - p_i) \cdot \tau_i}{\tau_i} + p_i = U,$$

where the first equality follows from (7.38), the second equality follows from (7.39), and the third inequality follows from (7.40).

For the newly constructed $\hat{\varphi}$, we have the following theorem:

Theorem 7.3
The constructed $\hat{\varphi}$ is a feasible transient cycle.

Proof. To prove that $\hat{\varphi}$ is a feasible initial transient cycle, we need to show that the newly constructed $\hat{\varphi}$ satisfies Criterion 7.1. By our assumption, $e_i(0) = E_{\max}$. At time $(\hat{a}_i + \hat{\tau}_i)$, we have

$$
\begin{aligned}
e_i(\hat{a}_i + \hat{\tau}_i) &= e_i(0) - \hat{p}_i \cdot \hat{a}_i + (u_i - \hat{p}_i) \cdot \hat{\tau}_i \\
&= E_{\max} - p_i \cdot (a_i - \tau) + (u_i - p_i) \cdot \tau_i \\
&= E_{\max} - p_i \cdot (a_i - \tau + \tau_i) + u_i \cdot \tau_i \\
&= E_{\max},
\end{aligned}
\tag{7.41}
$$

where the second equality follows since $e_i(0) = E_{\max}$, $\hat{p}_i = p_i$, $\hat{a}_i = a_i - \tau$, and $\hat{\tau}_i = \tau_i$, the last equality follows from (7.38) and (7.39). Therefore, the battery at node i is full when the WCV leaves it at $(\hat{a}_i + \hat{\tau}_i)$. At time τ, we have

$$
\begin{aligned}
e_i(\tau) &= e_i(\hat{a}_i + \hat{\tau}_i) - \hat{p}_i \cdot \left(\tau - (\hat{a}_i + \hat{\tau}_i)\right) \\
&= E_{\max} - p_i \cdot [\tau - (a_i - \tau + \tau_i)] \\
&= E_{\max} - p_i \cdot [2\tau - (a_i + \tau_i)] \\
&= e_i(2\tau) \\
&= E_i,
\end{aligned}
\tag{7.42}
$$

where the second equality follows from (7.41), and the fourth equality follows from (7.13). Therefore, Criterion 7.1(i) is met.

To show $e_i(t) \geq E_{\min}$ for $t \in [0, \tau]$, it is sufficient to show that $e_i(\hat{a}_i) \geq E_{\min}$ and $e_i(\tau) \geq E_{\min}$, since these two time instances are the local minimum for $e_i(t)$ during $t \in [0, \tau]$. We have $e_i(\hat{a}_i) = e_i(0) - \hat{p}_i \cdot \hat{a}_i = E_{\max} - p_i \cdot (a_i - \tau) \geq E_i - p_i \cdot (a_i - \tau) = e_i(a_i) \geq E_{\min}$. Also, by (7.42), $e_i(\tau) = E_i \geq E_{\min}$. Hence, $e_i(t) \geq E_{\min}$, for $t \in [0, \tau]$.

In summary, $\hat{\varphi}$ meets all the criteria of a feasible initial transient cycle. This completes the proof.   □

7.9 Numerical examples

In this section, we present some numerical examples to demonstrate how our solution can produce a renewable WSN and some interesting properties with such a network.

7.9.1 Simulation settings

We consider two randomly generated WSNs consisting of 50 and 100 nodes, respectively. The sensor nodes are deployed over a square area of 1 km × 1 km. The data rate (i.e., R_i, $i \in \mathcal{N}$) from each node is randomly generated

Table 7.2 Location and data rate R_i for each node in a 50-node network.

Node index	Location (m)	R_i (kb/s)	Node index	Location (m)	R_i (kb/s)
1	(815, 276)	1	26	(758, 350)	9
2	(906, 680)	8	27	(743, 197)	1
3	(127, 655)	4	28	(392, 251)	4
4	(913, 163)	6	29	(655, 616)	4
5	(632, 119)	3	30	(171, 473)	9
6	(98, 498)	7	31	(706, 352)	5
7	(278, 960)	3	32	(32, 831)	10
8	(547, 340)	7	33	(277, 585)	1
9	(958, 585)	6	34	(46, 550)	3
10	(965, 224)	8	35	(97, 917)	2
11	(158, 751)	5	36	(823, 286)	2
12	(971, 255)	1	37	(695, 757)	8
13	(957, 506)	4	38	(317, 754)	6
14	(485, 699)	10	39	(950, 380)	6
15	(800, 891)	2	40	(34, 568)	2
16	(142, 959)	9	41	(439, 76)	8
17	(422, 547)	5	42	(382, 54)	7
18	(916, 139)	10	43	(766, 531)	4
19	(792, 149)	1	44	(795, 779)	6
20	(959, 258)	5	45	(187, 934)	6
21	(656, 841)	3	46	(490, 130)	1
22	(36, 254)	10	47	(446, 569)	3
23	(849, 814)	1	48	(646, 469)	2
24	(934, 244)	8	49	(709, 12)	2
25	(679, 929)	8	50	(755, 337)	3

within [1, 10] kb/s. The power consumption coefficients are $\beta_1 = 50$ nJ/b, $\beta_2 = 0.0013$ pJ/(b · m^4), $\alpha = 4$, and $\rho = 50$ nJ/b [67]. The base station is assumed to be located at (500, 500) (in m) and the home service station for the WCV is assumed to be at the origin. The traveling speed of the WCV is $V = 5$ m/s.

For the battery at a sensor node, we choose a regular NiMH battery and its nominal cell voltage and the quantity of electricity is 1.2 V/2.5 Ah. We have $E_{max} = 1.2$ V $\times$ 2.5 A $\times$ 3600 sec $= 10.8$ KJ [94, Chapter 1]. We let $E_{min} = 0.05 \cdot E_{max} = 540$ J. We assume the wireless energy transfer rate $U = 5$ W, which is well within the feasible range [85].

We set the target $\epsilon = 0.01$ for the numerical results, i.e., our solution has an error no more than 0.01.

7.9.2 Results

50-node network   We first present complete results for the 50-node network. Table 7.2 gives the location of each node and its data rate for a 50-node network. The shortest Hamiltonian cycle is found by using the Concorde

Figure 7.8 An optimal traveling
path for the WCV for the
50-node sensor network,
assuming traveling direction is
counter clockwise.

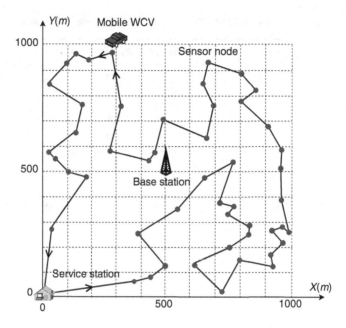

solver [28] and is shown in Fig. 7.8. For this optimal cycle, $D_{\text{TSP}} = 5821$ m
and $\tau_{\text{TSP}} = 1164.2$ s. For the target $\epsilon = 0.01$, by Theorem 7.2, we have

$$
m = \left\lceil \sqrt{\frac{U \cdot \tau_{\text{TSP}}}{4\epsilon(E_{\max} - E_{\min})}} \right\rceil = \left\lceil \sqrt{\frac{5 \times 1164.2}{4 \times 0.01 \times (10800 - 540)}} \right\rceil = 4 \,,
$$

which is a small number. In our solution, the cycle time $\tau = 30.73$ hours, the
vacation time $\tau_{\text{vac}} = 26.82$ hours, and the objective $\eta_{\text{vac}} = 87.27\%$.

In Corollary 7.2, the WCV can follow either direction of the shortest Hamiltonian cycle while achieving the same objective value $\eta_{\text{vac}} = 87.27\%$. Com-
paring the two solutions, the values for f_{ij}, f_{iB}, τ, τ_i, τ_{TSP}, τ_{vac} are identical,
while the values of a_i and E_i are different. This finding can be verified by
our simulation results in Table 7.3 (counter clockwise direction) and Table 7.4
(clockwise direction). As an example, Figs. 7.9(a) and (b) show the energy
cycle behavior of a sensor node (the 32nd node) under the two opposite traveling directions, respectively.

By Property 7.1, we find that there exists an energy bottleneck node in the
network with its energy dropping to $E_{\min}$ during a renewable energy cycle.
This property is confirmed in our numerical results. This bottleneck node is
the 48th node, whose energy behavior is shown in Fig. 7.10.

In Section 7.7.2, we showed that minimum energy routing may not be optimal for our problem (see Remark 7.1). This point is confirmed by our numerical results. In Fig. 7.11, we show that data routing in our solution differs from
the minimum energy routing for the 50-node network.

Table 7.3 The case of counter clockwise traveling direction: Node visited along the path, arrival time at each node, starting energy of each node in a renewable cycle, and charging time at each node for the 50-node network.

Node visited along the path	a_i (s)	E_i (J)	τ_i (s)	Node visited along the path	a_i (s)	E_i (J)	τ_i (s)
42	110702	10747	11	2	117778	10611	41
41	110725	10613	37	44	117848	10605	42
46	110777	9282	305	23	117903	10793	2
28	111113	7697	627	15	117923	10747	11
8	111776	7590	653	25	117960	10685	25
48	112461	714	2092	21	118002	10593	44
43	114579	10594	43	37	118065	8827	425
31	114660	6233	957	29	118519	8493	499
26	115627	10752	10	14	119056	10299	109
50	115639	9851	199	47	119192	10581	47
36	115855	10137	139	17	119246	9246	338
1	115997	9594	254	33	119614	4961	1287
27	116273	10551	53	38	120936	10059	164
5	116353	10646	33	7	121142	10754	10
49	116412	10610	40	45	121171	10658	31
19	116484	10660	29	16	121213	10738	14
18	116538	10622	38	35	121239	10259	120
4	116581	10329	100	32	121380	8628	483
10	116696	10596	43	11	121894	10010	176
24	116747	9648	245	3	122090	6697	924
20	116997	10773	6	40	123039	10790	2
12	117006	10794	1	34	123046	10747	12
39	117032	8565	477	6	123073	10519	63
13	117534	10020	167	30	123151	8319	563
9	117717	10613	40	22	123766	5722	1166

100-node network Table 7.5 gives the location of each node and its data rate for a 100-node network. The shortest hamiltonian cycle is shown in Fig. 7.12. For this optimal cycle, $D_{\text{TSP}} = 7687$ m and $\tau_{\text{TSP}} = 1537.4$ s. For the target $\epsilon = 0.01$, by Theorem 7.2, we have

$$m = \left\lceil \sqrt{\frac{U\tau_{\text{TSP}}}{4\epsilon(E_{\max} - E_{\min})}} \right\rceil = \left\lceil \sqrt{\frac{5 \times 1537.4}{4 \times 0.01 \times (10800 - 540)}} \right\rceil = 5.$$

The solution for the 100-node network includes the cycle time $\tau = 58.52$ hours, the vacation time $\tau_{\text{vac}} = 50.30$ hours, and the objective $\eta_{\text{vac}} = 85.95\%$. Additional results are shown in Table 7.6.

Table 7.4 The case of clockwise traveling direction: Node visited along the path, arrival time at each node, starting energy of each node in a renewable cycle, and charging time at each node for the 50-node network.

Node visited along the path	a_i (s)	E_i (J)	τ_i (s)	Node visited along the path	a_i (s)	E_i (J)	τ_i (s)
22	110676	5032	1166	9	117852	10613	40
30	111894	8032	563	13	117907	10023	167
6	112472	10489	63	39	118099	8588	477
34	112550	10741	12	12	118601	10795	1
40	112567	10789	2	20	118605	10773	6
3	112594	6301	924	24	118617	9668	245
11	113538	9944	176	10	118868	10600	43
32	113744	8461	483	4	118928	10339	100
35	114250	10222	120	18	119032	10626	38
16	114382	10734	14	19	119095	10664	29
45	114406	10648	31	49	119156	10615	40
7	114456	10751	10	5	119223	10650	33
38	114508	10012	164	27	119283	10558	53
33	114706	4676	1287	1	119357	9632	254
17	116024	9197	338	36	119613	10160	139
47	116368	10575	47	50	119770	9888	199
14	116443	10286	109	26	119972	10754	10
29	116590	8449	499	31	119992	6464	957
37	117118	8809	425	43	120986	10606	43
21	117561	10593	44	48	121056	1526	2092
25	117624	10685	25	8	123180	7927	653
15	117674	10747	11	28	123868	8059	627
23	117703	10793	2	46	124526	9472	305
44	117718	10605	42	41	124846	10637	37
2	117789	10611	41	42	124896	10754	11

7.10 Chapter summary

Linear approximation is a powerful approach to tackle certain nonlinear optimization problems. In this chapter, we showed how such an approach could be employed to solve a nonlinear programming problem in a WSN. In addition to the linear approximation technique, the problem in the case study is interesting on its own, and demonstrates how the so-called wireless energy transfer technology can be employed to address network lifetime problems in a WSN.

Figure 7.9 The energy behavior of a sensor node (the 32th) in the 50-node network during the initial transient cycle and the first two renewable cycles.

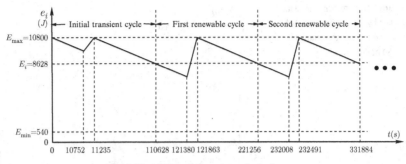

(a) Traveling path is counter clockwise.

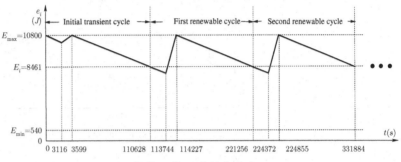

(b) Traveling path is clockwise.

Figure 7.10 The energy behavior of the bottleneck node (48th node) in the 50-node network. Traveling direction is counter clockwise.

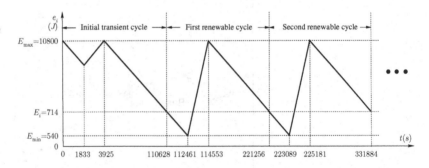

Figure 7.11 Comparison of data
routing by our solution and that
by minimum energy routing for
the 50-node network.

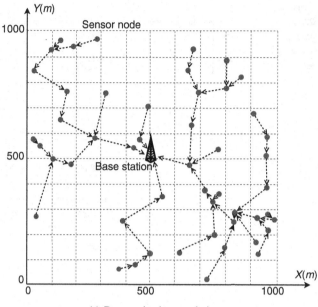

(a) Data routing in our solution.

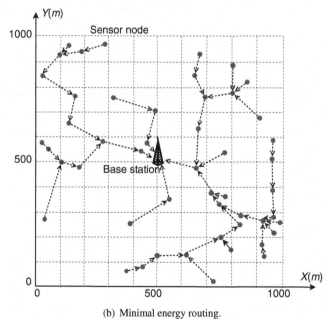

(b) Minimal energy routing.

Table 7.5 Location and data rate R_i for each node in a 100-node network.

Node index	Location (m)	R_i (kb/s)	Node index	Location (m)	R_i (kb/s)
1	(970, 383)	10	51	(295, 856)	7
2	(124, 85)	8	52	(306, 584)	10
3	(467, 734)	6	53	(106, 374)	8
4	(657, 332)	3	54	(594, 222)	7
5	(290, 840)	3	55	(283, 219)	10
6	(755, 372)	9	56	(155, 522)	1
7	(558, 828)	5	57	(1, 433)	3
8	(428, 177)	9	58	(284, 741)	10
9	(267, 130)	1	59	(551, 70)	8
10	(754, 880)	9	60	(871, 847)	5
11	(898, 44)	1	61	(42, 680)	7
12	(728, 687)	9	62	(905, 137)	5
13	(407, 734)	9	63	(131, 858)	4
14	(938, 437)	6	64	(834, 200)	3
15	(255, 380)	2	65	(800, 607)	4
16	(533, 980)	2	66	(918, 543)	1
17	(955, 399)	8	67	(137, 162)	5
18	(268, 440)	9	68	(505, 6)	4
19	(250, 157)	1	69	(405, 771)	10
20	(928, 326)	8	70	(174, 765)	6
21	(69, 314)	10	71	(575, 421)	9
22	(299, 895)	4	72	(606, 57)	5
23	(592, 247)	7	73	(214, 586)	5
24	(203, 311)	4	74	(520, 174)	9
25	(636, 409)	3	75	(989, 729)	10
26	(798, 708)	8	76	(490, 534)	6
27	(502, 144)	8	77	(695, 253)	10
28	(651, 871)	3	78	(411, 917)	8
29	(796, 83)	6	79	(35, 758)	6
30	(233, 462)	5	80	(293, 887)	10
31	(601, 30)	1	81	(801, 69)	9
32	(112, 753)	1	82	(347, 184)	10
33	(516, 700)	5	83	(83, 737)	7
34	(838, 215)	3	84	(511, 697)	4
35	(921, 680)	4	85	(367, 777)	5
36	(498, 557)	3	86	(739, 502)	10
37	(278, 851)	2	87	(525, 425)	1
38	(653, 559)	4	88	(805, 611)	2
39	(917, 902)	7	89	(817, 856)	5
40	(510, 420)	3	90	(189, 671)	6
41	(974, 358)	7	91	(124, 524)	1
42	(197, 489)	5	92	(821, 299)	7
43	(111, 256)	5	93	(638, 704)	9
44	(297, 929)	2	94	(16, 382)	2
45	(396, 467)	2	95	(896, 568)	5
46	(421, 254)	7	96	(515, 888)	3
47	(311, 431)	3	97	(545, 843)	10
48	(694, 703)	6	98	(606, 899)	7
49	(92, 402)	10	99	(760, 939)	3
50	(402, 182)	8	100	(855, 815)	2

Figure 7.12 An optimal traveling path for the WCV for the 100-node sensor network, assuming traveling direction is clockwise.

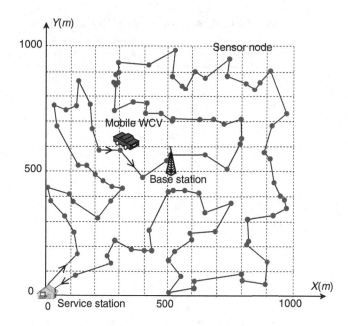

Table 7.6 The case of clockwise traveling direction: Node visited along the path, arrival time at each node, starting energy of each node in a renewable cycle, and charging time at each node for the 100-node network.

Node visited along the path	a_i (s)	E_i (J)	τ_i (s)	Node visited along the path	a_i (s)	E_i (J)	τ_i (s)
67	210727	10257	109	96	227822	10722	17
43	210855	9254	310	97	227850	9391	307
21	211179	10623	35	7	228162	7595	702
94	211232	10625	35	98	228880	10638	35
57	211278	10762	8	28	228926	9893	199
49	211305	10692	22	99	229151	10762	8
53	211333	9279	306	10	229171	10075	159
24	211661	8065	551	89	229344	10726	16
15	212230	8467	471	100	229371	9797	220
47	212716	4983	1181	60	229599	10613	41
18	213906	10698	21	39	229654	10687	25
30	213935	9012	364	75	229716	10581	48
42	214308	9784	207	35	229781	9415	305
56	214526	10787	3	95	230109	10561	53
91	214534	10778	4	66	230168	10790	2
61	214574	10682	24	14	230192	10720	18
79	214614	10681	24	17	230218	10723	17
83	214649	10274	107	1	230240	10287	113
32	214762	10365	89	41	230358	10716	19

Table 7.6 (Cont.)

Node visited along the path	a_i (s)	E_i (J)	τ_i (s)	Node visited along the path	a_i (s)	E_i (J)	τ_i (s)
63	214873	10640	33	20	230388	8701	464
70	214926	9839	196	92	230874	7791	668
90	215141	9489	268	34	231559	10732	15
73	215427	8932	383	64	231577	9786	225
52	215828	9431	281	62	231822	10614	41
45	216139	1745	1876	11	231882	10772	6
76	218038	10737	13	81	231908	10715	19
36	218056	7075	775	29	231930	9728	239
38	218862	10152	135	72	232207	10740	13
86	219018	4116	1402	31	232226	10786	3
65	220444	8386	508	68	232249	10724	17
88	220952	9402	294	59	232281	10261	120
26	221266	9845	201	27	232419	10221	129
12	221482	10546	54	74	232556	8179	587
48	221543	10725	16	54	233160	10733	15
93	221570	9030	374	23	233180	7167	817
33	221969	1095	2072	77	234018	10563	53
84	224042	1197	2072	6	234098	6861	890
3	226126	8828	427	4	235009	6073	1075
13	226565	9705	237	25	236100	5792	1146
69	226809	10424	82	71	237259	6167	1066
85	226899	9884	199	87	238335	4268	1516
58	227115	10534	58	40	239855	10743	13
5	227193	9991	176	46	239906	10677	28
37	227372	10781	4	8	239950	10321	111
51	227380	10397	88	50	240066	10481	74
80	227473	10691	24	82	240152	10643	37
22	227499	10752	11	55	240203	9350	338
44	227517	10766	7	19	240555	10785	3
78	227547	9805	217	9	240564	10775	6
16	227791	10742	13	2	240600	10658	33

7.11 Problems

7.1 Let us consider a continuous function $\theta(\mu) = \mu^2$, where $\mu \in [0, 1]$ (see Fig. 7.13). An approximating function $\hat{\theta}(\mu)$ is defined by dividing the interval $[0, 1]$ into four smaller intervals $[\mu_{k-1}, \mu_k]$, $1 \leq k \leq 4$, through the grid points $\mu_0 = 0$, $\mu_1 = \frac{1}{4}$, $\mu_2 = \frac{1}{2}$, $\mu_3 = \frac{3}{4}$, $\mu_4 = 1$. For any $\mu \in [0, 1]$, express its approximation $\hat{\theta}(\mu)$ using the piecewise linear approximation technique presented in Section 7.1. For $\mu = \frac{3}{8}$, find out $\hat{\theta}(\mu)$, z_k ($1 \leq k \leq 4$), and λ_k ($0 \leq k \leq 4$).

Figure 7.13 Piecewise linear approximation for a continuous function $\theta(\mu) = \mu^2$.

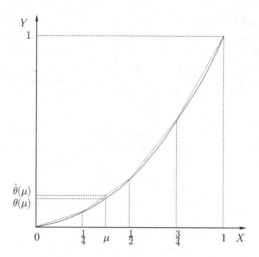

7.2   Discuss the pros and cons of energy-harvesting and wireless energy transfer for a sensor node.

7.3   Lemma 7.1 shows the equivalence (in terms of their optimal objective value) of a renewable energy cycle with a fully re-charged battery and a renewable energy cycle with partially re-charged battery. What's the purpose of giving this equivalence in Lemma 7.1?

7.4   Give the second part of the proof so as to complete the proof of Lemma 7.1.

7.5   How does Theorem 7.1 help us simplify the problem described in Section 7.4? Assuming Theorem 7.1 is not provided, is there any change in problem OPT?

7.6   Give a formal proof of Corollary 7.1.

7.7   In Corollary 7.2, we claim that two opposite directions of the shortest Hamiltonian cycle can achieve the same optimal objective. Explain why. In the two optimal solutions corresponding to the two opposite directions, which variables are changed and which are not? Justify your answer with the numerical results in Section 7.9.

7.8   Property 7.1 shows that for an optimal solution, there exists at least one "bottleneck" node in the network. Under a specific network, the WCV has two opposite outgoing directions starting from its home service station (both follow the shortest Hamiltonian cycle). In the two optimal solutions corresponding to the two opposite directions, do we have the same bottleneck node(s)? Justify your answer.

Table 7.7 Sensor node location in a ten-node network.

Node index	Location (m)	Node index	Location (m)
1	(60, 30)	6	(35,40)
2	(70, 65)	7	(45,55)
3	(90, 80)	8	(80,60)
4	(60, 40)	9	(20,60)
5	(45, 35)	10	(40,25)

Figure 7.14 A ten-node sensor network.

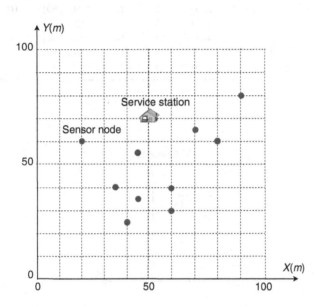

7.9 We consider a wireless sensor network consisting of ten nodes (see Fig. 7.14). The sensor nodes are deployed over a square area of 100 m × 100 m (see Table 7.7 for location of each sensor). The home service station for the WCV is assumed to be located at (50, 70) (in m). The traveling speed of the WCV is $V = 5$ m/s.

Find out the shortest Hamiltonian cycle for this network instance. For the optimal cycle, what is the total traveling distance D_{TSP} and the total traveling time τ_{TSP}?

7.10 What is the benefit of reformulating problem OPT to problem OPT-R?

7.11 Why do we put data routing as a part of the optimization problem? Why not simply use the minimum energy routing?

7.12 What is the purpose of using piecewise linear approximation for the quadratic terms η_i^2? How do we ensure that such an approximation can guarantee $(1 - \epsilon)$-optimality?

Table 7.8 Location and data rate R_i for each node in a 20-node network.

Node index	Location (m)	R_i (kb/s)	Node index	Location (m)	R_i (kb/s)
1	(600, 500)	8	11	(800, 700)	3
2	(700, 550)	9	12	(470, 700)	5
3	(560, 560)	1	13	(650, 250)	8
4	(400, 900)	9	14	(300, 100)	9
5	(560, 300)	6	15	(520, 540)	7
6	(100, 200)	4	16	(750, 150)	10
7	(590, 480)	3	17	(220, 200)	7
8	(800, 480)	6	18	(850, 300)	4
9	(200, 600)	5	19	(980, 150)	6
10	(400, 250)	10	20	(50, 950)	2

Figure 7.15 A three-node sensor network.

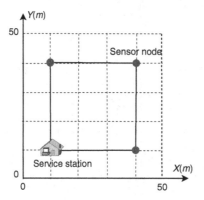

7.13 After solving problem OPT-L, why is it necessary to construct a feasible solution to problem OPT-R (and problem OPT)? Verify the feasibility of the constructed solution (in Section 7.7.3).

7.14 We consider a small WSN consisting of three nodes (see Fig. 7.15). The length of the shortest Hamiltonian cycle is 120 m for this network instance. The traveling speed of the WCV is 1.2 m/s. Each sensor node has a battery capacity of 10 KJ and is fully charged initially. The minimum energy at a sensor node battery for it to be operational is 500 J. The wireless energy transfer rate is 5 W. We are looking for a solution within 1% from optimum.

Answer the following questions:

(a) How can we set m, i.e., the number of segments that we use in the piecewise linear approximation such that 99% optimality is achieved?

(b) Use piecewise linear approximation and derive a relaxed linear formulation in the forms of problem OPT-L.

7.15 Why is it necessary to construct the initial transient cycle?

Table 7.9 The case of clockwise traveling direction: Node visited along the path, arrival time at each node, renewable cycle starting energy of each node, and charging time at each node for the 20-node network.

Node visited along the path	$a_i(s)$	E_i (J)	τ_i (s)	Node visited along the path	a_i (s)	E_i (J)	τ_i (s)
6	113337	10655	29	2	117427	9331	306
17	113390	10280	104	11	117768	10336	97
9	113574	5769	1018	8	117909	9400	293
20	114668	6280	923	18	118239	5765	1063
4	115661	7550	668	19	119342	9473	281
12	116372	8299	517	16	119669	8034	589
15	116922	10758	9	13	120287	10316	103
3	116940	10402	82	5	120411	1197	2091
7	117039	9608	247	10	122535	7361	754
1	117291	10252	114	14	123326	7823	657

Figure 7.16 An optimal traveling path for the 20-node sensor network. Only clockwise traveling direction is shown.

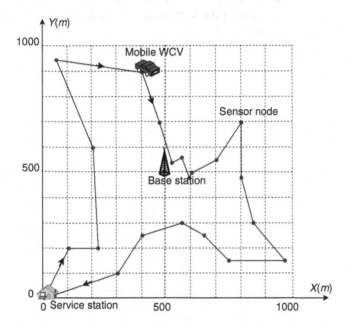

7.16 We consider a randomly generated WSN consisting of 20 nodes. Table 7.8 gives the location of each node and its data rate for a 20-node network. Refer to Section 7.9.1 for parameter setting.

The shortest Hamiltonian cycle is shown in Fig. 7.16. For this optimal cycle, $D_{\text{TSP}} = 4046$ m and $\tau_{\text{TSP}} = D_{\text{TSP}}/V = 809.2$ s. For the target $\epsilon = 0.01$, by Theorem 7.2, we have $m = 4$. In our solution, we have cycle time $\tau = 31.47$ hours, the vacation time $\tau_{\text{vac}} = 28.48$ hours, and the objective $\eta_{\text{vac}} = 90.50\%$. Table 7.9 shows nodes visited along the path of clockwise traveling direction, arrival time at each node, renewable cycle starting energy of each node, and charging time at each node for the 20-node network.

Assuming the WCV follows the clockwise traveling direction, answer the following questions:

(a) Based on Table 7.9, how to determine the energy consumption rate p_i at sensor node i? In particular, what are p_5, p_6, and p_{14}?

(b) In the initial transient cycle, the WCV re-charges sensor node i at the charging rate of u_i. How to determine the value of u_i for sensor node i? In particular, what are u_5, u_6, and u_{14}?

(c) Theorem 7.3 shows that the constructed initial cycle is a feasible transient cycle. Demonstrate this point at the 18th node. Draw the energy behavior of this specific node during the initial transient cycle and the first two renewable cycles.

(d) Recall that a bottleneck node in the network is defined as a node with its energy dropping to E_{min} during a renewable energy cycle. Which node is the bottleneck node in this network? Justify your answer and draw the energy behavior of the bottleneck node during the initial transient cycle plus the first two renewable cycles.

8 Approximation algorithm and its applications – Part 1

Do not follow where the path may lead. Go, instead, where there is no path and leave a trail.

Ralph Waldo Emerson

8.1 Review of approximation algorithms

Recall that Part II of this book focuses on methods for near-optimal and approximation solutions. Specifically, Chapters 5 to 7 follow the OR optimization approach to develop a $(1 - \varepsilon)$-optimal solution. In this and the following chapters, we will show how to develop the $(1 - \varepsilon)$-optimal solution by following the CS algorithm design approach.

We first give a brief overview of the so-called *approximation algorithms*. Such algorithms are designed to offer solutions that can approximate the unknown optimal solution according to certain benchmark performance criteria. In particular, in wireless network research, the most popular approximation algorithms can be classified as *constant-factor* approximation algorithms and $(1 - \varepsilon)$-*optimal* approximation algorithms.

- **Constant-factor approximation algorithms.** We use a maximization problem with a positive optimal objective value OPT (unknown) as an example. If we can prove that the obtained feasible solution achieves an objective value that is at least $c \cdot OPT$, where $c < 1$ is a constant, then the designed algorithm is a constant-factor approximation algorithm. Likewise, for a minimization problem, if we can prove that the obtained feasible solution achieves an objective value that is at most $c \cdot OPT$, where $c > 1$ is a constant, then the designed algorithm is a constant-factor approximation algorithm.

- **$(1-\varepsilon)$-optimal (or $(1+\varepsilon)$-optimal) approximation algorithms.** If the approximation factor can be set arbitrarily close to 1, then the approximation algorithm is a $(1-\varepsilon)$-optimal (or $(1+\varepsilon)$-optimal) approximation algorithm. That is, for a maximization problem and any small constant ε, $0 < \varepsilon \ll 1$, if the algorithm can find a solution with an objective value that is at least $(1-\varepsilon)OPT$, then this algorithm is $(1-\varepsilon)$-optimal. For a minimization problem and any small constant ε, if the algorithm can find a solution with an objective value that is at most $(1+\varepsilon)OPT$, then this algorithm is $(1+\varepsilon)$-optimal.

Both of the above two types of approximation algorithms are common in the literature to solve wireless network problems. The constant-factor approximation algorithms would only be useful if c is close to 1. But unfortunately, many results in the literature offer results that may be far away from 1. Although such results may still be of theoretical value, they are hardly useful in practice (in the sense of offering an accurate performance benchmark for the design of distributed networking algorithms and protocols).

Our preference is toward the design of $(1-\varepsilon)$-optimal approximation algorithms, which will be presented in a case study in this chapter. As expected, such algorithms are intellectually challenging and require much novelty in their design. Nevertheless, should we be able to design such an algorithm, then both its theoretical significance and practical value would be assured.

The case study in this chapter is a classic and fundamental problem on base station placement in a wireless sensor network (WSN). The setting of this problem is similar to that in Chapter 2. The problem here is to find the optimal location for the base station so that the network lifetime (until any sensor node runs out of energy) is maximized.

8.2 Case study: The base station placement problem

As discussed in Chapter 2, an important performance metric for a WSN is the so-called network lifetime. Given that energy expenditure at a node for transmitting data to another node directly depends on the physical distance between these two nodes, network lifetime is therefore highly dependent upon the physical topology of the network.

In this case study, we focus on base station location, which, as expected, has a significant impact on network lifetime performance. Specifically, we consider the following problem. Suppose each node i producing sensing data at a rate of r_i in a WSN, where all the sensing data are to be forwarded to the base station (via multi-hop if necessary): where should we place the base station in this WSN so that the network lifetime is maximized?

We will show that this problem can be formulated into a nonlinear programming (NLP), which is nonconvex and NP-hard in general [46]. One approach is to employ the branch-and-bound framework in Chapter 5. But that solution approach is tedious and not elegant. Instead, we will design an elegant approximation algorithm that can guarantee $(1 - \varepsilon)$-optimal network lifetime performance.

The $(1 - \varepsilon)$-optimal approximation algorithm that we are going to present is based on several techniques, which makes it possible to reduce an infinite search space to a finite-element search space for the base station location. The main idea of the approximation algorithm is to exploit a clever way of discretizing the cost parameter associated with energy consumption into a geometric sequence with tight upper and lower bounds. As a result, we can divide a continuous search space into a finite number of subareas. By further exploiting the cost property of each subarea, we can represent each subarea with a so-called "fictitious cost point" (FCP), which is an N-tuple cost vector with each element representing an upper bound on the cost for a corresponding sensor node in the network. Based on these ideas, we can successfully reduce an infinite search space for the base station location into a finite number of "points" for each of which we can solve an LP problem to find an achievable network lifetime and data routing solution. By comparing the achievable network lifetime obtained among all the FCPs, we show that the largest is $(1 - \varepsilon)$-optimal. We also show that placing the base station at *any physical point* in the subarea corresponding to the best FCP is $(1 - \varepsilon)$-optimal. We analyze the complexity of the approximation algorithm and show that it is polynomial.

The remainder of this chapter is organized as follows. Section 8.3 describes the network model used in the case study and formally states the base station placement problem. In Section 8.4, we show that for a given base station location, the maximum achievable network lifetime and the corresponding optimal routing can be found via a single LP problem. To search the best base station location, in Section 8.5, we narrow down the search space to the smallest enclosing disk (SED) that covers all the sensor nodes in the network. In Section 8.6, we divide the continuous search space of the SED into a finite number of subareas and represent each subarea by an FCP. Thus, we can find the best FCP corresponding to the maximum network lifetime. We then show that by placing the base station at *any point* in the subarea corresponding to the best FCP, we obtain a $(1 - \varepsilon)$-optimal network lifetime. In Section 8.7, we summarize all the steps as an algorithm. We also give an example for illustration. In Section 8.8, we prove the correctness of the algorithm and analyze its complexity. In Section 8.9, we give some additional numerical examples illustrating the efficacy of the algorithm. Section 8.10 summarizes this chapter.

8.3 Network model and problem description

8.3.1 Network model

We consider a static sensor network consisting of a set $\mathcal{N}$ of sensor nodes deployed over a two-dimensional area. The location of each sensor node is fixed and the initial energy at sensor node i is denoted as e_i. Each sensor node i generates data at a rate r_i. We assume that there is a single base station to be deployed in the area to collect sensing data.

In this chapter, we focus on the energy consumption due to communications (i.e., data transmission and reception). Suppose that sensor node i transmits data to sensor node j with a rate of f_{ij} (b/s). Then we model the transmission power at sensor node i as

$$u_{ij}^t = C_{ij} \cdot f_{ij}. \tag{8.1}$$

Here, C_{ij} is the cost associated with link $i \rightarrow j$ and can be modeled as

$$C_{ij} = \beta_1 + \beta_2 \cdot d_{ij}^\alpha, \tag{8.2}$$

where β_1 and β_2 are constant coefficients, d_{ij} is the physical distance between sensor nodes i and j, and α is the path-loss index, with $2 \leq \alpha \leq 4$.

The power consumption at the receiving sensor node i can be modeled as

$$u_i^r = \rho \cdot \sum_{\substack{k \in \mathcal{N} \\ k \neq i}} f_{ki}, \tag{8.3}$$

where f_{ki} (b/s) is the incoming bit-rate received by sensor i from sensor k and ρ is a constant coefficient.

In this chapter, we assume that the interference from different transmissions has been effectively avoided by appropriate MAC layer scheduling. For low bit rate and deterministic traffic pattern considered in this chapter, a contention-free MAC protocol is not hard to design and we omit its discussion in this chapter. Table 8.1 lists the notation used in this chapter.

8.3.2 Problem description

We aim to investigate how to optimally place a base station to collect data in a WSN so that the network lifetime can be maximized. Network lifetime is defined as the time until any sensor node first uses up its energy. To achieve optimality, we allow flow splitting, just as we did in previous chapters. Also, power control at a node is allowed, as modeled in (8.1) and (8.2).

Assume that base station B is located at a point p. Denote (x_B, y_B) as the position of point p and let T be the network lifetime. A feasible flow routing solution that achieves this network lifetime T should satisfy both flow balance and energy constraints at each sensor node. These constraints can be formally stated as follows. Denote f_{ij} and f_{iB} as the data rates from sensor node i to

Table 8.1 Notation.

Symbol	Definition
$\mathcal{A}$	The search space for the base station, which can be the smallest enclosing disk to cover all sensor nodes
$\mathcal{A}_m$	The mth subarea in the search space
B	The base station
$c_{iB}(p)$	Power consumption coefficient for transmitting data from sensor i to base station B at point p
C_{ij}	Power consumption coefficient for transmitting data from sensor i to sensor j
$C_{iB}^{\min}, C_{iB}^{\max}$	Lower and upper bounds of $c_{iB}(p)$
$C[h]$	The transmission cost for the hth circle
d_{ij} (or d_{iB})	Distance between sensor i and sensor j (or base station B)
e_i	Initial energy at sensor i
f_{ij} (or f_{iB})	Data rate from sensor i to sensor j (or base station B)
H_i	Total number of circles for discretization at sensor node i for a given ε
K	Total number of circles for discretization for a given ε
M	Total number of subareas for discretization for a given ε
$\mathcal{N}$	Set of sensor nodes in the network
N	$= |\mathcal{N}|$, number of sensor nodes in the network
$O_\mathcal{A}$	The center of the smallest enclosing disk $\mathcal{A}$
p_m	Fictitious cost point (FCP) representation for the mth subarea
p_{opt}	The best base station location
p^*	The best location among M FCPs
p_ε	A point in the subarea corresponding to p^*
r_i	Sensing data rate produced at sensor i
$R_\mathcal{A}$	The radius of the smallest enclosing disk $\mathcal{A}$
T_m	Maximum achievable network lifetime by placing the base station at p_m
T_{opt}	Optimal network lifetime achieved by placing the base station at p_{opt}
T^*	$= \max\{T_m : m = 1, 2, \ldots, M\}$
T_ε	$(1 - \varepsilon)$-optimal network lifetime achieved by p_ε
V_{ij} (or V_{iB})	Total data volume from sensor i to sensor j (or base station B)
(x_i, y_i)	Location of sensor node i
α	Path-loss index
β_1, β_2	Constant coefficients in transmission power modeling
ε	Desired approximation error, $0 < \varepsilon \ll 1$
ρ	Power consumption coefficient for receiving data
ψ_{opt}	The best flow routing solution when the base station is at p_{opt}
ψ^*	Flow routing solution when the base station is at p^*
ψ_ε	Flow routing solution when the base station is at p_ε

sensor node j and base station B, respectively. Then the flow balance for each sensor node i is

$$\sum_{\substack{k \in \mathcal{N} \\ k \neq i}} f_{ki} + r_i = \sum_{\substack{j \in \mathcal{N} \\ j \neq i}} f_{ij} + f_{iB},$$

i.e., the sum of total incoming flow rates plus the self-generated data rate is equal to the sum of the total outgoing flow rates. The energy constraint for each sensor node i is

$$\sum_{\substack{k \in \mathcal{N} \\ k \neq i}} \rho \cdot f_{ki} T + \sum_{\substack{j \in \mathcal{N} \\ j \neq i}} C_{ij} \cdot f_{ij} T + c_{iB}(p) \cdot f_{iB} T \leq e_i,$$

i.e., total consumed energy due to receiving and transmission over time T cannot exceed its initial energy e_i. By (8.2), we have

$$c_{iB}(p) = \beta_1 + \beta_2 \left[\sqrt{(x_B - x_i)^2 + (y_B - y_i)^2} \right]^\alpha,$$

which is a nonlinear function of base station location (x_B, y_B).

Our objective is to maximize the network lifetime T under the flow balance and energy constraints, i.e.,

Maximize T

$$\text{subject to} \quad \sum_{\substack{j \in \mathcal{N} \\ j \neq i}} f_{ij} + f_{iB} - \sum_{\substack{k \in \mathcal{N} \\ k \neq i}} f_{ki} = r_i \qquad (i \in \mathcal{N}) \quad (8.4)$$

$$\sum_{\substack{k \in \mathcal{N} \\ k \neq i}} \rho f_{ki} T + \sum_{\substack{j \in \mathcal{N} \\ j \neq i}} C_{ij} f_{ij} T + c_{iB}(p) f_{iB} T \leq e_i \qquad (i \in \mathcal{N}) \quad (8.5)$$

$$c_{iB}(p) - \beta_2 \left[\sqrt{(x_B - x_i)^2 + (y_B - y_i)^2} \right]^\alpha = \beta_1 \quad (i \in \mathcal{N})$$

$$(x_B, y_B) \in \mathcal{A}, T, f_{ij}, f_{iB} \geq 0 \qquad (i, j \in \mathcal{N}, i \neq j),$$

where $\mathcal{A}$ is an area containing all possible base station locations. This optimization problem is in the form of a *nonconvex programming*, which is NP-hard in general [46].

8.4 Optimal flow routing for a given base station location

As discussed earlier, the maximum network lifetime depends on both the base station location and data routing. To begin with, we show that for a *given* base station location, we can find the maximum network lifetime and optimal routing via a single LP problem. Here, the objective function is the network lifetime T and the constraints are given by (8.4) and (8.5). Multiplying both sides of (8.4) by T and denoting

$$V_{ij} = f_{ij} T \quad \text{and} \quad V_{iB} = f_{iB} T, \qquad (8.6)$$

where V_{ij} (or V_{iB}) can be interpreted as the total data volume from sensor node i to sensor node j (or base station B) over time T, we have the following formulation:

Maximize T

$$\text{subject to} \sum_{\substack{k\in\mathcal{N}\\k\neq i}} V_{ki} + r_i T - \sum_{\substack{j\in\mathcal{N}\\j\neq i}} V_{ij} - V_{iB} = 0 \qquad (i\in\mathcal{N})$$

$$\sum_{\substack{k\in\mathcal{N}\\k\neq i}} \rho V_{ki} + \sum_{\substack{j\in\mathcal{N}\\j\neq i}} C_{ij} V_{ij} + c_{iB}(p) V_{iB} \leq e_i \quad (i\in\mathcal{N})$$

$$T, V_{ij}, V_{iB} \geq 0 \qquad\qquad (i,j\in\mathcal{N}, i\neq j).$$

Note that for a given base station location, the $c_{iB}(p)$-values are constants. Therefore, the above formulation is in the form of an LP problem. Once we solve the above LP problem, we can obtain an optimal routing solution for f_{ij} and f_{iB} by $f_{ij} = \frac{V_{ij}}{T}$ and $f_{iB} = \frac{V_{iB}}{T}$.

8.5 Search space for base station location

Although for a given base station location we can find the corresponding maximum network lifetime via a single LP problem, it is not possible to examine all (infinite) points in the two-dimensional plane and select the point having the maximum network lifetime.

As a first step, we show that it is only necessary to consider points inside the so-called *smallest enclosing disk* (SED) [167],[1] which is a unique disk with the smallest radius that contains all the N sensor nodes in the network and can be found in $O(N)$ time [108]. This is formally stated in the following lemma:

Lemma 8.1

To maximize network lifetime, the base station location must be within the smallest enclosing disk $\mathcal{A}$ that covers all the N sensor nodes in the network.

Proof. The proof is based on contradiction. That is, if the base station location is not in the SED, then its corresponding network lifetime cannot be maximum. To see this, assume that the optimal base station location is at point p, which is outside the SED $\mathcal{A}$ (see Fig. 8.1). Denote $O_{\mathcal{A}}$ as the center of the SED. Let q be the intersecting point between the line segment $[p, O_{\mathcal{A}}]$ and the circle

---

[1] In fact, we can consider points in an even smaller area, i.e., the convex hull of all sensor nodes. However, using the convex hull does not reduce the order of complexity of our algorithm. On the other hand, the use of the SED simplifies the discussion.

Figure 8.1 A schematic diagram showing that an optimal base station location must be within the SED.

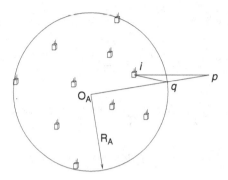

of the SED. Then for any sensor node i (all in $\mathcal{A}$), we have $d_{iq} < d_{ip}$. Consequently, $C_{iq} < C_{ip}$. As a result, we can save transmission energy for every sensor node $i \in \mathcal{N}$ by relocating p to q. This saving in energy at each node increases network lifetime, which shows that point p cannot be the optimal location to maximize network lifetime.   □

We have thus narrowed down the search space for base station B from a two-dimensional plane to the SED $\mathcal{A}$. However, the number of points in $\mathcal{A}$ remains infinite. It is tempting to partition $\mathcal{A}$ into small subareas (e.g., a grid-like structure), $\mathcal{A}_1, \mathcal{A}_2, \ldots$, up to say $\mathcal{A}_M$, i.e.,

$$\mathcal{A} = \bigcup_{m=1}^{M} \mathcal{A}_m.$$

When each subarea is sufficiently small (i.e., M is sufficiently large), we can use some point $q_m \in \mathcal{A}_m$ to represent $\mathcal{A}_m$, $m = 1, 2, \ldots, M$. By solving an LP problem on each of the M points, we can select the best location among all points and obtain a good solution for the base station placement. However, such an approach is a *heuristic* at best, and does not provide any *theoretical guarantee* on performance.

The key to providing a theoretical guarantee on performance is to partition each subarea in such a way that tight bounds on transmission cost can be guaranteed for any point in the subarea. If this is possible, then we may be able to exploit such a property and design an approximation algorithm that yields *provably* $(1 - \varepsilon)$-optimal network lifetime performance. The goal of this chapter is to develop such an algorithm having a guaranteed performance rather than merely a good heuristic. In the following section, we develop a novel technique to partition the SED $\mathcal{A}$ into subareas where each subarea can be represented by a special point with a set of tight bounds. Then we show how to design a $(1 - \varepsilon)$-optimal approximation algorithm based on these special points.

Figure 8.2 A sequence of circles with increasing costs with center at node 4.

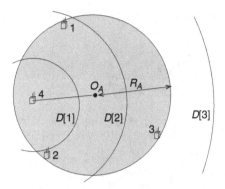

8.6 Subarea division and fictitious cost points

8.6.1 Subarea division

The subarea division (with guaranteed performance bounds) hinges upon a clever discretization of a cost parameter. A close look at the energy constraint in (8.5) suggests that the location of the base station is *embedded* in the cost parameters $c_{iB}(p)$. In other words, if we can discretize these cost parameters, we may also discretize the location for the base station.

Since the search space is narrowed down to the SED $\mathcal{A}$, we can limit the range for the distance between a sensor node i to the possible location for the base station. Denote $O_\mathcal{A}$ and $R_\mathcal{A}$ as the origin and radius of the SED $\mathcal{A}$. For each sensor node $i \in \mathcal{N}$, denote $D_{i,O_\mathcal{A}}$ as the distance from sensor node i to the origin of disk $\mathcal{A}$ (see node 4 in Fig. 8.2 as an example). Denote by $D_{iB}^{\min}$ and $D_{iB}^{\max}$ the minimum and maximum distances between sensor node i and any possible location for the base station B, respectively. Then we have

$$D_{iB}^{\min} = 0,$$
$$D_{iB}^{\max} = D_{i,O_\mathcal{A}} + R_\mathcal{A}.$$

Corresponding to $D_{iB}^{\min}$ and $D_{iB}^{\max}$, denote by $C_{iB}^{\min}$ and $C_{iB}^{\max}$ the minimum and maximum cost values between sensor node i and base station B, respectively. Then by (8.2), we have

$$C_{iB}^{\min} = \beta_1, \tag{8.7}$$
$$C_{iB}^{\max} = \beta_1 + \beta_2 (D_{iB}^{\max})^\alpha = \beta_1 + \beta_2 (D_{i,O_\mathcal{A}} + R_\mathcal{A})^\alpha. \tag{8.8}$$

Given the range of $d_{iB} \in [D_{iB}^{\min}, D_{iB}^{\max}] = [0, D_{i,O_\mathcal{A}} + R_\mathcal{A}]$ for each sensor node i, we now show how to partition the disk $\mathcal{A}$ into a finite number of subareas with the distance of each subarea to sensor node i meeting some tight bounds. Specifically, from a sensor node i, we draw a sequence of circles centered at this sensor node, each with increasing radius $D[1], D[2], \ldots, D[H_i]$

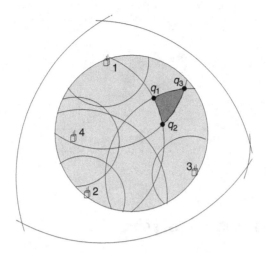

corresponding to costs $C[1], C[2], \ldots, C[H_i]$, which are defined with the following geometric property:

$$C[h] = C_{iB}^{\min}(1+\varepsilon)^h = \beta_1(1+\varepsilon)^h \quad (1 \le h \le H_i). \tag{8.9}$$

This geometric sequence $C[h]$ (with a factor of $(1+\varepsilon)$) is carefully chosen and will offer tight performance bounds for any point in a subarea (more on this later). The number of required circles H_i can be determined by having the last circle in the sequence (with radius $D[H_i]$) to completely contain disk $\mathcal{A}$, i.e., $D[H_i] \ge D_{iB}^{\max}$, or equivalently,

$$C[H_i] \ge C_{iB}^{\max}.$$

That is, we can determine H_i as follows:

$$H_i = \left\lceil \frac{\ln(C_{iB}^{\max}/C_{iB}^{\min})}{\ln(1+\varepsilon)} \right\rceil = \left\lceil \frac{\ln\left(1 + \frac{\beta_2}{\beta_1}(D_{i,O_{\mathcal{A}}} + R_{\mathcal{A}})^\alpha\right)}{\ln(1+\varepsilon)} \right\rceil. \tag{8.10}$$

For example, for node 4 in Fig. 8.2, we have $H_4 = 3$, i.e., $D[3]$ is the circle centered at node 4 that will completely contain the disk. As a result, with sensor node i as the center, we have a total of H_i circles, each with cost $C[h]$, $h = 1, 2, \ldots, H_i$.

The above partitioning of the SED $\mathcal{A}$ is with respect to a specific node i. We now perform the above process for *all* sensor nodes. These intersecting circles will cut disk $\mathcal{A}$ into a finite number of *irregular* subareas, with the boundaries of each subarea being either an arc (with a center at some sensor node i and some cost $C[h]$, $1 \le h < H_i$) or an arc segment of the SED $\mathcal{A}$. As an example, the SED $\mathcal{A}$ in Fig. 8.3 is now cut into 28 irregular subareas.

We now claim that under this subarea partitioning technique, for any point in a given subarea, its cost to each sensor node in the network can be *tightly* bounded quantitatively. This is because, with respect to each sensor node i, a subarea $\mathcal{A}_m$ must be enclosed within some arc centered at sensor node i. Denote the index of this arc (w.r.t. sensor node i) as $h_i(\mathcal{A}_m)$. So when the base station B is at any point $p \in \mathcal{A}_m$, we have

$$C[h_i(\mathcal{A}_m) - 1] \leq c_{iB}(p) \leq C[h_i(\mathcal{A}_m)], \tag{8.11}$$

where we define $C[0] = C_{iB}^{\min} = \beta_1$. Since $\frac{C[h_i(\mathcal{A}_m)]}{C[h_i(\mathcal{A}_m)-1]} = 1 + \varepsilon$ by (8.9), we have very tight lower and upper bounds for $c_{iB}(p)$. The reader may now have a better appreciation of the benefit of the discretization technique for costs and distances.

8.6.2 Fictitious cost point

We now introduce a novel concept called *fictitious cost point* (FCP). It will be used to represent an upper bound on the cost for any point in a subarea $\mathcal{A}_m$, $m = 1, 2, \ldots, M$.

Definition 8.1

Denote the FCP for subarea $\mathcal{A}_m$ ($m = 1, 2, \ldots, M$) as p_m, which is represented by an N-tuple vector with its ith element ($i = 1, 2, \ldots, N$) being an upper bound on the cost for any point in subarea $\mathcal{A}_m$ corresponding to the ith sensor node in the network.

That is, the N-tuple cost vector for FCP p_m is $[c_{1B}(p_m), c_{2B}(p_m), \ldots, c_{NB}(p_m)]$, with the ith element $c_{iB}(p_m)$ being

$$c_{iB}(p_m) = C[h_i(\mathcal{A}_m)], \tag{8.12}$$

where $h_i(\mathcal{A}_m)$ is determined by (8.11).

As an example, the FCP for subarea with corner points $\{q_1, q_2, q_3\}$ in Fig. 8.3 can be represented by a 4-tuple cost vector $[c_{1B}(p_m), c_{2B}(p_m), c_{3B}(p_m), c_{4B}(p_m)] = [C[2], C[3], C[2], C[3]]$, where the first component $C[2]$ represents an upper bound on the cost for any point in this subarea to sensor node 1, the second component $C[3]$ represents an upper bound on the cost (which is loose here) for any point in this subarea to sensor node 2, and so forth.

We emphasize that the reason we use the word "fictitious" is that an FCP p_m cannot be mapped to a *physical* point within the corresponding subarea $\mathcal{A}_m$. This is because there does not exist a physical point in subarea $\mathcal{A}_m$ that has its costs to all the N sensor nodes *equal* (one by one) to the respective N-tuple cost vector embodied by p_m *simultaneously*. As an example, any point within the dark subarea bounded by corner points q_1, q_2, and q_3 cannot have its costs correspond to the four sensor nodes in the network equal to the respective element in $[C[2], C[3], C[2], C[3]]$ simultaneously, where $[C[2], C[3], C[2], C[3]]$ is the cost vector for the FCP that represents this subarea.

The following important property for FCP p_m will be used in the proof of the $(1 - \varepsilon)$-optimality guarantee of the approximation algorithm.

Property 8.1

For any point $p \in \mathcal{A}_m$ and the corresponding FCP p_m, we have

$$c_{iB}(p_m) \leq (1 + \varepsilon)c_{iB}(p).$$

Proof. By the definitions of FCP p_m (see (8.12)) and $C[h]$ (see (8.9)), we have

$$c_{iB}(p_m) = C[h_i(\mathcal{A}_m)] = (1 + \varepsilon) \cdot C[h_i(\mathcal{A}_m) - 1] \leq (1 + \varepsilon) \cdot c_{iB}(p),$$

where the inequality follows from (8.11). $\qquad\square$

8.7 Summary of algorithm and example

By discretizing the cost parameters and the corresponding distances, we have partitioned the search space (SED $\mathcal{A}$) into a finite number of M subareas. By introducing the concept of FCPs, we can represent each subarea with a single point. As a result, we can now readily apply the LP formulation of Section 8.4 to examine each FCP and thereby select the FCP that offers the maximum network lifetime. The complete approximation algorithm is outlined in Algorithm 8.1. The correctness proof of its $(1 - \varepsilon)$-optimality guarantee is given in Section 8.8.

Example 8.1

We use a small three-node network to illustrate the steps of the approximation algorithm. The location, data rate, and initial energy for each sensor are specified in Table 8.2, where the units of distance, rate, and energy are all normalized. Also, we set $\alpha = 2$, $\beta_1 = 1$, $\beta_2 = 0.5$, and $\rho = 1$ under the normalized units. For illustration, we set the error bound to $\varepsilon = 0.2$.[2]

(1) We identify SED $\mathcal{A}$ with origin $O_{\mathcal{A}} = (0.61, 0.57)$ and radius $R_{\mathcal{A}} = 0.51$ (see Fig. 8.4).

(2) We first have $D_{i, O_{\mathcal{A}}} = R_{\mathcal{A}} = 0.51$ for each node i, $1 \leq i \leq 3$. We then find the lower and upper bounds on c_{iB} for each node i, $1 \leq i \leq 3$, as follows:

$$C_{iB}^{\min} = \beta_1 = 1,$$
$$C_{iB}^{\max} = \beta_1 + \beta_2 (D_{i, O_{\mathcal{A}}} + R_{\mathcal{A}})^{\alpha} = 1 + 0.5 \cdot (0.51 + 0.51)^2 = 1.52.$$

Algorithm 8.1 A $(1 - \varepsilon)$-approximation algorithm

---

1. Find the smallest enclosing disk $\mathcal{A}$ that covers all the N nodes.
2. Within $\mathcal{A}$, compute the lower and upper cost bounds $C_{iB}^{\min}$ and $C_{iB}^{\max}$ for each node $i \in \mathcal{N}$ by (8.7) and (8.8).
3. For a given $\varepsilon > 0$, define a sequence of costs $C[1], C[2], \ldots, C[H_i]$ by (8.9), where H_i is calculated by (8.10).
4. At each node i, draw a sequence of $(H_i - 1)$ circles centered at node i with increasing radius corresponding to cost $C[h]$, $h = 1, 2, \ldots, H_i - 1$. The intersection of these circles within disk $\mathcal{A}$ will partition $\mathcal{A}$ into M subareas $\mathcal{A}_1, \mathcal{A}_2, \ldots, \mathcal{A}_M$.
5. For each subarea $\mathcal{A}_m$, $1 \le m \le M$, define an FCP p_m by an N-tuple cost vector $[c_{1B}(p_m), c_{2B}(p_m), \ldots, c_{NB}(p_m)]$, where $c_{iB}(p_m)$ is defined in (8.12).
6. For each FCP p_m, $1 \le m \le M$, apply LP in Section 8.4 to obtain the achievable network lifetime T_m.
7. Select the FCP p^* that offers the maximum network lifetime among these M FCPs. The base station can be placed at any point p_ε within the subarea corresponding to p^*.
8. For the chosen point p_ε, apply LP in Section 8.4 and obtain a $(1 - \varepsilon)$-optimal network lifetime T_ε.

---

(3) For each node i, $1 \le i \le 3$, we find

$$H_i = \left\lceil \frac{\ln \left(1 + \frac{\beta_2}{\beta_1}(D_{i,O_\mathcal{A}} + R_\mathcal{A})^\alpha \right)}{\ln(1 + \varepsilon)} \right\rceil = \left\lceil \frac{\ln \left(1 + \frac{0.5}{1}(0.51 + 0.51)^2 \right)}{\ln(1 + 0.2)} \right\rceil = 3,$$

and

$$C[1] = \beta_1(1 + \varepsilon) = 1 \cdot (1 + 0.2) = 1.20,$$
$$C[2] = \beta_1(1 + \varepsilon)^2 = 1 \cdot (1 + 0.2)^2 = 1.44,$$
$$C[3] = \beta_1(1 + \varepsilon)^3 = 1 \cdot (1 + 0.2)^3 = 1.73.$$

(4) We draw circles centered at each node i, $1 \le i \le 3$, with cost $C[h]$, $1 \le h < H_i = 3$, to partition the original disk $\mathcal{A}$ into 16 subareas $\mathcal{A}_1, \mathcal{A}_2, \ldots, \mathcal{A}_{16}$.

(5) We define an FCP p_m for each subarea $\mathcal{A}_m$, $1 \le m \le 16$. For example, for FCP p_1, we define a 3-tuple cost vector as $[c_{1B}(p_1), c_{2B}(p_1), c_{3B}(p_1)] = [C[1], C[3], C[2]] = [1.20, 1.73, 1.44]$.

(6) We apply the LP formulation of Section 8.4 to these 16 FCPs and obtain the network lifetime for each FCP.

(7) Since the FCP $p^* = p_9$ has the maximum achievable network lifetime of 226.47 among all the 16 FCPs, we can place the base station at any point in the subarea $\mathcal{A}_9$, e.g., $p_\varepsilon = (0.6, 0.6)$.

Table 8.2 Sensor locations, data rate, and initial energy for the example sensor network.

Node index	Location	Data rate	Initial energy
1	$(0.1, 0.5)$	0.8	390
2	$(1.1, 0.7)$	1.0	400
3	$(0.4, 0.1)$	0.5	130

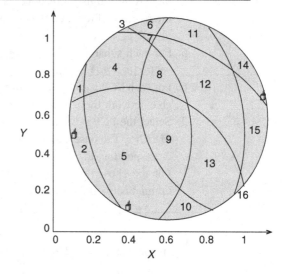

Figure 8.4 The SED is partitioned into 16 subareas for the three-node example.ff

(8) We apply the LP formulation of Section 8.4 to p_ε and obtain a $(1 - \varepsilon)$-optimal network lifetime $T_\varepsilon = 227.07$. This completes the algorithm.

8.8 Correctness proof and complexity analysis

In this section, we give a formal proof that the solution obtained by Algorithm 8.1 is $(1 - \varepsilon)$-optimal and analyze its complexity.

Denote p_{opt} as the optimal location for base station placement, T_{opt} and ψ_{opt} as the corresponding maximum network lifetime and flow routing solution, all of which are unknown.

Denote p^* as the best FCP among the M FCPs p_m, $m = 1, 2, \ldots, M$, based on their network lifetime performance. Denote by T^* and ψ^* the

---

[2] This ε is used here to simplify the illustration of each step. In Section 8.9, we use $\varepsilon = 0.05$ for all computations.

corresponding maximum network lifetime and flow routing solution, i.e., $T^* = \max\{T_m : m = 1, 2, \ldots, M\}$.

Based on Algorithm 8.1, Step 7, we choose a physical point p_ε in the sub-area corresponding to p^* for base station placement. For point p_ε, denote the network lifetime as T_ε and the corresponding flow routing solution as ψ_ε.

Our roadmap for the proof is as follows. In Theorem 8.1, we prove that T^* for the best FCP p^* is within $(1 - \varepsilon)$ of the optimum, i.e., $T^* \geq (1 - \varepsilon)T_{\text{opt}}$. Then, in Theorem 8.2, we show that for the physical point p_ε, its corresponding network lifetime T_ε is also within $(1 - \varepsilon)$ of the optimum, i.e., $T_\varepsilon \geq (1 - \varepsilon)T_{\text{opt}}$.

Theorem 8.1

$T^* \geq (1 - \varepsilon)T_{\text{opt}}$.

To prove Theorem 8.1, we first present the following lemma, which is a general case of the theorem:

Lemma 8.2

For any given base station location p and corresponding optimal routing solution φ and achievable network lifetime T (obtained via the stated LP formulation), denote $\mathcal{A}_m$ as the subarea that contains p for a given ε. Then for the corresponding FCP p_m, its achievable network lifetime T_m is at least $(1 - \varepsilon)$ of T, i.e., $T_m \geq (1 - \varepsilon)T$.

Proof. Instead of considering the optimal routing solution for FCP p_m, we use the same routing φ on p_m, which is clearly suboptimal. That is, denoting $\hat{T}_m$ as the network lifetime for FCP p_m under φ, we have $T_m \geq \hat{T}_m$. Then we only need to show that $\hat{T}_m \geq (1 - \varepsilon)T$.

To show $\hat{T}_m \geq (1 - \varepsilon)T$, we compute the total consumed energy on node $i \in \mathcal{N}$ under φ for FCP p_m and at time $(1 - \varepsilon)T$, which is

$$\sum_{\substack{k \in \mathcal{N} \\ k \neq i}} \rho f_{ki}(1 - \varepsilon)T + \sum_{\substack{j \in \mathcal{N} \\ j \neq i}} C_{ij} f_{ij}(1 - \varepsilon)T + c_{iB}(p_m)f_{iB}(1 - \varepsilon)T$$

$$< \sum_{\substack{k \in \mathcal{N} \\ k \neq i}} \rho f_{ki}T + \sum_{\substack{j \in \mathcal{N} \\ j \neq i}} C_{ij} f_{ij}T + (1 + \varepsilon)c_{iB}(p)f_{iB}(1 - \varepsilon)T$$

$$< \sum_{\substack{k \in \mathcal{N} \\ k \neq i}} \rho f_{ki}T + \sum_{\substack{j \in \mathcal{N} \\ j \neq i}} C_{ij} f_{ij}T + c_{iB}(p)f_{iB}T \leq e_i.$$

The first inequality holds via Property 8.1. The last inequality holds by the energy constraint corresponding to the routing solution φ for the point p. Thus, the network lifetime $\hat{T}_m$ for FCP p_m under the routing solution φ is at least $(1 - \varepsilon)T$. Thus, we have $T_m \geq \hat{T}_m \geq (1 - \varepsilon)T$. □

With Lemma 8.2, we are ready to prove Theorem 8.1 as follows:

Proof. Consider the special case of Lemma 8.2 where the given base station location p is the optimal location p_{opt}, with the corresponding optimal data routing solution ψ_{opt} and network lifetime T_{opt}. Following the same approach as in Lemma 8.2, we can find a corresponding subarea $\mathcal{A}_m$ that contains the point p_{opt} with the corresponding FCP p_m. As a result, for FCP p_m, we have $T_m \geq (1 - \varepsilon)T_{\text{opt}}$. Thus, for the best FCP p^* among all the FCPs, we have $T^* \geq T_m \geq (1 - \varepsilon)T_{\text{opt}}$. □

Theorem 8.1 guarantees that the best network lifetime among the M FCPs is at least $(1 - \varepsilon)$ of T_{opt}. Now, consider a point p_ε in the subarea represented by the best FCP p^*. We have the following theorem:

Theorem 8.2

$T_\varepsilon \geq (1 - \varepsilon)T_{\text{opt}}$.

Proof. Denote $\hat{T}_\varepsilon$ as the network lifetime for point p_ε under the same routing solution ψ^* for FCP p^*. Since ψ^* is a suboptimal routing for p_ε, we have $T_\varepsilon \geq \hat{T}_\varepsilon$. Thus, to show $T_\varepsilon \geq (1 - \varepsilon)T_{\text{opt}}$, we only need to show that $\hat{T}_\varepsilon \geq T^* \geq (1 - \varepsilon)T_{\text{opt}}$, where the second inequality follows from Theorem 8.1.

To establish $\hat{T}_\varepsilon \geq T^*$, we compute the total consumed energy on node $i \in \mathcal{N}$ under ψ^* for the point p_ε at time T^*, which is given by

$$\sum_{\substack{k \in \mathcal{N} \\ k \neq i}} \rho f_{ki} T^* + \sum_{\substack{j \in \mathcal{N} \\ j \neq i}} C_{ij} f_{ij} T^* + c_{iB}(p_\varepsilon) f_{iB} T^*$$

$$\leq \sum_{\substack{k \in \mathcal{N} \\ k \neq i}} \rho f_{ki} T^* + \sum_{\substack{j \in \mathcal{N} \\ j \neq i}} C_{ij} f_{ij} T^* + c_{iB}(p^*) f_{iB} T^* \leq e_i.$$

The first inequality holds by (8.11) and (8.12). The second inequality holds by the energy constraint on p^* under the routing solution ψ^*. Thus, the network lifetime $\hat{T}_\varepsilon$ for location p_ε under ψ^* is at least T^*. As a result, the maximum network lifetime T_ε for location p_ε is at least $\hat{T}_\varepsilon \geq T^* \geq (1 - \varepsilon)T_{\text{opt}}$. □

The complexity of Algorithm 8.1 can be measured by the number of LPs that need to be solved, which is equal to the total number of subareas M. Hence, we compute M.

The boundary segments of each subarea are defined by either an arc centered at some sensor node i (with some cost $C[h]$, $1 \leq h < H_i$, with H_i being given by (8.10)), or an arc of the disk $\mathcal{A}$. Since there are $H_i - 1$ circles radiating from each sensor node i and one circle for the disk $\mathcal{A}$, the total number of circles

Table 8.3 Each node's location, data generation rate, and initial energy for a ten-node network.

Location	Rate	Initial energy	Location	Rate	Initial energy
(0.81, 0.86)	0.7	390	(0.48, 0.22)	0.1	300
(0.25, 0.71)	0.4	400	(0.53, 0.16)	0.8	410
(0.47, 0.44)	1.0	440	(0.66, 0.52)	0.2	210
(0.28, 0.03)	0.6	330	(0.91, 0.86)	0.1	320
(0.25, 0.36)	0.2	440	(0.44, 0.21)	0.9	330

is $K = 1 + \sum_{i \in \mathcal{N}} (H_i - 1)$. The maximum number of subareas M that can be obtained by K circles is upper bounded by

$$M \le K^2 - K + 2. \tag{8.13}$$

The verification of (8.13) is left as a homework exercise. We hence have

$$M = O(K^2) = O\left(\left(\sum_{i \in \mathcal{N}} H_i\right)^2\right) = O\left(\left(\frac{N}{\varepsilon}\right)^2\right).$$

8.9 Numerical examples

In this section, we apply the approximation algorithm to various network topologies and use numerical results to demonstrate its efficacy. The units of distance, rate, and energy are all normalized appropriately. The normalized parameters in the energy consumption model are $\beta_1 = \beta_2 = \rho = 1$, and we set the path-loss index $\alpha = 2$.

We consider four randomly generated networks consisting of ten 20, 50, and 100 nodes deployed over a 1×1 square area. In all cases, the targeted accuracy for the approximation algorithm is 0.95-optimal, i.e., $\varepsilon = 0.05$.

The network setting (location, data rate, and initial energy for each node) for the ten-node network is specified in Table 8.3. By applying Algorithm 8.1, we find that the FCP with the cost vector [1.05, 1.28, 1.05, 1.22, 1.16, 1.05, 1.05, 1.05, 1.41, 1.05] has the maximum network lifetime $T^* = 357.49$, which is at least 95% of the optimal value. By placing the base station at a point in the corresponding subarea, e.g., at point (0.59, 0.31), the network lifetime is $T_\varepsilon = 359.17 > T^*$. This network lifetime is also at least 95% of the optimal value. The flow routing solution is displayed in Fig. 8.5, where a circle represents a sensor node and a star represents the location of the base station (0.59, 0.31).

The network setting for a 20-node network (with location, data rate, and initial energy for each of the 20 sensor nodes) is given in Table 8.4. By applying Algorithm 8.1, we find that the FCP with the cost vector [1.55, 1.05, 1.16, 1.05, 1.41, 1.22, 1.41, 1.28, 1.63, 1.05, 1.16, 1.98, 1.71, 1.16, 1.28,

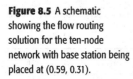

Figure 8.5 A schematic showing the flow routing solution for the ten-node network with base station being placed at (0.59, 0.31).

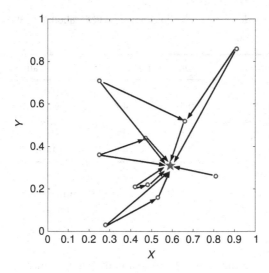

Table 8.4 Each node's location, data generation rate, and initial energy for a 20-node network.

Location	Rate	Initial energy	Location	Rate	Initial energy
(0.98, 0.49)	0.4	180	(0.09, 0.84)	0.7	60
(0.44, 0.67)	0.8	320	(0.65, 0.62)	0.1	100
(0.57, 0.52)	0.1	340	(0.92, 0.05)	0.1	310
(0.13, 0.19)	0.6	430	(1.00, 0.33)	0.6	280
(0.74, 0.73)	0.1	350	(0.63, 1.00)	0.2	210
(0.24, 0.19)	0.7	310	(0.11, 0.36)	0.3	70
(0.49, 0.38)	0.9	410	(0.89, 0.12)	0.7	420
(0.63, 0.33)	0.7	500	(0.52, 0.86)	0.3	270
(0.76, 0.63)	0.6	270	(0.24, 0.91)	0.9	160
(0.92, 0.33)	0.5	180	(0.40, 0.67)	1.0	180

1.80, 1.05, 1.05] has the maximum network lifetime $T^* = 82.86$ among all the FCPs. Subsequently, we place the base station at a point in the corresponding subarea, say at point (0.31, 0.79). The corresponding network lifetime is $T_\varepsilon = 82.91 > T^*$, which is also at least 95% of the optimal value. The flow routing solution is depicted in Fig. 8.6.

8.10 Chapter summary

This chapter showed how to design an approximation algorithm to provide a $(1 - \varepsilon)$-optimal solution to a nonconvex optimization problem. The case study focused on a classic base station placement problem in a WSN. The design of the $(1 - \varepsilon)$-approximation algorithm was based on several clever

Figure 8.6 A schematic showing the flow routing solution for the 20-node network with base station being placed at (0.31, 0.79).

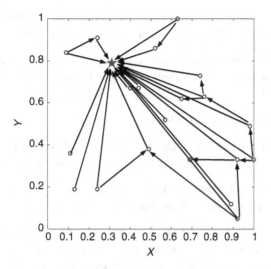

techniques, such as discretization of cost parameters (and distances), partitioning of the search space into a finite number of subareas, and representation of subareas with fictitious points (with tight bounds on costs). These three techniques could be exploited to develop approximation algorithms for other problems. We proved that the approximation algorithm is indeed $(1 - \varepsilon)$-optimal. The complexity of the algorithm involves solving a polynomial number of LPs, which is still polynomial.

8.11 Problems

8.1 (a) Explain the difference between constant-factor approximation algorithms and $(1 - \varepsilon)$-optimal (or $(1 + \varepsilon)$-optimal) approximation algorithms.
(b) Check research literature on wireless networking (from IEEE or ACM transactions/journals or conference proceedings) and identify one paper for the constant-factor approximation algorithms.
(c) Based on the paper that you found in Part (b), describe your preference between the two types of algorithms and give an explanation (in terms of performance and complexity issues).

8.2 Show that the formulation for the optimization problem in Section 8.3.2 is nonconvex.

8.3 Describe the two key techniques used in the design of the approximation algorithm in this chapter.

8.4 (a) By studying the reference given in the chapter, describe a linear time algorithm for the smallest enclosing disk problem.

(b) Apply your algorithm of Part (a) to an example having at least five randomly deployed nodes in a two-dimensional area.

8.5   Our discretization procedure in Section 8.6.1 partitions the SED into a number of irregular subareas. Why is such an irregular subarea division more attractive than a regular grid-type subarea division?

8.6   Derive (8.10) for H_i.

8.7   What is a fictitious cost point and how is it represented? What is the difference between a fictitious cost point and a physical point?

8.8   In Definition 8.1, why should we use upper bounds on transmission costs to define a fictitious cost point?

8.9   Consider the same setting as in Example 8.1, except that the three nodes 1, 2, and 3 are instead located at $(0, 0)$, $(0.93, 0.25)$, $(0.43, 0.75)$, respectively. Enumerate all the fictitious cost points in terms of their 3-tuple vectors.

8.10   If the transmission power at each node is upper bounded by a constant U $(U < \infty)$, how would you revise Algorithm 8.1?

8.11   Prove that (8.13) is correct.

8.12   In this chapter, we only considered how to optimally locate a single base station. Suppose now that we have two base stations and we want to determine where to place each of them optimally so as to maximize the network lifetime. How would you extend Algorithm 8.1 for this two base station problem? Note that in this case, each sensor can split its data and transmit them simultaneously to two base stations.

9

Approximation algorithm and its applications – Part 2

It is not length of life, but depth of life.

Ralph Waldo Emerson

9.1 Introduction

This chapter is a sequel to the last chapter. In the last chapter, we showed how to design a $(1 - \varepsilon)$-optimal approximation algorithm for a base station placement problem. The final solution gives a point (or any point within the corresponding fictitious cost point (FCP) subarea) for the base station. In this chapter, we extend the base station placement problem to the next level, where the base station is allowed to move around in the wireless sensor network (WSN).

The benefits of using a mobile base station to prolong sensor network life-time are easy to understand. Since the base station is the sink node for data collected by all the sensor nodes in the network, the set of sensor nodes near the base station would carry a considerable burden in relaying traffic from other sensor nodes to the base station. By allowing the base station to be mobile, we could alleviate the traffic burden from this fixed set of sensor nodes to other sensor nodes in the network, and thus extend network lifetime. Further, given new advances in unmanned autonomous vehicle (UAV) and customized robotics, having an unmanned vehicle carrying a base station for sensor data collection may not be too far from reality.

Although the potential benefit of using a mobile base station to prolong sensor network lifetime is significant, the technical difficulty of this optimization problem is enormous. There are two components that are tightly coupled in this problem. First, the location of the base station is time-dependent, i.e., at different time instances, the base station may be at different locations. Second, the multi-hop flow routing appears to be dependent on both time and location of the base station. As a result, an optimization problem with the objective of maximizing network lifetime needs to consider both base station location and flow routing, both of which are also time-dependent.

9.2 Case study: The mobile base station problem

In this chapter, we present theoretical results regarding the optimal movement of a mobile base station. We formulate an optimization problem that incorporates base station movement and multi-hop flow routing. The solution that we will present hinges upon two important intermediate results.

- The first result shows that as far as the network lifetime objective is concerned, we can *transform* the time-dependent problem to a location (space)-dependent problem. In particular, we show that flow routing only depends on the base station location, regardless of *when* the base station visits this location. Further, the specific time instances for the base station to visit a location is not important, as long as the total sojourn time for the base station to be present at this location is the same. This result allows us to focus on solving a location-dependent problem.
- The second result shows that to obtain a $(1 - \varepsilon)$-optimal solution to the location-dependent problem, we only need to consider a finite set of points within the smallest enclosing disk (SED) for the mobile base station's location. This result follows the same approach in Section 8.6. Then we can find the optimal sojourn time for the base station to stay at each FCP (as well as the corresponding flow routing solution) such that the overall network lifetime (i.e., sum of the sojourn times) is maximized via a single LP problem. We prove that the proposed solution can guarantee that the achieved network lifetime is at least $(1 - \varepsilon)$ of the maximum (unknown) network lifetime.

The rest of this case study is organized as follows. In Section 9.3, we describe the network model and formally state the base station movement problem. In Section 9.4, we transform the time-dependent problem to a location-dependent problem. In Section 9.5, we first develop an optimal solution for a constrained mobile base station (C-MB) problem, where the base station is allowed to be present among a set of given locations. Then we present the solution for the unconstrained mobile base station (U-MB) problem, where the base station is allowed to roam anywhere in the two-dimensional plane. Here, we give a formal proof of $(1 - \varepsilon)$-optimality of the proposed algorithm.

In Section 9.6, we present some numerical examples illustrating the efficacy of the proposed algorithm.

9.3 Problem and its formulation

We consider a set of $\mathcal{N}$ sensor nodes deployed over a two-dimensional area, with the location of each sensor node $i \in \mathcal{N}$ being at a fixed point (x_i, y_i). We assume that each node i generates data at a fixed rate of r_i. There is a base station B for the sensor network and it serves as the sink node for all data collected by the sensor nodes. Data generated by each sensor node should be transmitted to the base station via single or multi-hop.

Communication energy is assumed to be the dominant source of energy consumption at a node, which has been modeled in Section 8.3.1. We assume that each sensor node $i \in \mathcal{N}$ is initially provisioned with an amount of energy e_i. The base station is not constrained with energy and is free to roam in the two-dimensional plane. Again, network lifetime is defined as the first time instance when any of the sensor nodes runs out of energy. From (8.1), (8.2), and (8.3), it is not hard to realize that the location of the base station and the corresponding multi-hop flow routing among the nodes will affect energy consumption behavior at each node and thus the network lifetime. Table 9.1 lists the notation used in this chapter.

Our goal is to find how to optimally move a mobile base station to collect data in a sensor network so that the network lifetime can be maximized. Denote $(x, y)(t)$ as the position of base station B at time t and T the network lifetime (which is the objective function of our optimization problem). Then a feasible flow routing solution realizing this network lifetime T must satisfy both flow conservation and energy constraint at each sensor node. These constraints can be formally stated as follows. Denote $g_{ij}(t)$ and $g_{iB}(t)$ as the data rates from node i to node j and base station B at time t, respectively. Under multi-hop multi-path routing, the flow conservation for each node $i \in \mathcal{N}$ at any time $t \in [0, T]$ is

$$\sum_{\substack{k \in \mathcal{N} \\ k \neq i}} g_{ki}(t) + r_i = \sum_{\substack{j \in \mathcal{N} \\ j \neq i}} g_{ij}(t) + g_{iB}(t),$$

i.e., for node i, the sum of total incoming flow rates plus self-generated data rate is equal to the total outgoing flow rates at time t. Note that in our problem, data generated at each node should be transmitted to the base station in real time.

Table 9.1 Notation.

General notation

$\mathcal{A}$	The movement region for the base station
$c_{iB}(p)$	Transmission energy cost from sensor i to base station B at point p
$c_{iB}(t)$	Transmission energy cost from sensor i to base station B at time t
C_{ij}	Transmission energy cost from sensor i to sensor j
e_i	Initial energy at sensor node i
$f_{iB}(p)$	Flow rate from sensor i to base station B when B is at point p
$f_{ij}(p)$	Flow rate from sensor i to sensor j when base station B is at point p
$g_{iB}(t)$	Flow rate from sensor i to base station B at time t
$g_{ij}(t)$	Flow rate from sensor i to sensor j at time t
$\mathcal{N}$	The set of sensor nodes in the network
N	$= |\mathcal{N}|$, the number of sensor nodes in the network
$p(s)$	The arrived point when base station B traverses a distance s along $\mathcal{P}$
$\mathcal{P}$	The traveling path for base station B
r_i	Bit rate generated at sensor node i
$s(t)$	Cumulative distance traversed by base station B up to time t
S	The total traversed distance at the end of network lifetime
$u(s)$	$= \frac{1}{\|v(t(s))\|}$ when the base station traverses point $p(s)$ at time t
$U(s)$	Sojourn time for base station B at distance s
$v(t)$	The velocity of base station B at time t
$w(p)$	$= \sum_{s \in \mathcal{Z}(p)} u(s)$ when base station B traverses point p but never dwells
$W(p)$	Sojourn time for base station B at point p
$(x, y)(t)$	Location of base station B at time t
(x_i, y_i)	Location of sensor node i
$\mathcal{Z}(p)$	Set of distance s with $p(s) = p$
α	Path-loss index, $2 \le \alpha \le 4$
β_1, β_2	Two constant terms in power consumption model for transmission
ρ	Power consumption coefficient for receiving data

C-MB problem-specific notation

M	The number of predetermined locations
p_m	The mth location
$T^*_{\text{C-MB}}$	The maximum network lifetime achieved by $\psi^*_{\text{C-MB}}$
$\psi^*_{\text{C-MB}}$	An optimal solution to the C-MB problem

U-MB problem-specific notation

$\mathcal{A}_m$	The mth subarea in $\mathcal{A}$
$C^{\min}_{iB}, C^{\max}_{iB}$	Lower and upper bounds of $c_{iB}(p)$ for $p \in \mathcal{A}$
$C[h]$	$= \beta_1(1+\varepsilon)^h$, the transmission energy cost for the hth circle
H_i	The required number of circles at sensor node i
M	The number of subareas under a given ε
$O_{\mathcal{A}}, R_{\mathcal{A}}$	The center and radius of $\mathcal{A}$
p_m	The defined FCP for $\mathcal{A}_m$
S_m	$= \{s : p(s) \in \mathcal{A}_m, 0 \le s \le S\}$
$T_{\text{U-MB}}$	$(1 - \varepsilon)$-optimal network lifetime achieved by $\psi_{\text{U-MB}}$
$W(\mathcal{A}_m)$	Sojourn time for base station B in subarea $\mathcal{A}_m$
$W(p_m)$	Sojourn time for the base station at FCP p_m
ε	Targeted approximation error, $\varepsilon > 0$ and $\varepsilon \ll 1$
$\psi_{\text{U-MB}}$	A $(1 - \varepsilon)$-optimal solution to the U-MB problem

The energy constraint for each node $i \in \mathcal{N}$ is

$$\int_0^T \left[\sum_{\substack{k \in \mathcal{N} \\ k \neq i}} \rho \cdot g_{ki}(t) + \sum_{\substack{j \in \mathcal{N} \\ j \neq i}} C_{ij} \cdot g_{ij}(t) + c_{iB}(t) \cdot g_{iB}(t) \right] dt \leq e_i,$$

i.e., total consumed energy due to reception and transmission over time T cannot exceed its initial energy e_i. We have

$$c_{iB}(t) = \beta_1 + \beta_2 \left[\sqrt{(x(t) - x_i)^2 + (y(t) - y_i)^2} \right]^\alpha,$$

by (8.2), where (x_i, y_i) is the location of node i.

Denote by $\mathcal{A}$ the movement region for the base station, which can be narrowed down to the SED for all nodes in the network, as we have shown in the last chapter. Note that the SED can be found in polynomial time [167]. The optimization problem that we are interested in can be formulated as follows:

Maximize T

subject to $\sum_{\substack{k \in \mathcal{N} \\ k \neq i}} g_{ki}(t) + r_i = \sum_{\substack{j \in \mathcal{N} \\ j \neq i}} g_{ij}(t) + g_{iB}(t)$ $(i \in \mathcal{N}, t \in [0, T])$

$\int_0^T \left[\sum_{\substack{k \in \mathcal{N} \\ k \neq i}} \rho \cdot g_{ki}(t) + \sum_{\substack{j \in \mathcal{N} \\ j \neq i}} C_{ij} \cdot g_{ij}(t) + c_{iB}(t) \cdot g_{iB}(t) \right] dt \leq e_i$ $(i \in \mathcal{N})$

$c_{iB}(t) = \beta_1 + \beta_2 \left[\sqrt{(x(t) - x_i)^2 + (y(t) - y_i)^2} \right]^\alpha$ $(i \in \mathcal{N}, t \in [0, T])$

$(x, y)(t) \in \mathcal{A}$ $(t \in [0, T])$

$T, g_{ij}(t), g_{iB}(t) \geq 0$ $(i, j \in \mathcal{N}, i \neq j, t \in [0, T])$,

where the base station location (i.e., $(x, y)(t)$ for $t \in [0, T]$) and the corresponding flow routing (i.e., $g_{ij}(t)$ and $g_{iB}(t)$ for $t \in [0, T]$) form a joint optimization space for the objective T. Since the left-hand-side in the second constraint is not a polynomial function of optimization variables, this formulation is in the form of a *nonpolynomial program*.

9.4 From time domain to space domain

The difficulty in the problem formulation lies in that the base station location $(x, y)(t)$ and flow routing $g_{ij}(t)$ and $g_{iB}(t)$ are all functions of time. This adds considerable difficulty in the optimization problem. In this section, we show that as far as network lifetime performance is concerned, such dependency on time can be relaxed. Specifically, we will show (Lemma 9.1) that flow routing only needs to be dependent on the location of the base station and can be independent of *when* the base station is present at this location. Further, as long as the total sojourn time for the base station to be present at this location is the same, the specific time instance (i.e., "when") the base station visits this location is not important (Lemma 9.2). These results effectively transform the problem to a location-dependent problem.

We first give the following definition for a time-dependent solution:

Definition 9.1

A time-dependent solution consists of a network lifetime T, a path $\mathcal{P} = \{(x, y)(t) : t \in [0, T]\}$ for the base station, and a flow routing $g_{ij}(t)$ and $g_{iB}(t)$ at time t, $i, j \in \mathcal{N}, i \neq j$, where $(x, y)(t)$, $g_{ij}(t)$ and $g_{iB}(t)$ are all functions of time t.

For such a solution, denote $v(t)$ as the base station velocity at time t (and thus $\|v(t)\|$ is the base station speed at time t). Denote $s(t) = \int_{\tau=0}^{t} \|v(\tau)\| d\tau$ as the distance traversed by the base station up to time t. Suppose the total traversed distance at the end of network lifetime T is $s(T) = S$. Then we have $s(t) \in [0, S]$ for $t \in [0, T]$. Note that for a given path $\mathcal{P}$, we can identify the corresponding base station location for any s, which we denote as $p(s)$. Denote $U(s)$ as the sojourn time at distance s. The base station may visit the same point multiple times. Then multiple distances may correspond to the same point. Denote $\mathcal{Z}(p)$ as the set of such distances that correspond to the same point p. Then the total sojourn time at a point p is

$$W(p) = \sum_{s \in \mathcal{Z}(p)} U(s). \tag{9.1}$$

Now we give the following definition:

Definition 9.2

A location-dependent solution consists of a path $\mathcal{P}$ for the base station, $W(p)$ at each point $p \in \mathcal{P}$, flow routing $f_{ij}(p)$ and $f_{iB}(p)$, $i, j \in \mathcal{N}, i \neq j$, when the base station is at point p, and a network lifetime T, where $W(p)$, $f_{ij}(p)$, and $f_{iB}(p)$ are all functions of location p.

The following theorem shows that for the objective of network lifetime maximization, it is sufficient to consider location-dependent solutions.

Theorem 9.1

The optimal location-dependent solution can achieve the same maximum network lifetime as the optimal time-dependent solution.

The proof of Theorem 9.1 is based on the following two lemmas:

Lemma 9.1

Given a feasible time-dependent solution, we can construct a location-dependent solution with the same network lifetime.

Proof. The proof is based on the following construction. For a given time-dependent solution φ, it consists of a network lifetime T, a path $\mathcal{P}$ for the base station, and a flow routing $g_{ij}(t)$ and $g_{iB}(t)$, $i, j \in \mathcal{N}, i \neq j$. To construct a location-dependent solution $\bar{\varphi}$, we let the base station follow the same path $\mathcal{P}$ and for each point $p \in \mathcal{P}$, we compute $W(p)$ by (9.1). For $\bar{\varphi}$, we define location-dependent flow rates $f_{ij}(p)$ and $f_{iB}(p)$ for each point $p \in \mathcal{P}$ by the average of $g_{ij}(t)$ and $g_{iB}(t)$ over all visits to p during $[0, T]$ as follows:

- If the base station dwells at p at least once (with $W(p) > 0$), we define

$$f_{ij}(p) = \frac{\int_{t \in [0,T]}^{(x,y)(t)=p} g_{ij}(t)dt}{W(p)}, \tag{9.2}$$

$$f_{iB}(p) = \frac{\int_{t \in [0,T]}^{(x,y)(t)=p} g_{iB}(t)dt}{W(p)}. \tag{9.3}$$

- If the base station traverses p (maybe multiple times) but never dwells, then there is a unique time corresponding to each $s \in \mathcal{Z}(p)$. Denote such time as $t(s)$. Define

$$u(s) = \frac{1}{\|v(t(s))\|}, \tag{9.4}$$

$$w(p) = \sum_{s \in \mathcal{Z}(p)} u(s). \tag{9.5}$$

Based on $u(s)$ and $w(p)$, we can define

$$f_{ij}(p) = \frac{\sum_{s \in \mathcal{Z}(p)} g_{ij}(t(s)) \cdot u(s)}{w(p)}, \tag{9.6}$$

$$f_{iB}(p) = \frac{\sum_{s \in \mathcal{Z}(p)} g_{iB}(t(s)) \cdot u(s)}{w(p)}. \tag{9.7}$$

To show the data routing scheme with $f_{ij}(p)$ and $f_{iB}(p)$ is feasible and $\bar{\varphi}$ has the same network lifetime T, we need to prove that (i) when the base station visits each point $p \in \mathcal{P}$, flow conservation holds at this node, and (ii) at time T, the energy consumption at each node is the same as that in solution φ. Both (i) and (ii) can be intuitively explained by noting that $f_{ij}(p)$ and $f_{iB}(p)$ are defined by the average of $g_{ij}(t)$ and $g_{iB}(t)$, respectively.

Now we formally prove (i) and (ii).

(i) For flow conservation at point p, if the base station dwells at p at least once, then we have the following flow conservation:

$$\sum_{\substack{k \in \mathcal{N} \\ k \neq i}} f_{ki}(p) + r_i = \sum_{\substack{k \in \mathcal{N} \\ k \neq i}} \frac{\int_{t \in [0,T]}^{(x,y)(t)=p} g_{ki}(t) dt}{W(p)} + \frac{\int_{t \in [0,T]}^{(x,y)(t)=p} r_i dt}{W(p)}$$

$$= \frac{\int_{t \in [0,T]}^{(x,y)(t)=p} \left[\sum_{k \in \mathcal{N}}^{k \neq i} g_{ki}(t) + r_i \right] dt}{W(p)}$$

$$= \frac{\int_{t \in [0,T]}^{(x,y)(t)=p} \left[\sum_{j \in \mathcal{N}}^{j \neq i} g_{ij}(t) + g_{iB}(t) \right] dt}{W(p)}$$

$$= \sum_{\substack{j \in \mathcal{N} \\ j \neq i}} \frac{\int_{t \in [0,T]}^{(x,y)(t)=p} g_{ij}(t) dt}{W(p)} + \frac{\int_{t \in [0,T]}^{(x,y)(t)=p} g_{iB}(t) dt}{W(p)}$$

$$= \sum_{\substack{j \in \mathcal{N} \\ j \neq i}} f_{ij}(p) + f_{iB}(p).$$

The first equality holds by (9.2) and the fact that $W(p) = \int_{t \in [0,T]}^{(x,y)(t)=p} 1 dt$. The third equality holds by the flow conservation in solution φ. The last equality holds by (9.2) and (9.3).

If the base station traverses (but never dwells at) p, then we have the following flow conservation:

$$\sum_{\substack{k \in \mathcal{N} \\ k \neq i}} f_{ki}(p) + r_i = \sum_{\substack{k \in \mathcal{N} \\ k \neq i}} \frac{\sum_{s \in \mathcal{Z}(p)} g_{ki}(t(s)) u(s)}{w(p)} + \frac{\sum_{s \in \mathcal{Z}(p)} r_i \cdot u(s)}{w(p)}$$

$$= \frac{\sum_{s \in \mathcal{Z}(p)} \left[\sum_{k \in \mathcal{N}}^{k \neq i} g_{ki}(t(s)) + r_i \right] u(s)}{w(p)}$$

$$= \frac{\sum_{s \in \mathcal{Z}(p)} \left[\sum_{k \in \mathcal{N}}^{k \neq i} g_{ij}(t(s)) + g_{iB}(t(s)) \right] u(s)}{w(p)}$$

$$= \sum_{\substack{j \in \mathcal{N} \\ j \neq i}} \frac{\sum_{s \in \mathcal{Z}(p)} g_{ij}(t(s)) u(s)}{w(p)} + \frac{\sum_{s \in \mathcal{Z}(p)} g_{iB}(t(s)) u(s)}{w(p)}$$

$$= \sum_{\substack{j \in \mathcal{N} \\ j \neq i}} f_{ij}(p) + f_{iB}(p).$$

The first equality holds by (9.6) and (9.5). The third equality holds by the flow conservation in solution φ. The last equality holds by (9.6) and (9.7).

(ii) For energy consumption at time T, we want to show that the energy consumption at each node i in the constructed location-dependent solution $\bar{\varphi}$

is the same as that in the given time-dependent solution φ, i.e.,

$$
\sum_{\substack{k\in\mathcal{N}}}^{k\neq i} \rho \left[\sum_{\substack{s\in[0,S]}}^{U(s)>0} f_{ki}(p(s))U(s) + \int_{\substack{s\in[0,S]}}^{U(s)=0} f_{ki}(p(s))u(s)ds \right]
$$

$$
+ \sum_{\substack{j\in\mathcal{N}}}^{j\neq i} C_{ij} \left[\sum_{\substack{s\in[0,S]}}^{U(s)>0} f_{ij}(p(s))U(s) + \int_{\substack{s\in[0,S]}}^{U(s)=0} f_{ij}(p(s))u(s)ds \right]
$$

$$
+ \sum_{\substack{s\in[0,S]}}^{U(s)>0} c_{iB}(p(s))f_{iB}(p(s))U(s) + \int_{\substack{s\in[0,S]}}^{U(s)=0} c_{iB}(p(s))f_{iB}(p(s))u(s)ds
$$

$$
= \sum_{\substack{k\in\mathcal{N}}}^{k\neq i} \rho \int_0^T g_{ki}(t)dt + \sum_{\substack{j\in\mathcal{N}}}^{j\neq i} C_{ij} \int_0^T g_{ij}(t)dt + \int_0^T c_{iB}(t)g_{iB}(t)dt,
$$

where $c_{iB}(p)$ is the energy cost from sensor i to the base station B when the base station is at point p. To show that the above equality holds, it is sufficient to show that the following three equalities hold:

$$
\sum_{\substack{s\in[0,S]}}^{U(s)>0} f_{ki}(p(s))U(s) + \int_{\substack{s\in[0,S]}}^{U(s)=0} f_{ki}(p(s))u(s)ds = \int_0^T g_{ki}(t)dt
$$
$$
(k,i \in \mathcal{N}, k \neq i), \tag{9.8}
$$

$$
\sum_{\substack{s\in[0,S]}}^{U(s)>0} f_{ij}(p(s))U(s) + \int_{\substack{s\in[0,S]}}^{U(s)=0} f_{ij}(p(s))u(s)ds = \int_0^T g_{ij}(t)dt
$$
$$
(i,j \in \mathcal{N}, j \neq i), \tag{9.9}
$$

$$
\sum_{\substack{s\in[0,S]}}^{U(s)>0} c_{iB}(p(s))f_{iB}(p(s))U(s) + \int_{\substack{s\in[0,S]}}^{U(s)=0} c_{iB}(p(s))f_{iB}(p(s))u(s)ds
$$
$$
= \int_0^T c_{iB}(t)g_{iB}(t)dt \quad (i \in \mathcal{N}). \tag{9.10}
$$

We now prove (9.10). The proofs for (9.8) and (9.9) are very similar (but simpler) and are left as homework problems.

On the left-hand-side (LHS) of (9.10), we observe that the summation and integration are over distances in $[0, S]$. Recall that $\mathcal{Z}(p)$ denotes the set of total traversed distances when the base station visits $p \in \mathcal{P}$. If the base station visits p multiple times, then $\mathcal{Z}(p)$ has multiple elements. However, for each $s \in \mathcal{Z}(p)$, $c_{iB}(p(s))$ is the same since $p(s)$ is the same point p. Further, based on the definitions in (9.3) and (9.7) for $f_{iB}(\cdot)$, we have that $f_{iB}(p(s))$ is also the same for each $s \in \mathcal{Z}(p)$. Thus, for a point $p \in \mathcal{P}$, we can group these distances in $\mathcal{Z}(p)$ together in the summation as well as integration on the LHS of (9.10).

Now we select one distance from each group $\mathcal{Z}(p)$, $p \in \mathcal{P}$, to represent this group. In particular, we can select the smallest distance in $\mathcal{Z}(p)$. Denote $\mathcal{Y}(\mathcal{P})$

as the set of these representatives for each $\mathcal{Z}(p)$. Then $\mathcal{Y}(\mathcal{P})$ is a subset of $[0, S]$.

For the summation (first term) on the LHS of (9.10), we have

$$
\sum_{s \in [0,S]}^{U(s)>0} c_{iB}(p(s)) f_{iB}(p(s)) U(s) = \sum_{s \in \mathcal{Y}(\mathcal{P})}^{W(p(s))>0} \sum_{z \in \mathcal{Z}(p(s))}^{U(z)>0} c_{iB}(p(z)) f_{iB}(p(z)) U(z)
$$

$$
= \sum_{s \in \mathcal{Y}(\mathcal{P})}^{W(p(s))>0} c_{iB}(p(s)) f_{iB}(p(s)) \sum_{z \in \mathcal{Z}(p(s))}^{U(z)>0} U(z)
$$

$$
= \sum_{s \in \mathcal{Y}(\mathcal{P})}^{W(p(s))>0} c_{iB}(p(s)) f_{iB}(p(s)) \sum_{z \in \mathcal{Z}(p(s))} U(z)
$$

$$
= \sum_{s \in \mathcal{Y}(\mathcal{P})}^{W(p(s))>0} c_{iB}(p(s)) f_{iB}(p(s)) W(p(s)).
$$

$$(9.11)$$

The first equality holds by grouping those distances corresponding to the same point $p(s)$ that the base station dwells. The second equality holds by $c_{iB}(z) = c_{iB}(s)$ and $f_{iB}(p(z)) = f_{iB}(p(s))$ for each $z \in \mathcal{Z}(p(s))$. The last equality holds by (9.1).

Following the same token, for the integration (second term) on the LHS of (9.10), it can be shown that

$$
\int_{s \in [0,S]}^{U(s)=0} c_{iB}(p(s)) f_{iB}(p(s)) u(s) ds
$$

$$
= \int_{s \in \mathcal{Y}(\mathcal{P})}^{W(p(s))=0} c_{iB}(p(s)) f_{iB}(p(s)) w(p(s)) ds.
$$

$$(9.12)$$

Thus by (9.11) and (9.12), we have

$$
\text{LHS of } (9.10) = \sum_{s \in \mathcal{Y}(\mathcal{P})}^{W(p(s))>0} c_{iB}(p(s)) f_{iB}(p(s)) W(p(s))
$$

$$
+ \int_{s \in \mathcal{Y}(\mathcal{P})}^{W(p(s))=0} c_{iB}(p(s)) f_{iB}(p(s)) w(p(s)) ds
$$

$$
= \sum_{s \in \mathcal{Y}(\mathcal{P})}^{W(p(s))>0} c_{iB}(p(s)) \frac{\int_{t \in [0,T]}^{(x,y)(t)=p(s)} g_{iB}(t) dt}{W(p(s))} W(p(s))
$$

$$
+ \int_{s \in \mathcal{Y}(\mathcal{P})}^{W(p(s))=0} c_{iB}(p(s)) \frac{\sum_{\hat{s} \in \mathcal{Z}(p(s))} g_{iB}(t(\hat{s})) u(\hat{s})}{w(p(s))} w(p(s)) ds
$$

$$= \sum_{s \in \mathcal{Y}(\mathcal{P})}^{W(p(s))>0} c_{iB}(p(s)) \int_{t \in [0,T]}^{(x,y)(t)=p(s)} g_{iB}(t)dt$$

$$+ \int_{s \in \mathcal{Y}(\mathcal{P})}^{W(p(s))=0} c_{iB}(p(s)) \sum_{\hat{s} \in \mathcal{Z}(p(s))} g_{iB}(t(\hat{s}))u(\hat{s})ds, \qquad (9.13)$$

where the second equality holds by (9.3) and (9.7).

For the right-hand-side (RHS) of (9.10), we have

$$\text{RHS of (9.10)} = \int_{t \in [0,T]}^{\|v(t)\|>0} c_{iB}(t)g_{iB}(t)dt + \int_{t \in [0,T]}^{\|v(t)\|=0} c_{iB}(t)g_{iB}(t)dt. \quad (9.14)$$

We now transform each of the above integrations from t-domain to s-domain.

- For the case of $\|v(t)\| = 0$ (i.e., the base station dwells at the current point). Denote $t_1(s)$ as the time when the base station has traversed a distance s and starts this dwelling period. Denote $t_2(s)$ as the time when the base station completes this dwelling period. Then $U(s) > 0$ and for any time t during $[t_1(s), t_2(s)]$, the base station dwells and $\|v(t)\| = 0$. Thus,

$$\int_{t \in [0,T]}^{\|v(t)\|=0} c_{iB}(t)g_{iB}(t)dt = \sum_{s \in [0,S]}^{U(s)>0} \int_{t=t_1(s)}^{t_2(s)} c_{iB}(t)g_{iB}(t)dt$$

$$= \sum_{s \in [0,S]}^{U(s)>0} c_{iB}(p(s)) \int_{t=t_1(s)}^{t_2(s)} g_{iB}(t)dt, \qquad (9.15)$$

where the first equality holds as the integration $\int_{t \in [0,T]}^{\|v(t)\|=0}$ is limited to those dwelling periods, each corresponding to a distance $s \in [0, S]$ with $U(s) > 0$. The second equality holds by $c_{iB}(t) = c_{iB}(p(s))$ for $t \in [t_1(s), t_2(s)]$.

- For the case of $\|v(t)\| > 0$ (i.e., the base station is traversing at the current point $p(s)$). We have

$$\int_{t \in [0,T]}^{\|v(t)\|>0} c_{iB}(t)g_{iB}(t)dt = \int_{s \in [0,S]}^{U(s)=0} c_{iB}(p(s))g_{iB}(s) \cdot \frac{1}{\|v(t(s))\|}ds$$

$$= \int_{s \in [0,S]}^{U(s)=0} c_{iB}(p(s))g_{iB}(s)u(s)ds, \qquad (9.16)$$

where $g_{iB}(s)$ is the flow rate from sensor i to the base station B when the base station is traversing at point $p(s)$. The first equality holds by $ds = \|v(t(s))\|dt$ and the second equality holds by (9.4).

Therefore, by (9.14), (9.15), and (9.16), we have

RHS of (9.10)

$$= \sum_{s\in[0,S]}^{U(s)>0} c_{iB}(p(s)) \int_{t=t_1(s)}^{t_2(s)} g_{iB}(t)dt + \int_{s\in[0,S]}^{U(s)=0} c_{iB}(p(s))g_{iB}(s)u(s)ds.$$

(9.17)

For the summation (first term) on the RHS of (9.17), we have

$$\sum_{s\in[0,S]}^{U(s)>0} c_{iB}(p(s)) \int_{t=t_1(s)}^{t_2(s)} g_{iB}(t)dt$$

$$= \sum_{s\in\mathcal{Y}(\mathcal{P})}^{W(p(s))>0} \sum_{z\in\mathcal{Z}(p(s))}^{U(z)>0} c_{iB}(p(z)) \int_{t=t_1(z)}^{t_2(z)} g_{iB}(t)dt$$

$$= \sum_{s\in\mathcal{Y}(\mathcal{P})}^{W(p(s))>0} c_{iB}(p(s)) \sum_{z\in\mathcal{Z}(p(s))}^{U(z)>0} \int_{t=t_1(z)}^{t_2(z)} g_{iB}(t)dt$$

$$= \sum_{s\in\mathcal{Y}(\mathcal{P})}^{W(p(s))>0} c_{iB}(p(s)) \int_{t\in[0,T]}^{(x,y)(t)=p(s)} g_{iB}(t)dt,$$

(9.18)

where the first equality holds by grouping those distances corresponding to the same point $p(s)$ where the base station dwells. The second equality holds by $c_{iB}(p(z)) = c_{iB}(p(s))$ for each $z \in \mathcal{Z}(p(s))$.

Following the same token, for the integration (second term) on the RHS of (9.17), it can be shown that

$$\int_{s\in[0,S]}^{U(s)=0} c_{iB}(p(s))g_{iB}(s)u(s)ds$$

$$= \int_{s\in\mathcal{Y}(\mathcal{P})}^{W(p(s))=0} c_{iB}(p(s)) \sum_{\hat{s}\in\mathcal{Z}(p(s))} g_{iB}(t(\hat{s}))u(\hat{s})ds.$$

(9.19)

Therefore, by (9.17), (9.18), and (9.19), we have

$$\text{RHS of (9.10)} = \sum_{s\in\mathcal{Y}(\mathcal{P})}^{W(p(s))>0} c_{iB}(p(s)) \int_{t\in[0,T]}^{(x,y)(t)=p(s)} g_{iB}(t)dt$$

$$+ \int_{s\in\mathcal{Y}(\mathcal{P})}^{W(p(s))=0} c_{iB}(p(s)) \sum_{\hat{s}\in\mathcal{Z}(p(s))} g_{iB}(t(\hat{s}))u(\hat{s})ds.$$

(9.20)

By (9.13) and (9.20), (9.10) is proved.

Based on our results in (i) and (ii), the constructed location-dependent solution $\bar{\varphi}$ with $\mathcal{P}$, $W(p)$ (and $w(p)$), $f_{ij}(p)$, and $f_{iB}(p)$ is feasible and has the same network lifetime T as that achieved by time-dependent solution φ. This completes the proof. $\qquad\square$

The following lemma further extends Lemma 9.1 and says that the ordering and specific time instances for the base station to visit a particular point p is not important.

Lemma 9.2
Under a location-dependent solution, as long as $W(p)$ (and $w(p)$) at each point p remains the same, the network lifetime T will remain unchanged regardless of the ordering and frequency of the base station's presence at each point.

Lemma 9.2 can be easily proved by analyzing the energy consumption behavior at each node over time T. We leave its proof as a homework problem. Combining Lemmas 9.1 and 9.2, and considering the special case that φ is optimal, we have Theorem 9.1.

Based on Theorem 9.1, we conclude that as far as the network lifetime is concerned, it is sufficient for us to study location-dependent solutions. This result allows us to develop a provable approximation algorithm in the space domain, which we will present in the following section.

9.5 A (1 − ε)-optimal algorithm

Note that in the location-dependent problem formulation, there are an infinite number of points in $\mathcal{P}$. In this section, we first consider the case when the base station is only allowed to be present at a finite set of M positions. We call this problem the *constrained mobile base station* (C-MB) problem. Based on this intermediate result, we then devise a solution to the general problem where the base station is allowed to roam anywhere on the two-dimensional plane. We term the latter problem the *unconstrained mobile base station* (U-MB) problem.

9.5.1 Optimal sojourn time computation for the C-MB problem

We now show that the C-MB problem can be formulated as an LP, which can be solved in polynomial time. Recall that in the C-MB problem, the location of the base station is limited to a finite set of M locations p_m, $m = 1, 2, \ldots, M$. Thus if the base station dwells at p_m at least once, then we have $W(p_m) > 0$ otherwise (i.e., base station never dwells at p_m, we have $W(p_m) = 0$. We need to find optimal $W(p_m)$ for all points and the corresponding flow routing $f_{ij}(p_m)$ and $f_{iB}(p_m)$ for each point p_m with $W(p_m) > 0$.

When the base station is at point p_m, $1 \le m \le M$, the flow conservation for node $i \in \mathcal{N}$ is

$$\sum_{\substack{k \in \mathcal{N} \\ k \neq i}} f_{ki}(p_m) + r_i = \sum_{\substack{j \in \mathcal{N} \\ j \neq i}} f_{ij}(p_m) + f_{iB}(p_m). \tag{9.21}$$

The energy constraint for node $i \in \mathcal{N}$, at the end of network lifetime T is

$$\sum_{m=1}^{M} \left[\sum_{\substack{k \in \mathcal{N} \\ k \neq i}} \rho \cdot f_{ki}(p_m) + \sum_{\substack{j \in \mathcal{N} \\ j \neq i}} C_{ij} \cdot f_{ij}(p_m) + c_{iB}(p_m) \cdot f_{iB}(p_m) \right] W(p_m) \le e_i. \tag{9.22}$$

Note that for each i and p_m, $c_{iB}(p_m)$ is a constant.

We can formulate C-MB problem as an LP problem by letting $V_{ij}(p_m) = f_{ij}(p_m) \cdot W(p_m)$ and $V_{iB}(p_m) = f_{iB}(p_m) \cdot W(p_m)$, where $V_{ij}(p_m)$ (or $V_{iB}(p_m)$) can be interpreted as the total data volume from sensor node i to sensor node j (or base station B) when the base station is at p_m. We have

LP(C-MB)   Maximize T

$$\text{subject to} \qquad \sum_{m=1}^{M} W(p_m) - T = 0$$

$$\sum_{\substack{k \in \mathcal{N} \\ k \neq i}} V_{ki}(p_m) + r_i \cdot W(p_m) - \sum_{\substack{j \in \mathcal{N} \\ j \neq i}} V_{ij}(p_m) - V_{iB}(p_m) = 0$$

$$(i \in \mathcal{N}, 1 \le m \le M) \tag{9.23}$$

$$\sum_{m=1}^{M} \left[\sum_{\substack{k \in \mathcal{N} \\ k \neq i}} \rho \cdot V_{ki}(p_m) + \sum_{\substack{j \in \mathcal{N} \\ j \neq i}} C_{ij} \cdot V_{ij}(p_m) + c_{iB}(p_m) \cdot V_{iB}(p_m) \right] \le e_i$$

$$(i \in \mathcal{N}) \tag{9.24}$$

$$T, W(p_m), V_{ij}(p_m), V_{iB}(p_m) \ge 0 \qquad (i, j \in \mathcal{N}, i \neq j, 1 \le m \le M),$$

where (9.23) and (9.24) follow from (9.21) and (9.22), respectively. Once we solve the above LP problem, we have $W(p_m)$ for $1 \le m \le M$. For each point p_m with $W(p_m) > 0$, we can obtain $f_{ij}(p_m)$ and $f_{iB}(p_m)$ by $f_{ij}(p_m) = \frac{V_{ij}(p_m)}{W(p_m)}$ and $f_{iB}(p_m) = \frac{V_{iB}(p_m)}{W(p_m)}$. Recall that for those points with $W(p_m) = 0$, it means that the base station will not visit those points in this solution.

We summarize the result in this section with the following proposition:

Proposition 9.1
The C-MB problem can be solved via a single LP in polynomial time.

The solution to the above LP problem yields the sojourn time for the base station at each location p_m, $m = 1, 2, \ldots, M$, and the optimal flow routing when the base station is at p_m. So far, we assume that base station B can move from one point to another in zero time. We will discuss how to relax this assumption in Section 9.5.4.

9.5.2 Solution to the U-MB problem and proof of (1 − ϵ)-optimality

We now convert the U-MB problem to a C-MB problem with the $(1 - \varepsilon)$ network lifetime performance guarantee. Our approach is to exploit the energy cost function and how the location of the base station affects the energy cost. Note that the location of the base station is *embedded* in the cost parameter c_{iB} and that these cost parameters directly affect network lifetime. Thus, to design a $(1 - \varepsilon)$-optimal algorithm, we divide disk $\mathcal{A}$ into subareas (see Section 8.6), with each subarea to be associated with some nice properties for c_{iB}s that can be used to prove $(1 - \varepsilon)$-optimality.

Once we have M nonuniform subareas and represent them by M FCPs (see Section 8.6), we can readily apply the LP approach discussed in Section 9.5.1 to formulate an optimization problem on these M FCPs. In the following, we will show how to construct a $(1 - \varepsilon)$-optimal solution to the U-MB problem by solving the C-MB problem on the FCPs.

Denote $\psi^*_{\text{U-MB}}$ as an optimal solution to the U-MB problem and $T^*_{\text{U-MB}}$ as the maximum network lifetime, both of which are unknown. Our objective is to find a solution to the U-MB problem that has provable $(1 - \varepsilon)$-optimal network lifetime. Denote $\psi^*_{\text{C-MB}}$ as an optimal solution to the C-MB problem obtained by solving an LP problem for the M FCP p_m, $m = 1, 2, \ldots, M$, and $T^*_{\text{C-MB}}$ the corresponding network lifetime.

Our roadmap to construct a solution to the U-MB problem and to prove its $(1 - \varepsilon)$-optimality is as follows. In Theorem 9.2, we will prove that $T^*_{\text{C-MB}} \geq (1 - \varepsilon)T^*_{\text{U-MB}}$ (see Fig. 9.1). Since the optimal solution $\psi^*_{\text{C-MB}}$ corresponding to $T^*_{\text{C-MB}}$ is based on the M FCPs instead of physical points, in Theorem 9.3 we will further show how to construct a solution $\psi_{\text{U-MB}}$ to the U-MB problem based on $\psi^*_{\text{C-MB}}$ and prove that the corresponding network lifetime is $(1 - \varepsilon)$-optimal, i.e., $T_{\text{U-MB}} \geq (1 - \varepsilon)T^*_{\text{U-MB}}$ (see Fig. 9.1).

Theorem 9.2

For a given $\varepsilon > 0$, define subareas $\mathcal{A}_m$ and FCPs p_m, $m = 1, 2, \ldots, M$, as in Section 9.5.2. Then we have $T^*_{\text{C-MB}} \geq (1 - \varepsilon) \cdot T^*_{\text{U-MB}}$.

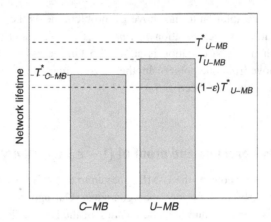

To prove that Theorem 9.2 is true, we need the following lemma:

Lemma 9.3

Given a feasible solution $\pi_{\text{U-MB}}$ to the U-MB problem with a network life-time $T_{\text{U-MB}}$, we can construct a solution $\pi_{\text{C-MB}}$ to the C-MB problem with a network lifetime $T_{\text{C-MB}} \geq (1 - \varepsilon) \cdot T_{\text{U-MB}}$.

The proof of this lemma is based on the following construction. Solution $\pi_{\text{U-MB}}$ consists of a specific path $\mathcal{P}$ for the base station, $W(p)$ (and $w(p)$), $f_{ij}(p)$, $f_{iB}(p)$ values for each point $p \in \mathcal{P}$, and a network lifetime $T_{\text{U-MB}}$. For subareas $\mathcal{A}_m$, $m = 1, 2, \ldots, M$, denote $W(\mathcal{A}_m)$ as the total sojourn time during $[0, T_{\text{U-MB}}]$ when the base station B is within each subarea $\mathcal{A}_m$. We have

$$W(\mathcal{A}_m) = \sum_{s \in \mathcal{S}_m}^{U(s)>0} U(s) + \int_{s \in \mathcal{S}_m}^{U(s)=0} u(s)ds, \qquad (9.25)$$

where $\mathcal{S}_m = \{s : p(s) \in \mathcal{A}_m, 0 \leq s \leq S\}$. To construct solution $\pi_{\text{C-MB}}$, we can let the base station spend $W(p_m)$ amount of time on FCP p_m, $m = 1, 2, \ldots, M$, where

$$W(p_m) = (1 - \varepsilon) \cdot W(\mathcal{A}_m),$$

and, for each point p_m with $W(p_m) > 0$, set the flow routing when the base station is at p_m as

$$f_{ij}(p_m) = \frac{1}{W(\mathcal{A}_m)} \left(\sum_{s \in \mathcal{S}_m}^{U(s)>0} f_{ij}(p(s))U(s) + \int_{s \in \mathcal{S}_m}^{U(s)=0} f_{ij}(p(s))u(s)ds \right),$$

$$f_{iB}(p_m) = \frac{1}{W(\mathcal{A}_m)} \left(\sum_{s \in \mathcal{S}_m}^{U(s)>0} f_{iB}(p(s))U(s) + \int_{s \in \mathcal{S}_m}^{U(s)=0} f_{iB}(p(s))u(s)ds \right).$$

The details of the proof of Lemma 9.3 are similar to the proof of Lemma 9.1 and are left in a homework problem.

Lemma 9.3 is a powerful result. With this lemma, we are now ready to prove Theorem 9.2.

Proof. Consider the special case of Lemma 9.3 where the given solution to the U-MB problem is an optimal solution $\pi^*_{\text{U-MB}}$ with network lifetime $T^*_{\text{U-MB}}$. By Lemma 9.3, we can transform it into a solution to the C-MB problem with a network lifetime at least $(1 - \varepsilon)T^*_{\text{U-MB}}$, i.e., there is a solution to the C-MB problem on the FCPs with a network lifetime at least $(1 - \varepsilon)T^*_{\text{U-MB}}$. As a result, the optimal solution $\psi^*_{\text{C-MB}}$ to the C-MB problem must have a network lifetime $T^*_{\text{C-MB}} \geq (1 - \varepsilon)T^*_{\text{U-MB}}$. $\qquad\qquad\square$

Theorem 9.2 guarantees that the network lifetime obtained by the LP solution on the M FCPs is at least $(1 - \varepsilon)$ of $T^*_{\text{U-MB}}$. However, a FCP may not be mapped to a physical point, which is required in the final solution. In the following theorem, we show how to construct a solution with each point being physically realizable. Further, the network lifetime for this constructed solution is greater than or equal to the maximum network lifetime for the C-MB problem, i.e., $T_{\text{U-MB}} \geq T^*_{\text{C-MB}}$. As a result, this new solution is $(1 - \varepsilon)$-optimal.

Theorem 9.3

For a given $\varepsilon > 0$, define subareas $\mathcal{A}_m$ and FCPs p_m, $m = 1, 2, \ldots, M$, as discussed in Section 9.5.2. Given an optimal solution $\psi^*_{\text{C-MB}}$ on these M FCPs with $W^*(p_m)$, $f^*_{ij}(p_m)$, $f^*_{iB}(p_m)$, and a network lifetime $T^*_{\text{C-MB}}$, a $(1 - \varepsilon)$-optimal solution $\psi_{\text{U-MB}}$ to the U-MB problem can be constructed by having the base station stay in $\mathcal{A}_m$ for

$$W(\mathcal{A}_m) = W^*(p_m) \tag{9.26}$$

amount of time and by having a corresponding flow routing for any physical point $p \in \mathcal{A}_m$ as

$$f_{ij}(p) = f^*_{ij}(p_m), \tag{9.27}$$

$$f_{iB}(p) = f^*_{iB}(p_m). \tag{9.28}$$

In Theorem 9.3, note that in the constructed solution to the U-MB problem, as long as the base station is within $\mathcal{A}_m$ (any point in this subarea), the flow routing is the same.

Proof. We will show that the constructed solution $\psi_{\text{U-MB}}$ is feasible, i.e., (i) flow conservation holds at any point, and (ii) the network lifetime of $\psi_{\text{U-MB}}$ is at least $T^*_{\text{C-MB}}$. Based on (i) and (ii), as well as the fact that $T^*_{\text{C-MB}} \geq (1 - \varepsilon)T^*_{\text{U-MB}}$ (see Theorem 9.2), we know that $T_{\text{U-MB}}$ must be at least $(1 - \varepsilon)T^*_{\text{U-MB}}$, i.e., $\psi_{\text{U-MB}}$ is $(1 - \varepsilon)$-optimal.

(i) For flow conservation, when the base station location p is in $\mathcal{A}_m$, we have

$$\sum_{\substack{k \in \mathcal{N} \\ k \neq i}} f_{ki}(p) + r_i = \sum_{\substack{k \in \mathcal{N} \\ k \neq i}} f_{ki}^*(p_m) + r_i$$

$$= \sum_{\substack{j \in \mathcal{N} \\ j \neq i}} f_{ij}^*(p_m) + f_{iB}^*(p_m)$$

$$= \sum_{\substack{j \in \mathcal{N} \\ j \neq i}} f_{ij}(p) + f_{iB}(p).$$

The first equality holds by (9.27). The second equality holds by the flow conservation in solution $\psi_{\text{C-MB}}^*$. The third equality holds by (9.27) and (9.28). Thus, solution $\psi_{\text{U-MB}}$ is feasible.

(ii) The total energy consumption at node $i \in \mathcal{N}$ by time $T_{\text{C-MB}}^*$ in solution $\psi_{\text{U-MB}}$ is

$$\sum_{\substack{k \in \mathcal{N} \\ k \neq i}} \rho \left[\sum_{\substack{s \in [0,S] \\ U(s)>0}} f_{ki}(p(s))U(s) + \int_{\substack{s \in [0,S] \\ }}^{U(s)=0} f_{ki}(p(s))u(s)ds \right]$$

$$+ \sum_{\substack{j \in \mathcal{N} \\ j \neq i}} c_{ij} \left[\sum_{\substack{s \in [0,S] \\ U(s)=0}} f_{ij}(p(s))U(s) + \int_{\substack{s \in [0,S] \\ }}^{U(s)=0} f_{ij}(p(s))u(s)ds \right]$$

$$+ \sum_{\substack{s \in [0,S] \\ U(s)>0}} c_{iB}(p(s))f_{iB}(p(s))U(s) + \int_{\substack{s \in [0,S] \\ }}^{U(s)=0} c_{iB}(p(s))f_{iB}(p(s))u(s)ds.$$

$$(9.29)$$

For the last two terms in (9.29), we have

$$\sum_{\substack{s \in [0,S] \\ U(s)>0}} c_{iB}(p(s))f_{iB}(p(s))U(s) + \int_{\substack{s \in [0,S] \\ }}^{U(s)=0} c_{iB}(p(s))f_{iB}(p(s))u(s)ds$$

$$= \sum_{m=1}^{M} \left[\sum_{\substack{s \in \mathcal{S}_m \\ U(s)>0}} c_{iB}(p(s))f_{iB}(p(s))U(s) + \int_{\substack{s \in \mathcal{S}_m \\ }}^{U(s)=0} c_{iB}(p(s))f_{iB}(p(s))u(s)ds \right]$$

$$\leq \sum_{m=1}^{M} \left[\sum_{\substack{s \in \mathcal{S}_m \\ U(s)>0}} c_{iB}(p_m)f_{iB}^*(p_m)U(s) + \int_{\substack{s \in \mathcal{S}_m \\ }}^{U(s)=0} c_{iB}(p_m)f_{iB}^*(p_m)u(s)ds \right]$$

$$= \sum_{m=1}^{M} c_{iB}(p_m)f_{iB}^*(p_m) \left[\sum_{\substack{s \in \mathcal{S}_m \\ U(s)>0}} U(s) + \int_{\substack{s \in \mathcal{S}_m \\ }}^{U(s)=0} u(s)ds \right]$$

$$= \sum_{m=1}^{M} c_{iB}(p_m)f_{iB}^*(p_m)W(\mathcal{A}_m)$$

$$= \sum_{m=1}^{M} c_{iB}(p_m)f_{iB}^*(p_m)W^*(p_m) \qquad (9.30)$$

Algorithm 9.1 A $(1 - \varepsilon)$-optimal algorithm

1. Within $\mathcal{A}$, compute $C_{iB}^{\min}$ and $C_{iB}^{\max}$ for each node $i \in \mathcal{N}$ by (8.7) and (8.8).
2. For a given $\varepsilon > 0$, define a sequence of costs $C[1], C[2], \ldots, C[H_i]$ by (8.9), where H_i is defined by (8.10).
3. For each node $i \in \mathcal{N}$, draw a sequence of $(H_i - 1)$ circles corresponding to cost $C[h]$ centered at node i, $1 \leq h < H_i$. The intersection of these circles within disk $\mathcal{A}$ will divide $\mathcal{A}$ into M subareas $\mathcal{A}_1, \mathcal{A}_2, \ldots, \mathcal{A}_M$.
4. For each subarea $\mathcal{A}_m$, $1 \leq m \leq M$, define a FCP p_m, which is represented by an N-tuple cost vector $[c_{1B}(p_m), c_{2B}(p_m), \ldots, c_{NB}(p_m)]$, where $c_{iB}(p_m)$ is defined by (8.12).
5. For the C-MB problem on these M FCPs, apply the LP formulation in Section 9.5.1 and obtain an optimal solution $\psi_{\text{C-MB}}^*$ with $W^*(p_m)$, $f_{ij}^*(p_m)$, and $f_{iB}^*(p_m)$.
6. Construct a $(1 - \varepsilon)$-optimal solution $\psi_{\text{U-MB}}$ to the U-MB problem based on $\psi_{\text{C-MB}}^*$ using the procedure in Theorem 9.3.

---

for $i \in \mathcal{N}$. The second inequality holds by $c_{iB}(p(s)) \leq c_{iB}(p_m)$ in Property 8.1 and (9.28). The fourth equality holds by (9.25). The last equality holds by (9.26).

Following the same token, it can be shown that

$$
\sum_{s \in [0,S]}^{U(s)>0} f_{ij}(p(s))U(s) + \int_{s \in [0,S]}^{U(s)=0} f_{ij}(p(s))u(s)ds = \sum_{m=1}^{M} f_{ij}^*(p_m)W^*(p_m).
$$

(9.31)

for $i, j \in \mathcal{N}$ and $i \neq j$. Thus, by (9.30) and (9.31), the total energy consumption at node i by time $T_{\text{C-MB}}^*$, which is shown in (9.29), is no more than

$$
\sum_{k \in \mathcal{N}}^{k \neq i} \rho \sum_{m=1}^{M} f_{ki}^*(p_m)W^*(p_m) + \sum_{j \in \mathcal{N}}^{j \neq i} C_{ij} \sum_{m=1}^{M} f_{ij}^*(p_m)W^*(p_m)
$$

$$
+ \sum_{m=1}^{M} c_{iB}(p_m)f_{iB}^*(p_m)W^*(p_m) \leq e_i,
$$

where the inequality holds by the energy constraint in solution $\psi_{\text{C-MB}}^*$. Thus, the network lifetime of solution $\psi_{\text{U-MB}}$ is at least $T_{\text{C-MB}}^* \geq (1 - \varepsilon)T_{\text{U-MB}}^*$. This completes the proof. $\qquad\square$

9.5.3 Summary of algorithm and example

We now summarize the design of the $(1 - \varepsilon)$-optimal algorithm in Algorithm 9.1. In Algorithm 9.1, Step 5 has the highest complexity (solving an LP problem) among all steps. Since there are $(H_i - 1)$ circles radiating from

Table 9.2 Sensor location, data rate, and initial energy of the example sensor network.

Node index	(x_i, y_i)	r_i	e_i
1	(0.2, 0.9)	0.6	170
2	(0.4, 0.6)	1.0	420
3	(0.6, 0.3)	0.8	460
4	(1.0, 0.2)	0.4	230

Figure 9.2 The SED for the example sensor network.

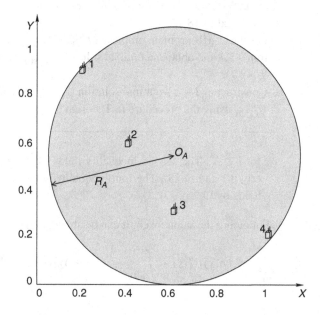

sensor node $i \in \mathcal{N}$ and one circle for disk $\mathcal{A}$, the total number of subareas M obtained through the intersection of these circles is upper bounded by $O([1 + \sum_{i=1}^{N}(H_i - 1)]^2) = O((N/\varepsilon)^2)$. Thus, the LP problem in Step 5 has polynomial size and the complexity of the overall algorithm is polynomial.

Example 9.1

To illustrate the steps in Algorithm 9.1, we solve a small four-node WSN problem as an example. The location, data rate, and initial energy for each sensor node are shown in Table 9.2, where the units of distance, rate, and energy are all normalized with appropriate dimensions. We use $\alpha = 2$ in this example and set $\beta_1 = 1$, $\beta_2 = 0.5$ and $\rho = 1$ under normalized units. For illustration, we set $\varepsilon = 0.2$.[1]

In Step 1, we first identify the SED $\mathcal{A}$ with origin $O_{\mathcal{A}} = (0.60, 0.55)$ and radius $R_{\mathcal{A}} = 0.53$ (see Fig. 9.2). Then we have $D_{1,O_{\mathcal{A}}} = 0.53$, $D_{2,O_{\mathcal{A}}} = 0.21$, $D_{3,O_{\mathcal{A}}} = 0.25$, and $D_{4,O_{\mathcal{A}}} = 0.53$. We then find the lower and upper bounds of c_{iB} for each node i as follows:

$$C_{iB}^{\min} = \beta_1 = 1,$$
$$C_{iB}^{\max} = \beta_1 + \beta_2(D_{i,O_{\mathcal{A}}} + R_{\mathcal{A}})^{\alpha}.$$

Thus, we have

$$C_{1B}^{\max} = 1 + 0.5 \cdot (0.53 + 0.53)^2 = 1.56,$$
$$C_{2B}^{\max} = 1 + 0.5 \cdot (0.21 + 0.53)^2 = 1.27,$$
$$C_{3B}^{\max} = 1 + 0.5 \cdot (0.25 + 0.53)^2 = 1.30,$$
$$C_{4B}^{\max} = 1 + 0.5 \cdot (0.53 + 0.53)^2 = 1.56.$$

In Step 2, for $\varepsilon = 0.2$, since

$$H_i = \left\lceil \frac{\ln(1 + \frac{\beta_2}{\beta_1}(D_{i,O_{\mathcal{A}}} + R_{\mathcal{A}})^{\alpha})}{\ln(1 + \varepsilon)} \right\rceil,$$

we have

$$H_1 = \left\lceil \frac{\ln(1 + \frac{0.5}{1}(0.53 + 0.53)^2)}{\ln(1 + 0.2)} \right\rceil = 3,$$

$$H_2 = \left\lceil \frac{\ln(1 + \frac{0.5}{1}(0.21 + 0.53)^2)}{\ln(1 + 0.2)} \right\rceil = 2,$$

$$H_3 = \left\lceil \frac{\ln(1 + \frac{0.5}{1}(0.25 + 0.53)^2)}{\ln(1 + 0.2)} \right\rceil = 2,$$

$$H_4 = \left\lceil \frac{\ln(1 + \frac{0.5}{1}(0.53 + 0.53)^2)}{\ln(1 + 0.2)} \right\rceil = 3,$$

and

$$C[1] = \beta_1(1 + \varepsilon) = 1 \cdot (1 + 0.2) = 1.20,$$
$$C[2] = \beta_1(1 + \varepsilon)^2 = 1 \cdot (1 + 0.2)^2 = 1.44,$$
$$C[3] = \beta_1(1 + \varepsilon)^3 = 1 \cdot (1 + 0.2)^3 = 1.73.$$

In Step 3, we draw a sequence of circles centered at each node i, $1 \leq i \leq 4$, and with cost $C[h]$, $1 \leq h < H_i$, to divide the SED $\mathcal{A}$ into 16 subareas $\mathcal{A}_1, \mathcal{A}_2, \ldots, \mathcal{A}_{16}$ (see Fig. 9.3).

In Step 4, we define a FCP p_m for each subarea $\mathcal{A}_m$, $1 \leq m \leq 16$. For example, for FCP p_1, we define the 4-tuple cost vector as $[c_{1B}(p_m), c_{2B}(p_m), c_{3B}(p_m), c_{4B}(p_m)] = [C[1], C[1], C[2], C[3]] = [1.20, 1.20, 1.44, 1.73]$.

In Step 5, we obtain an optimal solution $\psi_{\text{C-MB}}^*$ to C-MB problem on these 16 FCPs by the LP approach discussed in Section 9.5.1. We obtain the network lifetime $T_{\text{C-MB}}^* = 247.76$, $W^*(p_7) = 144.22$, $W^*(p_{12}) = 82.50$, $W^*(p_{16}) = 21.04$, and for all other 13 FCPs, we have $W^*(p_m) = 0$ (meaning the base station will not visit these 13 subareas). When the base station is at FCP p_7, the routing is $f_{1B}^*(p_7) = 0.60$, $f_{2B}^*(p_7) = 0.51$, $f_{23}^*(p_7) = 0.49$, $f_{3B}^*(p_7) = 1.29$, and $f_{4B}^*(p_7) = 0.40$. When the base station is at FCP p_{12}, the routing is $f_{12}^*(p_{12}) = 0.60$, $f_{2B}^*(p_{12}) = 1.60$, $f_{3B}^*(p_{12}) = 0.80$, and $f_{4B}^*(p_{12}) = 0.40$. When the base station is at FCP p_{16}, the routing is $f_{12}^*(p_{16}) = 0.60$, $f_{23}^*(p_{16}) = 1.60$, $f_{34}^*(p_{16}) = 2.40$, and $f_{4B}^*(p_{16}) = 2.80$.

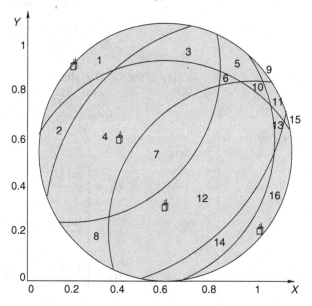

Figure 9.3 The subareas for the example sensor network.ff

In Step 6, we obtain a $(1 - \varepsilon)$-optimal solution $\psi_{\text{U-MB}}$ to U-MB problem as follows. Let the base station stay at any point in subarea $\mathcal{A}_7$ for 144.22 unit of time, stay at any point in subarea $\mathcal{A}_{12}$ for 82.50 unit of time, and stay at any point in subarea $\mathcal{A}_{16}$ for 21.04 unit of time. When the base station is at a point p in subarea $\mathcal{A}_7$, the routing is $f_{1B}(p) = 0.60$, $f_{2B}(p) = 0.51$, $f_{23}(p) = 0.49$, $f_{3B}(p) = 1.29$, and $f_{4B}(p) = 0.40$. When the base station is at a point p in subarea $\mathcal{A}_{12}$, the routing is $f_{12}(p) = 0.60$, $f_{2B}(p) = 1.60$, $f_{3B}(p) = 0.80$, and $f_{4B}(p) = 0.40$. When the base station is at a point p in subarea $\mathcal{A}_{16}$, the routing is $f_{12}(p) = 0.60$, $f_{23}(p) = 1.60$, $f_{34}(p) = 2.40$, and $f_{4B}(p) = 2.80$. The network lifetime for $\psi_{\text{U-MB}}$ is greater than or equal to 247.76 and is $(1 - \varepsilon)$-optimal.

9.5.4 Discussions

We now discuss the design of a path $\mathcal{P}$ based on $W^*(p_m)$ values. Such a path is certainly not unique. In Example 9.1, the base station can move from

Table 9.3 Each node's location, data generation rate and initial energy for a ten-node network.

Location	Data rate	Initial energy	Location	Data rate	Initial energy
(0.0, 0.8)	0.8	150	(0.6, 0.7)	0.6	370
(1.0, 1.0)	1.0	200	(0.4, 0.6)	0.2	420
(0.3, 0.4)	0.6	130	(0.2, 0.9)	0.6	100
(0.8, 0.5)	0.8	460	(0.9, 0.1)	0.4	80
(0.7, 0.3)	0.3	170	(0.5, 0.2)	1.0	150

subarea 7 to 12 and to 16 (denote as $(\mathcal{A}_7, \mathcal{A}_{12}, \mathcal{A}_{16})$) or, another path can be $(\mathcal{A}_{16}, \mathcal{A}_{12}, \mathcal{A}_7)$. Note that for any path, as long as the total sojourn time at each subarea $\mathcal{A}_m$ is $W^*(p_m)$, the achieved network lifetime is $(1 - \varepsilon)$-optimal. Thus, all of these paths are equally good under network lifetime objective. It may be arguable that one path is better than another under some other objective, e.g., minimizing the total traveled distance. However, such objective can be formulated as a separate problem and its discussion is beyond the scope of this chapter.

Along a path $\mathcal{P}$, it is possible that one subarea and the next subarea that the base station visits are not adjacent. We argue that the traveling time between two subareas (e.g., minutes) is likely on a much smaller time scale than network lifetime (e.g., months). It can be shown that if buffering is available at sensor nodes when base station is in transition from one subarea to the next subarea, then the $(1 - \varepsilon)$-optimal network lifetime can still be achieved. In this case, a node only needs to slightly delay its transmission until the base station arrives at the next subarea and then empties the buffer with a higher rate for a brief period of time.

9.6 Numerical examples

Now we apply the $(1 - \varepsilon)$-optimal algorithm for different sized networks and use numerical results to demonstrate the efficacy of the algorithm. We consider four randomly generated networks consisting of ten, 20, 50, and 100 nodes deployed over a unit square area, respectively. The data rate and initial energy for each node are randomly generated between [0.1, 1] and [50, 500], respectively. The units of distance, rate, and energy are all normalized appropriately. The normalized parameters in energy consumption model are $\beta_1 = \beta_2 = \rho = 1$. We assume the path-loss index $\alpha = 2$ and set $\varepsilon = 0.05$.

The network setting (location, data rate, and initial energy for each node) for the ten-node network is given in Table 9.3. By applying Algorithm 9.1, we obtain a $(1 - \varepsilon)$-optimal network lifetime 142.86, which is guaranteed to be at least 95% of the optimum. In Table 9.4, we have seven subareas that will

Table 9.4 Sojourn time at each optimal location for the ten-node network.

$\mathcal{A}_m(x, y)$	$W(\mathcal{A}_m)$
(0.93, 0.96)	4.28
(0.88, 0.19)	3.54
(0.68, 0.90)	0.04
(0.94, 0.89)	50.36
(0.69, 0.22)	44.00
(0.90, 0.40)	27.15
(0.23, 0.86)	13.49

Figure 9.4 Network topology and optimal locations for base station movement for the ten-node network.

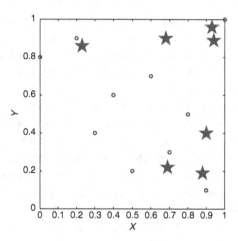

be visited by the base station in the $(1 - \varepsilon)$-optimal solution (also shown in Fig. 9.4). For illustration purpose, we use a star to represent the corresponding subarea that the base station will visit in the solution. For example, we put a star on location $(0.93, 0.96)$ to represent the subarea that contains this point. Table 9.4 also lists the corresponding sojourn time for the base station to stay in each of these seven subareas. The flow routing solution when the base station is in each of the seven subareas is different, as expected. Fig. 9.5 shows a possible path for the ten-node network. Note that as we discussed in Section 9.5.4, such a path is not unique.

It is worth noting that for 95% optimality, only seven subareas need to be visited by the base station. It turns out that for 20-, 50-, and 100-node networks, the number of subareas that needs to be visited by the base station is also very small (six subareas for 20-node network, eight subareas for 50-node network, and 12 subareas for 100-node network). This observation is not obvious. But it is a good news as it hints that the base station may not need to move frequently to many different locations to achieve near-optimal solution.

Table 9.5 Each node's location, data generation rate and initial energy for a 20-node network.

Location	Data rate	Initial energy	Location	Data rate	Initial energy
(0.52, 0.02)	0.6	480	(0.29, 0.14)	0.6	120
(0.74, 0.76)	0.3	310	(0.05, 0.99)	0.4	60
(0.95, 0.03)	0.8	150	(0.84, 0.06)	1.0	180
(0.53, 0.63)	0.6	220	(0.99, 0.37)	0.4	340
(0.58, 1.00)	0.4	230	(0.73, 0.67)	0.8	220
(0.48, 0.84)	0.7	160	(0.53, 0.27)	0.5	380
(0.17, 0.83)	0.1	380	(0.57, 0.05)	0.7	250
(0.73, 0.39)	0.1	500	(0.88, 0.84)	0.2	240
(0.36, 0.98)	0.1	430	(0.26, 0.12)	0.9	440
(0.76, 0.02)	0.7	500	(0.71, 0.21)	0.3	70

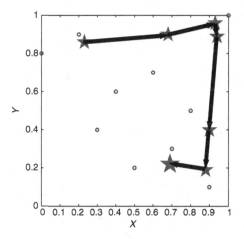

Figure 9.5 A possible base station moving path for the ten-node network.

The network setting for a small 20-node network (with location, data rate, and initial energy for each of the 20 sensor nodes) is given in Table 9.5. By applying Algorithm 9.1, we obtain a $(1 - \varepsilon)$-optimal network lifetime 144.23. Again, we use a star to represent the subarea that base station will visit in the solution. For this particular 20-node network setting, we have six subareas (see Fig. 9.6) that the base station will visit in the final solution, with the corresponding sojourn time in each subarea shown in Table 9.6. Again, we show a possible path for the 20-node network in Fig. 9.7.

The network setting for the 50-node network (with location, data rate, and initial energy for each of the 50 sensor nodes) is given in Table 9.8. By applying Algorithm 9.1, we obtain a $(1 - \varepsilon)$-optimal network lifetime 122.30. In Table 9.7, we have eight subareas (see Fig. 9.8) that the base station will visit in the $(1 - \varepsilon)$-optimal solution, as well as the sojourn time for the base station

Table 9.6 Sojourn time at each optimal location for the 20-node network.

$\mathcal{A}_m(x, y)$	$W(\mathcal{A}_m)$
(0.28, 0.33)	2.86
(0.27, 0.47)	8.44
(0.95, 0.86)	9.62
(0.80, 0.06)	5.34
(0.70, 0.11)	108.05
(0.88, 0.05)	9.92

Figure 9.6 Network topology and optimal locations for base station movement for the 20-node network.

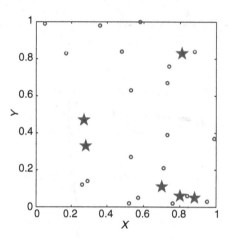

Figure 9.7 A possible base station moving path for the 20-node network.

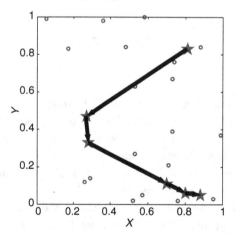

in each of these eight subareas. Fig. 9.9 shows a possible path for the 50-node network.

Finally, we consider a 100-node network shown in Table 9.9. We omit to list the each node's coordinates, data rate, and initial energy to conserve

Table 9.7 Sojourn time at each optimal location for the 50-node network.

$\mathcal{A}_m(x, y)$	$W(\mathcal{A}_m)$
(0.13, 0.11)	0.39
(0.17, 0.64)	1.47
(0.32, 0.90)	26.23
(0.25, 0.61)	27.72
(0.49, 0.12)	3.43
(0.94, 0.10)	9.66
(0.34, 0.26)	8.37
(0.16, 0.30)	45.03

Figure 9.8 Network topology and optimal locations for base station movement for the 50-node network.

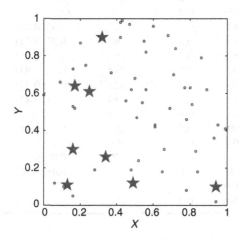

Figure 9.9 A possible base station moving path for the 50-node network.

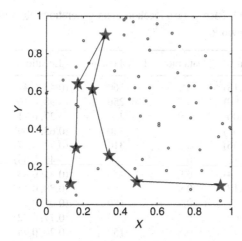

Table 9.8 Each node's location, data generation rate and initial energy for a 50-node network.

Location	Data rate	Initial energy	Location	Data rate	Initial energy
(0.52, 0.24)	1.0	290	(0.80, 0.63)	0.8	260
(0.68, 0.91)	0.8	160	(0.00, 0.59)	0.3	50
(0.32, 0.90)	0.4	480	(0.81, 0.54)	0.7	150
(0.54, 0.55)	0.1	500	(0.78, 0.46)	0.8	150
(0.78, 0.08)	0.1	140	(0.84, 0.18)	0.7	160
(0.94, 0.02)	0.8	300	(0.61, 0.43)	0.7	400
(0.06, 0.12)	0.1	220	(0.11, 0.10)	0.9	300
(0.45, 0.56)	0.4	370	(0.20, 0.87)	0.5	470
(0.83, 0.29)	0.4	400	(0.16, 0.05)	0.9	140
(0.17, 0.52)	0.7	160	(0.61, 0.42)	0.2	450
(0.42, 0.98)	0.8	280	(0.72, 0.84)	0.5	480
(0.55, 0.88)	0.4	320	(0.56, 0.82)	0.9	240
(0.95, 0.43)	0.6	390	(0.42, 0.90)	0.3	490
(0.99, 0.41)	0.8	180	(0.37, 0.71)	0.1	470
(0.16, 0.53)	0.8	190	(0.71, 0.70)	0.6	220
(0.89, 0.62)	0.7	340	(0.47, 0.97)	0.6	140
(0.69, 0.52)	0.7	220	(0.16, 0.73)	0.1	150
(0.86, 0.79)	0.4	50	(0.51, 0.47)	0.1	90
(0.23, 0.75)	0.6	150	(0.77, 0.63)	1.0	390
(0.43, 0.99)	0.5	290	(0.65, 0.18)	0.4	340
(0.60, 0.96)	0.3	500	(0.48, 0.19)	0.5	70
(0.56, 0.62)	0.4	420	(0.09, 0.66)	0.8	140
(0.50, 0.68)	1.0	170	(0.48, 0.62)	0.6	300
(0.17, 0.66)	1.0	250	(0.93, 0.36)	0.5	270
(0.66, 0.30)	0.1	100	(0.28, 0.19)	0.8	160

Table 9.9 Each node's location, data generation rate and initial energy for a 100-node network.

Location	Data rate	Initial energy	Location	Data rate	Initial energy
(0.85, 0.13)	0.1	160	(0.43, 0.44)	0.5	290
(0.74, 0.99)	0.5	250	(0.98, 0.16)	0.8	160
(0.31, 0.06)	0.9	430	(0.27, 0.71)	0.1	190
(0.93, 0.67)	0.7	240	(0.68, 0.09)	0.2	130
(0.65, 0.56)	0.3	310	(0.77, 0.77)	0.2	320
(0.75, 0.12)	0.4	90	(0.49, 0.44)	0.4	400
(0.63, 0.74)	0.9	470	(0.35, 0.83)	0.5	190
(0.88, 0.17)	0.9	440	(0.88, 0.12)	0.4	460
(0.91, 0.48)	0.1	180	(0.08, 0.27)	0.8	210
(0.94, 0.02)	0.9	220	(0.77, 0.72)	0.1	380
(0.26, 0.35)	0.8	150	(0.20, 0.74)	1.0	470

Table 9.9 (Cont.)

Location	Data rate	Initial energy	Location	Data rate	Initial energy
(0.77, 0.35)	0.5	160	(0.20, 0.56)	0.7	320
(0.42, 0.73)	0.3	100	(0.66, 0.12)	1.0	190
(0.65, 0.58)	0.5	320	(0.44, 0.06)	0.5	450
(0.93, 0.15)	0.7	150	(0.61, 0.05)	0.7	210
(0.32, 0.04)	0.2	230	(0.10, 0.26)	0.7	480
(0.63, 0.96)	0.6	110	(0.52, 0.85)	0.1	440
(0.51, 0.75)	0.5	360	(0.06, 0.65)	0.4	320
(0.03, 0.27)	0.1	350	(0.49, 0.16)	0.8	380
(0.64, 0.43)	0.1	360	(0.79, 0.15)	0.4	480
(0.67, 0.40)	0.6	230	(0.27, 0.50)	0.4	140
(0.84, 0.38)	0.1	210	(0.22, 0.81)	0.6	290
(0.21, 0.16)	0.7	430	(0.77, 0.46)	0.6	370
(0.37, 0.93)	0.9	460	(0.06, 0.04)	0.8	230
(0.86, 0.96)	0.1	90	(0.95, 0.95)	0.3	80
(0.77, 0.48)	0.6	270	(0.10, 0.42)	0.4	250
(0.56, 0.09)	0.9	270	(0.91, 0.86)	0.7	280
(0.72, 0.26)	1.0	260	(0.33, 0.55)	0.7	490
(0.88, 0.51)	0.3	110	(0.38, 0.33)	0.1	490
(0.66, 0.49)	0.1	90	(0.01, 0.88)	0.9	420
(0.80, 0.92)	0.8	170	(0.07, 0.19)	0.3	230
(0.48, 0.78)	0.1	70	(0.12, 0.31)	0.9	470
(0.87, 0.38)	1.0	150	(0.95, 0.72)	0.8	370
(0.92, 0.68)	0.1	60	(0.85, 0.65)	1.0	210
(0.71, 0.12)	0.8	50	(0.51, 0.94)	0.2	170
(0.69, 0.79)	0.1	200	(0.24, 0.78)	0.6	250
(0.38, 0.90)	0.2	180	(0.26, 0.48)	0.2	70
(0.77, 1.00)	0.2	120	(0.87, 0.74)	0.4	150
(0.44, 0.34)	0.9	200	(0.80, 0.47)	0.6	440
(0.54, 0.21)	1.0	340	(0.71, 0.06)	0.8	380
(0.44, 0.48)	0.1	450	(0.13, 0.68)	0.2	460
(0.54, 0.05)	0.6	130	(0.60, 0.98)	0.2	460
(0.91, 0.04)	0.7	220	(0.89, 0.01)	0.7	430
(0.57, 0.21)	0.2	220	(0.95, 0.33)	0.2	430
(0.61, 0.77)	0.4	160	(0.09, 0.09)	0.3	260
(0.75, 0.55)	0.2	170	(0.33, 0.86)	0.5	490
(0.39, 0.41)	0.8	190	(0.01, 0.72)	0.1	170
(0.26, 0.32)	0.8	230	(0.15, 0.42)	0.1	160
(0.25, 0.97)	0.9	310	(0.03, 0.69)	0.5	190
(0.35, 0.92)	0.3	290	(0.01, 0.27)	0.6	120

space. They are all randomly generated as we described early in this section. By applying Algorithm 9.1, we obtain a $(1 - \varepsilon)$-optimal network lifetime 149.45. For this particular 100-node network setting, we have 12 subareas (see Fig. 9.10) that the base station will visit, with the corresponding sojourn time in each subarea shown in Table 9.10. Fig. 9.11 shows a possible path for the 100-node network.

Table 9.10 Sojourn time at each optimal location for the 100-node network.

$\mathcal{A}_m(x, y)$	$W(\mathcal{A}_m)$
(0.75, 0.83)	0.15
(0.86, 0.05)	0.38
(0.69, 0.28)	50.14
(0.24, 0.04)	2.10
(0.59, 0.88)	0.57
(0.81, 0.82)	0.09
(0.20, 0.53)	21.21
(0.91, 0.09)	0.05
(0.40, 0.68)	41.64
(0.89, 0.19)	3.16
(0.61, 0.79)	23.82
(0.19, 0.02)	6.14

Figure 9.10 Network topology and optimal locations for base station movement for the 100-node network.

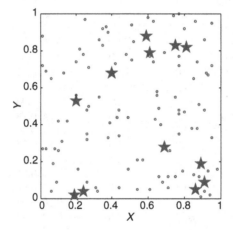

9.7 Chapter summary

This chapter is a sequel to the last chapter. Again, our interest is in the design of a $(1 - \varepsilon)$-optimal approximation algorithm. But the problem is much harder than that in the last chapter. By allowing the base station to be mobile, both the location of the base station and the multi-hop flow routing in the network are time-dependent.

To address this problem, we showed that as far as network lifetime objective is concerned, we can *transform* the time-dependent problem to a location (space)-dependent problem. In particular, we showed that flow routing only depends on the base station location, regardless of *when* the base station visits this location. Further, the specific time instances for the base station to visit a

Figure 9.11 A possible base station moving path for the 100-node network.

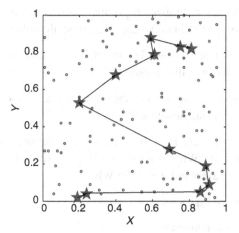

location are not important, as long as the total sojourn time for the base station to be present at this location is the same. This result allowed us to focus on solving a location-dependent problem.

Based on the above result, we further showed that to obtain a $(1 - \varepsilon)$-optimal solution to the location-dependent problem, we only need to consider a finite set of points within the SED for the mobile base station's location. Here, we followed the same approach as that in Section 8.6, i.e., discretization of energy cost through a geometric sequence, division of a disk into a finite number of subareas, and representation of each subarea with a FCP. Then we can find the optimal sojourn time for the base station to stay at each FCP (as well as the corresponding flow routing solution) so that the overall network lifetime (i.e., sum of the sojourn times) is maximized via a single LP problem. We proved that the proposed solution can guarantee that the achieved network lifetime is at least $(1 - \varepsilon)$ of the maximum (unknown) network lifetime.

This chapter offers some excellent examples on how to transform a problem from time domain to space domain and how to prove results through construction. Students are encouraged to gain a deep understanding of these techniques, which should transcend to other optimization problems in wireless networks.

9.8 Problems

9.1 What is the benefit of using mobile base station? Compared to the static base station problem in the last chapter, what are the challenges in employing a mobile base station?

9.2 Show that the formulation for the optimization problem in Section 9.3 is nonconvex.

9.3 For the optimization problem formulated in Section 9.3, can we discretize time t to obtain an approximation algorithm?

9.4 Describe the three key techniques used in the design of the approximation algorithm in the chapter.

9.5 What is the relationship between a time-dependent solution and a location-dependent solution?

9.6 Can we use Eqs. (9.2) and (9.3) to define $f_{ij}(p)$ and $f_{iB}(p)$ for every point $p \in \mathcal{P}$? Why? Can we use Eqs. (9.6) and (9.7) to define $f_{ij}(p)$ and $f_{iB}(p)$ for every point $p \in \mathcal{P}$? Why?

9.7 Prove Eq. (9.8).

9.8 Prove Lemma 9.2.

9.9 Complete the proof of Lemma 9.3.

9.10 In Section 9.5.4, it is said that the design of a path $\mathcal{P}$ based on $W^*(p_m)$ values is not unique. We may use an additional objective to select the best path, e.g., minimizing the total traveled distance. Discuss other possible objective(s) for path selection.

Methods for Efficient Heuristic Solutions

10 An efficient technique for mixed-integer optimization

Remember, happiness doesn't depend upon who you are or what you have; it depends solely upon what you think.

Dale Carnegie

10.1 Sequential fixing: an introduction

In Chapter 5, we presented a branch-and-bound framework for solving mixed-integer nonlinear programs (MINLPs), which can be used to solve a broad class of problems. Furthermore, it can guarantee a $(1 - \varepsilon)$-optimal solution for any given $\varepsilon \geq 0$ but its worst-case complexity is exponential. In this chapter, we present another technique called *sequential fixing* (SF) to solve certain mixed-integer optimization problems. SF is designed to iteratively determine (fix) binary integer variables. Unlike branch-and-bound, it is a heuristic procedure and has polynomial-time complexity. Nevertheless, it is a very efficient technique, and, in a number of problem instances, we found that this technique can offer highly competitive solutions.

The basic idea behind SF is as follows. For some mixed-integer optimization problems, such as certain MINLP and any mixed-integer linear programming (MILP), if we were able to set the optimal values for all the integer variables and thereby reduce the original problem to an LP, then we can solve the reduced problem optimally in polynomial time. Thus, the key challenge in such contexts is how to determine the values for all the integer variables. This can be done by examining the linear or continuous relaxation of the original problem, which is obtained by relaxing all the integer variables to continuous variables. Although the solution to this relaxation may not have an integer value for each integer variable, we can set (i.e., fix) the values of one or more integer variables based on the *closeness* in the relaxed solution to certain integer values. Instead of determining all integer variable values via a single relaxation, we can fix only one or a few integer variables in each iteration. For the remaining (unfixed) integer variables, we can solve a new relaxation (with some integer variables' values being already fixed) and then fix one or more integer variables. This SF procedure terminates after we fix all the integer variables. The values of other variables in the original problem can be obtained by solving the resulting LP problem.

Unlike most techniques in other chapters, SF is a heuristic procedure. To measure its performance, we can compare its solution value to some performance bound, e.g., a lower bound for a minimization problem, or an upper bound for a maximization problem. Such a bound can be obtained via the initial relaxation to the original problem. Note that the optimal objective value lies between this bound and the solution obtained by the SF algorithm. Therefore, if we can show (e.g., via a large number of simulations) that the SF solution has a value very close to the computed relaxation-based bound, then we can claim that the SF solution must be even closer to the optimum, thus validating its performance.

In the case study of this chapter, we show how SF can be employed to solve a spectrum sharing problem in a multi-hop cognitive radio network (CRN).

10.2 Case study: Spectrum sharing for cognitive radio networks

Consider a multi-hop CRN. For such a network, each node senses a set of spectrum bands that it can use for communication. Due to the unequal size of spectrum bands, it may be necessary to further divide a large band into subbands (likely of unequal size) to schedule transmission and reception. There are many fundamental problems that can be posed for such a wireless network in the context of rates and capacity. In this chapter, we consider the following problem. Suppose there exists a set of user sessions in the network that is characterized by several source–destination pairs, each having a certain rate requirement. Then, how can we perform spectrum allocation, scheduling and interference avoidance, and multi-hop flow routing such that the required network-wide radio spectrum resource is minimized?

To formulate the problem mathematically, we characterize the structure and constraints of a CRN at multiple layers. Special attention is given to the modeling of spectrum sharing and unequal (nonuniform) subband division, scheduling and interference modeling, and multi-hop routing. We formulate an optimization problem with the objective of minimizing the required network-wide radio spectrum resource for a set of session rate requirements. Since such a problem formulation is a mixed-integer nonlinear programming (MINLP), which is NP-hard in general [46], we aim to derive a near-optimal solution.

We present a near-optimal algorithm for the formulated MINLP problem. First, we develop a lower bound for the objective function by relaxing the integer variables and employing a linearization technique. This lower bound will be used as a measure for the quality of any solution. Then we present an SF solution procedure where the determination of integer variables is performed iteratively through a sequence of LPs. After fixing all the integer variables, the other variables in the optimization problem can be derived by solving the residual LP problem. Since the solution obtained by the proposed SF algorithm represents an upper bound for the minimization problem, we compare it to the lower bound developed earlier. Simulations show that the results obtained by the SF algorithm are very close to the lower bound, thus suggesting that (1) the lower bound is very tight; and (2) the solution obtained by the SF algorithm is very close to the optimum, and thus is near-optimal.

The remainder of this chapter is organized as follows. In Section 10.3, we characterize a CRN based on multiple layers and formulate its structure in terms of mathematical constraints. We also describe the optimal spectrum sharing problem and formulate it as an MINLP problem. In Section 10.4, we develop a lower bound for this MINLP problem by relaxing integer variables and using linearization. In Section 10.5, we describe an SF algorithm. Section 10.6 presents simulation results and demonstrates the near-optimal performance of the SF algorithm. Section 10.7 summarizes this chapter.

10.3 Mathematical modeling and problem formulation

We consider an ad hoc network consisting of a set $\mathcal{N}$ of nodes. There is a set $\mathcal{L}$ of unicast communication sessions. Denote $s(l)$ and $d(l)$ as the source and destination nodes of session $l \in \mathcal{L}$, and $r(l)$ as the rate requirement (in b/s) of session l. Table 10.1 lists the notation used in this chapter.

10.3.1 Modeling of multi-layer characteristics

Spectrum sharing and subband division   In a multi-hop CRN, the available spectrum bands at one node may be different from another node in the network. Given a set of available frequency bands at a node, the size (or bandwidth) of each band may differ drastically. For example, among the least-utilized spectrum bands found in [106], the bandwidth between [1240, 1300] MHz (allocated to amateur radio) is 60 MHz, while the bandwidth between [1525, 1710] MHz (allocated to mobile satellites, GPS systems, and meteorological applications) is 185 MHz. Such a large difference in bandwidths among the available bands suggests the need to further divide a large band into smaller subbands for a more flexible and efficient frequency allocation. Since an equal subband division of the available spectrum band is likely to yield suboptimal performance, an unequal division may be necessary.

Table 10.1 Notation.

Symbol	Definition
$d(l)$	Destination node of session l
d_{ij}	Distance between nodes i and j
$f_{ij}(l)$	Data rate that is attributed to session l on link $i \to j$
g_{ij}	Propagation gain from node i to node j
$\mathcal{I}_j^m$	The set of nodes that can use band m and are within the interference range of node j
$K^{(m)}$	The maximum number for subband divisions in band m
$\mathcal{L}$	The set of active user sessions in the network
M	$= \|\mathcal{M}\|$, the number of available bands in the network
$\mathcal{M}$	$= \bigcup_{i \in \mathcal{N}} \mathcal{M}_i$, the set of available bands in the network
$\mathcal{M}_i$	The set of available bands at node $i \in \mathcal{N}$
$\mathcal{M}_{ij}$	$= \mathcal{M}_i \bigcap \mathcal{M}_j$, the set of available bands for link $i \to j$
$\mathcal{N}$	The set of nodes in the network
$r(l)$	Rate of session $l \in \mathcal{L}$
R_T, R_I	Transmission range and interference range, respectively
$s(l)$	Source node of session l
$\mathcal{T}_i^m$	The set of nodes that can use band m and are within the transmission range of node i
$\mathcal{T}_i$	$= \bigcup_{m \in \mathcal{M}_i} \mathcal{T}_i^m$, the set of nodes within the transmission range of node i
$u^{(m,k)}$	The fraction of bandwidth for the kth subband in band m
$W^{(m)}$	Bandwidth of band $m \in \mathcal{M}$
$x_{ij}^{(m,k)}$	Binary indicator to mark whether or not subband (m, k) is used for link $i \to j$.
α	Path-loss index
η	Ambient Gaussian noise density
ρ	Transmission power spectral density at a transmitter
ρ_T	The minimum threshold of power spectral density to decode a transmission at a receiver
ρ_I	The maximum threshold of power spectral density for interference to be negligible at a receiver

More formally, we model the union of the available frequency bands among all the nodes in the network as a set of M unequally sized bands (see Fig. 10.1). Denote $\mathcal{M}$ as the set of these bands and let $\mathcal{M}_i \subseteq \mathcal{M}$ be the set of available bands at node $i \in \mathcal{N}$, which is likely to be different from that at another node, say $j \in \mathcal{N}$, i.e., possibly, $\mathcal{M}_i \neq \mathcal{M}_j$. For example, at node i, $\mathcal{M}_i$ may consist of bands 1, 3, and 5, whereas at node j, $\mathcal{M}_j$ may consist of bands 1, 4, and 6. Denote $W^{(m)}$ as the bandwidth of band $m \in \mathcal{M}$. For more flexible and efficient bandwidth allocation and to overcome the disparity in the bandwidth size among the spectrum bands, we assume that band m can be further divided into up to $K^{(m)}$ subbands, each of which may be of unequal bandwidth. Denote

Figure 10.1 A schematic illustration of bands and subbands in spectrum sharing.

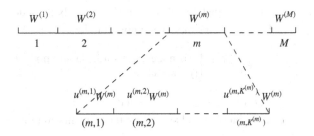

$u^{(m,k)}$ as the fraction of bandwidth for the kth subband in band m, which is part of our set of optimization decision variables. Then we have

$$\sum_{k=1}^{K^{(m)}} u^{(m,k)} = 1.$$

Note that some $u^{(m,k)}$-variables can be 0 in the final optimization solution, in which case we will have a fewer number of subbands than $K^{(m)}$. As an example, Fig. 10.1 shows M bands in the network, and for a specific band m, it displays a further division into $K^{(m)}$ subbands. Thus, the M bands in the network are effectively divided into $\sum_{m=1}^{M} K^{(m)}$ subbands, each of which may be of a different size.

Transmission range and interference range We assume that the power spectral density from the transmitter of a CR node is ρ. In this chapter, we assume that all nodes use the same power density for transmission. A widely used model for power propagation gain is [56]

$$g_{ij} = \beta \cdot d_{ij}^{-\alpha}, \tag{10.1}$$

where β is an antenna related constant, α is the path-loss index, and d_{ij} is the distance between nodes i and j. We assume that data transmission is successful only if the received power spectral density at the receiver exceeds a threshold ρ_T. Likewise, we assume that an interference will become nonnegligible only if it produces a power spectral density over a threshold of ρ_I at a receiver. Based on the threshold ρ_T, the transmission range for a node is thus given by $R_T = (\beta\rho/\rho_T)^{1/\alpha}$, which is derived from $\beta \cdot (R_T)^{-\alpha} \cdot \rho = \rho_T$. Similarly, based on the interference threshold ρ_I ($< \rho_T$), the interference range for a node is given by $R_I = (\beta\rho/\rho_I)^{1/\alpha}$. Because $\rho_I < \rho_T$, we have $R_I > R_T$.

Scheduling and interference constraints Scheduling can be done either in time domain or frequency domain. In this chapter, we consider a frequency domain subband assignment, i.e., how to assign subbands at a node for transmission and reception. A feasible scheduling on frequency bands must ensure that there is no interference *at the same node* and *among the neighboring nodes*.

Suppose that band m is available at both node i and node j, i.e., $m \in \mathcal{M}_i \cap \mathcal{M}_j$. To simplify the notation, let $\mathcal{M}_{ij} = \mathcal{M}_i \cap \mathcal{M}_j$. Denote

$$x_{ij}^{(m,k)} = \begin{cases} 1 & \text{if node } i \text{ transmits data to node } j \text{ on subband } (m,k), \\ 0 & \text{otherwise.} \end{cases}$$

For a node $i \in \mathcal{N}$ and a band $m \in \mathcal{M}_i$, denote $\mathcal{T}_i^m$ as the set of nodes that can use band m and are within the transmission range to node i, i.e.,

$$\mathcal{T}_i^m = \{j : d_{ij} \le R_T, j \ne i, m \in \mathcal{M}_j\}.$$

We assume that node i cannot transmit to multiple nodes on the same frequency subband. We have

$$\sum_{q \in \mathcal{T}_i^m} x_{iq}^{(m,k)} \le 1. \tag{10.2}$$

For a frequency subband (m,k), if node i uses this subband for transmitting data to a node $j \in \mathcal{T}_i^m$, then any other node that can produce interference on node j should not use this subband. Note that the so-called "hidden terminal" problem is a special case under this constraint. To model this constraint, we denote $\mathcal{I}_j^m$ as the set of nodes that can produce interference at node j on band m, i.e.,

$$\mathcal{I}_j^m = \{p : d_{pj} \le R_I, \mathcal{T}_p^m \ne \emptyset\}.$$

The physical meaning of $\mathcal{T}_p^m \ne \emptyset$ in the above definition is that node p may use band m for a valid transmission to some node in $\mathcal{T}_p^m$, which then causes interference to node j. Thus, we have

$$x_{ij}^{(m,k)} + \sum_{q \in \mathcal{T}_p^m} x_{pq}^{(m,k)} \le 1 \, (p \in \mathcal{I}_j^m, p \ne i). \tag{10.3}$$

In (10.3), if $x_{ij}^{(m,k)} = 1$, i.e., node i uses frequency subband (m,k) to transmit to node j, then any node p that can cause interference on node j should not transmit on this subband, i.e., $\sum_{q \in \mathcal{T}_p^m} x_{pq}^{(m,k)} = 0$. On the other hand, if $x_{ij}^{(m,k)} = 0$, (10.3) degenerates into (10.2), i.e., node p may transmit on subband (m,k) to at most one node $q \in \mathcal{T}_p^m$, i.e., $\sum_{q \in \mathcal{T}_p^m} x_{pq}^{(m,k)} \le 1$.

It is important to understand that in the interference constraint (10.3), if $x_{ij}^{(m,k)} = 0$, two nodes that can produce interference at node j but are far apart and outside each other's interference range can use the same subband (m,k) for transmission. We use an example to illustrate this point. In Fig. 10.2, suppose that node 1 transmits to node 2 on subband (m,k). Then any node that can cause interference at node 2 (i.e., node 3 or 5) cannot use the same subband for transmission. On the other hand, if node 1 does not use subband (m,k) to transmit to node 2, then node 3 may use this subband to transmit (to node 4) as stated in (10.3). Likewise, node 5 may also use this subband to transmit

Figure 10.2 An example illustrating interference among the links.

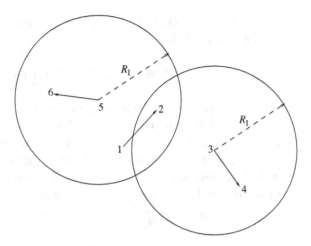

(to node 6) as stated in (10.3). That is, both nodes 3 and 5 may use the same subband for transmission.

Flow routing At the network level, a source node may need a number of relay nodes to route the data stream toward its destination node. Clearly, a route having only a single path may be overly restrictive and may not be able to take advantage of load balancing. A set of paths (with flow splitting) is more flexible to route the traffic from a source node to its destination. Mathematically, this can be modeled as follows. Denote $f_{ij}(l)$ as the data rate on link $i \to j$ that is attributed to session l, where $i \in \mathcal{N}$, $j \in \bigcup_{m \in \mathcal{M}_i} \mathcal{T}_i^m$, and $l \in \mathcal{L}$. To simplify the notation, let $\mathcal{T}_i = \bigcup_{m \in \mathcal{M}_i} \mathcal{T}_i^m$. If node i is the source node of session l, i.e., $i = s(l)$, then

$$\sum_{j \in \mathcal{T}_i} f_{ij}(l) = r(l). \tag{10.4}$$

If node i is an intermediate relay node for session l, i.e., $i \neq s(l)$ and $i \neq d(l)$, then

$$\sum_{j \in \mathcal{T}_i, j \neq s(l)} f_{ij}(l) = \sum_{p \in \mathcal{T}_i, p \neq d(l)} f_{pi}(l). \tag{10.5}$$

If node i is the destination node of session l, i.e., $i = d(l)$, then

$$\sum_{p \in \mathcal{T}_i} f_{pi}(l) = r(l). \tag{10.6}$$

It can be easily verified that if (10.4) and (10.5) are satisfied, then (10.6) must be satisfied. As a result, it is sufficient to list only (10.4) and (10.5) in the formulation.

Link capacity constraint In addition to the above flow balance equations at each node i for each session l, the aggregate flow rate on each radio link cannot exceed this link's capacity. To model this mathematically, we need to first find the capacity on link $i \to j$ and subband (m, k). If node i sends data to node j on subband (m, k), i.e., $x_{ij}^{(m,k)} = 1$, then the capacity on link $i \to j$ and subband (m, k) is given by

$$c_{ij}^{(m,k)} = u^{(m,k)} \cdot W^{(m)} \cdot \log_2 \left(1 + \frac{g_{ij}\rho}{\eta} \right),$$

where η is the ambient Gaussian noise density. Note that the denominator inside the log function contains only η. This is due to one of our interference constraints stated earlier, i.e., when node i transmits to node j on subband (m, k), then all the other neighbors of node j within its interference range are prohibited from using this subband. This interference constraint significantly helps simplify the calculation of the link capacity $c_{ij}^{(m,k)}$. When $x_{ij}^{(m,k)} = 0$, we have $c_{ij}^{(m,k)} = 0$. Thus, $c_{ij}^{(m,k)}$ can be written in the following compact form:

$$c_{ij}^{(m,k)} = x_{ij}^{(m,k)} \cdot u^{(m,k)} \cdot W^{(m)} \log_2 \left(1 + \frac{g_{ij}\rho}{\eta} \right). \tag{10.7}$$

Now, returning to our earlier requirement that the aggregate data rate on each link $i \to j$ cannot exceed the link's capacity, we have

$$\sum_{l \in \mathcal{L}, s(l) \neq j, d(l) \neq i} f_{ij}(l) \leq \sum_{m \in \mathcal{M}_{ij}} \sum_{k=1}^{K^{(m)}} c_{ij}^{(m,k)}$$

$$= \sum_{m \in \mathcal{M}_{ij}} \sum_{k=1}^{K^{(m)}} x_{ij}^{(m,k)} \cdot u^{(m,k)} W^{(m)} \log_2 \left(1 + \frac{g_{ij}\rho}{\eta} \right).$$

10.3.2 Problem formulation

For a multi-hop CRN of the type that we are investigating, various performance objectives can be used. In this chapter, we use the same objective function as in Chapter 5, i.e., the bandwidth-footprint-product (BFP). Since the footprint is identical for all nodes, it can be removed without any loss of generality. It is not hard to see that the solution procedure in this chapter can be likewise applied for several other performance objectives as well.

Mathematically, we have the following optimization problem:

$$\text{Minimize} \sum_{i \in \mathcal{N}} \sum_{m \in \mathcal{M}_i} \sum_{j \in \mathcal{T}_i^m} \sum_{k=1}^{K^{(m)}} W^{(m)} x_{ij}^{(m,k)} u^{(m,k)}$$

$$\text{subject to} \sum_{k=1}^{K^{(m)}} u^{(m,k)} = 1 \qquad (m \in \mathcal{M})$$

$$\sum_{q \in \mathcal{T}_i^m} x_{iq}^{(m,k)} \leq 1 \qquad (i \in \mathcal{N}, m \in \mathcal{M}_i, 1 \leq k \leq K^{(m)}) \tag{10.8}$$

$$x_{ij}^{(m,k)} + \sum_{q \in \mathcal{T}_p^m} x_{pq}^{(m,k)} \leq 1 (i \in \mathcal{N}, m \in \mathcal{M}_i, j \in \mathcal{T}_i^m,$$

$$1 \leq k \leq K^{(m)}, p \in \mathcal{I}_j^m, p \neq i) \qquad (10.9)$$

$$\sum_{\substack{l \in \mathcal{L} \\ s(l) \neq j, d(l) \neq i}} f_{ij}(l) - \sum_{m \in \mathcal{M}_{ij}} \sum_{k=1}^{K^{(m)}} W^{(m)} \log_2 \left(1 + \frac{g_{ij}\rho}{\eta}\right) x_{ij}^{(m,k)} u^{(m,k)} \leq 0$$

$$(i \in \mathcal{N}, j \in \mathcal{T}_i)$$

$$\sum_{\substack{j \in \mathcal{T}_i \\ j \neq s(l)}} f_{ij}(l) = r(l) \qquad (l \in \mathcal{L}, i = s(l))$$

$$\sum_{\substack{j \in \mathcal{T}_i \\ j \neq s(l)}} f_{ij}(l) - \sum_{p \in \mathcal{T}_i}^{p \neq d(l)} f_{pi}(l) = 0 \qquad (l \in \mathcal{L}, i \in \mathcal{N}, i \neq s(l), d(l))$$

$$x_{ij}^{(m,k)} = 0 \text{ or } 1, u^{(m,k)} \geq 0 \qquad (i \in \mathcal{N}, m \in \mathcal{M}_i, j \in \mathcal{T}_i^m, 1 \leq k \leq K^{(m)})$$

$$f_{ij}(l) \geq 0 \qquad (l \in \mathcal{L}, i \in \mathcal{N}, i \neq d(l), j \in \mathcal{T}_i, j \neq s(l)),$$

where $W^{(m)}$, g_{ij}, ρ, η, and $r(l)$ are all constants, and $x_{ij}^{(m,k)}$, $u^{(m,k)}$, and $f_{ij}(l)$ are the optimization decision variables.

The above optimization problem is a *mixed-integer nonlinear programming* (MINLP) problem, which is NP-hard in general [46]. Our approach to solve this problem is as follows. In Section 10.4, we first propose the computation of a lower bound for the problem, which can be obtained by relaxing the integer variables and using a linearization technique. Using this lower bound as a performance benchmark, we develop in Section 10.5 a highly effective algorithm based on the SF procedure that we discussed in Section 10.1. Using extensive simulation results, we show that the SF algorithm can offer solutions with objective values very close to the computed lower bounds. Since the optimal objective value lies between the lower bound and the solution obtained by the SF algorithm, the solution produced by the SF algorithm must be even closer to the optimum.

10.4 Deriving a lower bound

The complexity of the problem formulated in Section 10.3.2 arises from the binary $x_{ij}^{(m,k)}$-variables and the product of variables $x_{ij}^{(m,k)} u^{(m,k)}$. To derive a lower bound for the problem, we first multiply (10.8) and (10.9) by the corresponding $u^{(m,k)}$, so that $x_{ij}^{(m,k)}$ appears throughout as a product with $u^{(m,k)}$. We then relax the integer (binary) requirement on $x_{ij}^{(m,k)}$ with $0 \leq x_{ij}^{(m,k)} \leq 1$ and replace $x_{ij}^{(m,k)} u^{(m,k)}$ with a single variable, say $s_{ij}^{(m,k)}$, i.e., $s_{ij}^{(m,k)} = x_{ij}^{(m,k)} u^{(m,k)} \leq u^{(m,k)}$. Such a relaxation leads to the following lower-bounding problem formulation:

$$\text{Minimize} \sum_{i \in \mathcal{N}} \sum_{m \in \mathcal{M}_i} \sum_{j \in \mathcal{T}_i^m} \sum_{k=1}^{K^{(m)}} W^{(m)} s_{ij}^{(m,k)}$$

$$\text{subject to} \sum_{k=1}^{K^{(m)}} u^{(m,k)} = 1 \quad (m \in \mathcal{M})$$

$$\sum_{q \in \mathcal{T}_i^m} s_{iq}^{(m,k)} - u^{(m,k)} \le 0 \quad (i \in \mathcal{N}, m \in \mathcal{M}_i, 1 \le k \le K^{(m)}) \qquad (10.10)$$

$$s_{ij}^{(m,k)} + \sum_{q \in \mathcal{T}_p^m} s_{pq}^{(m,k)} - u^{(m,k)} \le 0 \left(i \in \mathcal{N}, m \in \mathcal{M}_i, j \in \mathcal{T}_i^m, \right.$$

$$\left. 1 \le k \le K^{(m)}, p \in \mathcal{I}_j^m, p \ne i \right) \quad (10.11)$$

$$\sum_{\substack{s(l) \ne j, d(l) \ne i}}^{l \in \mathcal{L}} f_{ij}(l) - \sum_{m \in \mathcal{M}_{ij}} \sum_{k=1}^{K^{(m)}} W^{(m)} \log_2\left(1 + \frac{g_{ij}\rho}{\eta}\right) s_{ij}^{(m,k)} \le 0$$

$$(i \in \mathcal{N}, j \in \mathcal{T}_i)$$

$$\sum_{\substack{j \in \mathcal{T}_i \\ j \ne s(l)}} f_{ij}(l) = r(l) \qquad (l \in \mathcal{L}, i = s(l))$$

$$\sum_{\substack{j \in \mathcal{T}_i \\ j \ne s(l)}} f_{ij}(l) - \sum_{\substack{p \in \mathcal{T}_i \\ p \ne d(l)}} f_{pi}(l) = 0 \qquad (l \in \mathcal{L}, i \in \mathcal{N}, i \ne s(l), d(l))$$

$$u^{(m,k)}, s_{ij}^{(m,k)} \ge 0 \qquad (i \in \mathcal{N}, m \in \mathcal{M}_i, j \in \mathcal{T}_i^m, 1 \le k \le K^{(m)})$$

$$f_{ij}(l) \ge 0 \qquad (l \in \mathcal{L}, i \in \mathcal{N}, i \ne d(l), j \in \mathcal{T}_i, j \ne s(l)).$$

This new (relaxed) formulation is a standard LP problem, the solution of which can be obtained in polynomial time. Due to the relaxation (and thus enlarged optimization space), the solution value to this LP problem yields a lower bound for the original problem of Section 10.3.2.

10.5　A near-optimal algorithm based on sequential fixing

10.5.1　Basic algorithm

We now take a closer look at the original MINLP problem formulation in Section 10.3.2. Observe that once the binary values for all the x-variables are determined, i.e., we have ascertained whether or not a node will indeed use a particular subband to send data to another node, then this MINLP reduces to an LP, which can be solved in polynomial time. Thus, the key obstacle in solving this MINLP problem lies in the determination of the binary values for all the x-variables. To this end, we propose the following two-step solution procedure:

1. Fix the binary values for the x-variables iteratively through a sequence of LPs.

2. Once all the x-variables are fixed, find a solution (to determine how to divide subbands and the flow routing) based on the x-variable values obtained in Step 1.

Such a two-step approach will yield a suboptimal (upper bounding) solution to the original MINLP problem. The quality of this algorithm can be assessed by comparing the resulting objective value to the lower bound that we derived in the previous section.

As said, the key to the two-step approach resides in the determination of the binary values for all the $x_{ij}^{(m,k)}$-variables. Our main idea is to fix the values of the $x_{ij}^{(m,k)}$-variables *sequentially* by solving a series of relaxed LP problems and setting at least one binary value for some $x_{ij}^{(m,k)}$-variable during each iteration. Specifically, during the first iteration, we relax all the binary variables $x_{ij}^{(m,k)}$ to satisfy $0 \leq x_{ij}^{(m,k)} \leq 1$, as in Section 10.4, to derive an LP problem. Upon solving this LP problem, we obtain a solution with each $x_{ij}^{(m,k)} = s_{ij}^{(m,k)}/u^{(m,k)}$ being a value between zero and one. Among all these x-variable values, we select some $x_{ij}^{(m,k)}$ that has the largest value. Then we fix (set) this particular $x_{ij}^{(m,k)}$ to 1. As a result of this fixing, by (10.8), we also need to fix $x_{iq}^{(m,k)} = 0$ for $q \in \mathcal{T}_i^m$ and $q \neq j$. Further, by (10.9), we can fix $x_{pq}^{(m,k)}$ to 0 for $p \in \mathcal{I}_j^m$, $p \neq i$, and $q \in \mathcal{T}_p^m$.

Algorithm 10.1 An SF algorithm

1. Set up and solve the initial relaxed LP problem, as shown in Section 10.4.

2. Suppose $x_{ij}^{(m,k)}$ has the largest value among all the x-variables that remain to be fixed; fix this $x_{ij}^{(m,k)} = 1$.
   Also, fix $x_{iq}^{(m,k)} = 0$ (for $q \in \mathcal{T}_i^m$ and $q \neq j$) and $x_{pq}^{(m,k)} = 0$ (for $p \in \mathcal{I}_j^m$, $p \neq i$, and $q \in \mathcal{T}_p^m$).

3. If all the $x_{ij}^{(m,k)}$-variables are fixed, go to Step 5.

4. Reformulate and solve a new relaxed LP problem with the newly fixed x-variables and go to Step 2.

5. Formulate and solve the LP problem based on the fixed $x_{ij}^{(m,k)}$-variable values.

Now, having fixed some x-variables in the first iteration, we update the problem to obtain a new LP problem for the second iteration as follows. For those $x_{ij}^{(m,k)}$-variables that are already fixed to one, since $s_{ij}^{(m,k)} = x_{ij}^{(m,k)} u^{(m,k)} = u^{(m,k)}$, we can replace the corresponding $s_{ij}^{(m,k)}$ by $u^{(m,k)}$. For those $x_{iq}^{(m,k)}$ and $x_{pq}^{(m,k)}$ that are fixed to zero, we can set $s_{iq}^{(m,k)} = 0$ and $s_{pq}^{(m,k)} = 0$. As a result,

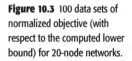

Figure 10.3 100 data sets of normalized objective (with respect to the computed lower bound) for 20-node networks.

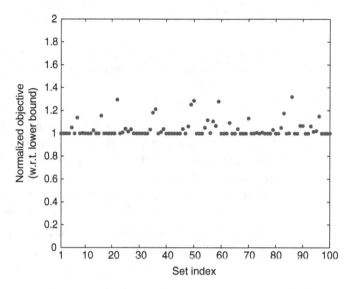

all the terms in the LP problem involving these s-variables can be removed and the corresponding constraints in (10.10) and (10.11) can also be removed.

In the second iteration, we solve this new LP problem and then fix some additional x-variables based on the same process (now the ordering of the x-values is done only for the remaining not-yet-fixed or free x-variables). The iteration continues and eventually we fix all the x-variables to either zero and one.

Upon fixing all the x-variable values, the original MINLP reduces to an LP problem, which can be solved in polynomial time. The complete SF algorithm is displayed in Fig. 10.3.

10.5.2 A speedup technique

In the SF algorithm, we need to solve a polynomial sequence of LPs, which yields polynomial-time complexity. By exploiting the spatial and spectral dimensions involved in radio resource allocation, we may decrease the number of LPs by fixing more x-variables during each iteration in Fig. 10.3. As a result, the number of iterations can be further decreased. From the spatial dimension, a subband usage will only have an impact within the interference range, and the same subband can be used by other links outside this range. Thus, for the same subband (m, k), we may fix *multiple* links that have nonoverlapping interference ranges within a single iteration of the SF algorithm. From the spectral dimension, the transmission on one subband will not interfere with the transmission on a different subband. Thus, for the same link $i \rightarrow j$, we may fix multiple subbands within a single iteration of the SF algorithm. Specifically, we can use a threshold $\alpha > 0.5$ in this fixing process and fix all the x-variables that exceed α to one in a single iteration. Note that in (10.8) and (10.9), it is

Table 10.2 Available bands $\mathcal{M}$ in the network in the simulation study.

Band index	Spectrum range (MHz)	Bandwidth (MHz)
I	[1240, 1300]	60.0
II	[1525, 1710]	185.0
III	[902, 928]	26.0
IV	[2400, 2483.5]	83.5
V	[5725, 5850]	125.0

required that at most one binary x-variable equals one, whereas in the relaxed problem, there is at most one fraction $s_{iq}^{(m,k)}/u^{(m,k)}$ that exceeds 0.5. Thus, $\alpha > 0.5$ ensures that both the constraints (10.8) and (10.9) (interference constraints at each node and among the nodes) will hold during the SF procedure. We used $\alpha = 0.85$ in our numerical results in the next section. In the case that none of the x-variables exceeds α, we will fall back to the basic algorithm in Fig. 10.3 and simply choose the largest valued x-variable.

10.6 Numerical examples

In this section, we present numerical results for our SF algorithm and compare it to the lower bound that we obtained in Section 10.4. We consider $|\mathcal{N}| = 20, 30$, or 40 nodes in a 500×500 area (in meters). Among these nodes, there are $|\mathcal{L}| = 5$ active sessions, each having a rate that is randomly generated within [10, 100] Mb/s.

We assume that there are $M = 5$ bands that can be used for the entire network (see Table 10.2). Bands I and II are among the least-utilized spectrum bands found in [106] (less than 2%), and bands III, IV, and V are unlicensed ISM bands used for 802.11. Recall that the set of available bands at each CR node is a subset of these five bands based on the node's location, and that the set of available bands at any two nodes in the network may not be identical. In the simulation, this is done by randomly selecting a subset of bands from the pool of five bands for each node. Further, we assume that bands I to V can be divided into $3, 5, 2, 4$, and 4 subbands, although other desirable divisions can be used. Note that the size of each subband may be unequal and is part of the optimization problem.

We assume that the transmission range at each node is 100 m and that the interference range is 150 m, although other settings can be used. The path-loss index α is assumed to be 4 and we set $\beta = 62.5$. The threshold ρ_T is assumed to be 10η. Thus, we have $\rho_I = \left(\frac{100}{150}\right)^{\alpha} \rho_T$, and the transmission power spectral density is given by $\rho = (100)^{\alpha} \rho_T / \beta = 1.6 \cdot 10^7 \eta$.

Note that it is possible that there is no feasible solution for a specific data set. This could be attributed to the loss of connectivity in certain areas in the network (due to random network topology), resource bottleneck in a hot area, etc. Thus, we only report results based on those data sets that have feasible solutions.

We first present simulation results for 100 data sets for 20-node networks that can produce feasible solutions. For each data set, the network topology, the source–destination pairs and bit rates of all sessions, and the available frequency bands at each node are randomly generated. We used the SF algorithm to determine a feasible solution, and we compared its objective value with the lower bound derived as in Section 10.4. The run time consumed by each simulation was less than 10 seconds on a Pentium 3.4 GHz machine.

Fig. 10.3 depicts the normalized objective values obtained by the SF algorithm with respect to the computed lower bounds for 100 data sets. The average normalized objective value obtained among the 100 simulations is 1.04 and the standard derivation is 0.07. There are two observations that can be made from this figure. First, since the ratio of the solution value obtained by SF (upper bound on the optimal solution value) to the lower bound value is close to 1 (in many cases, they coincide with each other), the lower bound must be very tight. Second, since the actual (unknown) optimal solution value lies between the solution value obtained by the SF algorithm and the lower bound, the SF solution value must be even closer to the optimal value than that indicated by the foregoing ratio.

To get a sense of how the actual (rather than the normalized) numerical results appear in the simulations, we list the 100 sets of results in Table 10.3. Note that in many cases, the result obtained by the SF algorithm is identical to the respective lower bound obtained via the underlying relaxation, which indicates that the solution found by SF in such cases is optimal.

Simulation results for the 100 random data sets corresponding to the 30-node and the 40-node networks that produce feasible solutions are shown in Figs. 10.4 and 10.5, respectively. For the 30-node networks, the average normalized value obtained was 1.10, with a standard derivation of 0.16. For the 40-node networks, the average normalized value obtained was 1.18, with a standard derivation of 0.16. Thus, the derived SF solutions are again close to optimality for these data sets.

10.7 Chapter summary

In this chapter, we presented an effective approach to address a class of mixed-integer optimization problems. The technique, called sequential fixing is designed to iteratively determine (fix) binary integer variables. It is a heuristic procedure and has a polynomial-time complexity. Its performance is

Table 10.3 Simulation results (in MHz) of 100 data sets for 20-node networks.

Set index	Lower bound	Result by SF	Set index	Lower bound	Result by SF	Set index	Lower bound	Result by SF
1	138.33	138.33	35	270.76	319.71	68	214.82	214.82
2	156.12	156.12	36	325.59	394.43	69	177.85	177.85
3	173.53	173.53	37	288.72	288.72	70	233.50	264.12
4	189.70	189.70	38	244.77	247.74	71	298.32	298.32
5	203.05	213.18	39	215.72	223.83	72	215.45	215.45
6	184.37	184.37	40	126.05	126.05	73	246.76	248.23
7	160.45	182.33	41	191.79	191.79	74	270.04	270.04
8	232.23	232.23	42	171.11	171.11	75	163.74	165.16
9	223.00	223.53	43	157.95	157.95	76	213.42	213.42
10	182.13	182.13	44	238.02	238.02	77	151.64	151.64
11	220.20	220.20	45	129.96	219.96	78	369.62	369.62
12	277.83	277.83	46	244.70	254.07	79	110.70	114.29
13	130.54	134.05	47	293.82	293.82	80	180.11	180.11
14	172.62	172.62	48	372.99	396.06	81	162.15	162.15
15	256.96	256.96	49	240.53	301.21	82	183.67	192.64
16	178.73	178.73	50	367.20	472.34	83	366.16	430.61
17	152.08	152.08	51	128.39	128.39	84	130.76	130.76
18	359.03	359.03	52	238.61	238.61	85	420.90	421.74
19	150.61	150.61	53	127.82	127.82	86	177.20	234.00
20	164.97	164.97	54	195.19	205.00	87	205.77	205.77
21	156.43	156.43	55	208.53	232.90	88	211.40	211.40
22	238.41	308.51	56	188.86	189.51	89	226.85	242.22
23	184.78	184.78	57	170.53	188.55	90	164.47	175.27
24	241.42	243.22	58	183.51	195.68	91	203.80	203.80
25	135.39	140.96	59	376.52	481.59	92	267.63	267.63
26	247.30	251.18	60	265.62	265.62	93	157.91	167.85
27	280.80	290.85	61	209.20	209.20	94	303.65	308.71
28	353.98	354.17	62	139.66	139.73	95	225.93	230.63
29	260.56	260.56	63	168.52	184.07	96	194.93	224.04
30	127.06	127.06	64	111.29	111.29	97	160.25	160.25
31	170.35	170.35	65	118.07	118.07	98	165.02	165.02
32	207.74	207.74	66	179.47	186.34	99	99.79	99.79
33	183.59	183.59	67	100.05	100.05	100	168.68	168.68
34	138.33	143.00						

typically measured by comparing its solution value to some performance bound, e.g., a lower bound for a minimization problem, or an upper bound for a maximization problem. Based on our own experience, we found that this SF technique is very efficient and can offer highly competitive solutions.

As a case study, we studied an optimization problem in a multi-hop CRN. Since the problem formulation is an MINLP, we developed a lower bound to estimate the optimal objective value. Subsequently, we presented an SF algorithm for this optimization problem. Numerical examples showed that the

Figure 10.4 100 data sets of normalized objective (with respect to the computed lower bound) for 30-node networks.

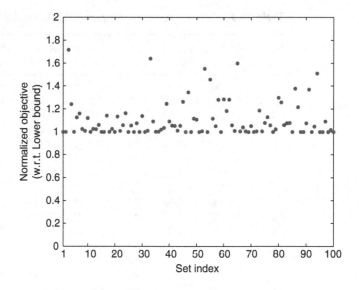

Figure 10.5 100 data sets of normalized objective (with respect to the computed lower bound) for 40-node networks.

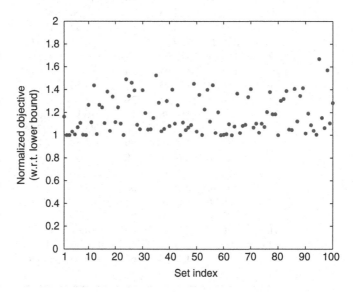

solutions produced by this SF algorithm can offer objective values that are very close to the computed lower bounds, thus confirming their near-optimality.

10.8 Problems

10.1 For the problem considered in this chapter, what is the purpose of developing a lower bound? If a solution obtained by the SF algorithm is very close to the lower bound, what can we claim?

10.2 Why is it necessary to divide a large band into smaller subbands in a CRN?

10.3 Show that, for a session, if flow balance holds at the source node (Eq. (10.4)) and at the relay nodes (Eq. (10.5)), then flow balance also holds at the session's destination node (Eq. (10.6)).

10.4 In this chapter, we used the so-called network-wide bandwidth-footprint-product (BFP) as the objective function. However, it is possible that two or more neighboring nodes that transmit on the same frequency band have some partial overlap of their footprints. As a result, the overlapped area will be counted multiple times in the objective function. Provide a justification on why this objective function still makes sense.

10.5 The capacity $c_{ij}^{(m,k)}$ on link $i \to j$ and subband (m, k) can be expressed as

$$c_{ij}^{(m,k)} = u^{(m,k)} W^m \log_2 \left(1 + \frac{g_{ij}\rho}{\eta}\right).$$

Referring to the RHS, explain why there is no interference term in the denominator inside the log function when calculating the SINR.

10.6 For the formulation of the original problem presented in this chapter, classify the constraints based on the different layers. Which constraints couple multiple layers?

10.7 In the SF algorithm, we fix the $x_{ij}^{(m,k)}$-variable having the largest value among all undetermined x-variables to 1 during each iteration. Comment on the alternative strategy of fixing the $x_{ij}^{(m,k)}$-variable having the smallest value among all undetermined x-variables to 0. Which is a more effective technique and why?

10.8 To speed up variable fixing, a threshold α is used to fix several undetermined x-variables to 1. Discuss the pros and cons of setting a large vs. a small value of α.

10.9 In the numerical examples, the solution obtained by the SF algorithm coincides with the lower bound solution in many cases. What can we infer from this observation and why? Also, if the SF algorithm fails to find a solution, can we claim that the original problem is infeasible? Why, or why not?

10.10 Explain why the SF algorithm is of polynomial-time complexity. Derive an order of complexity for this algorithm.

10.11 Consider a mixed-integer optimization problem where the goal is to maximize an objective function rather than to minimize it. Describe how to design an SF algorithm for this maximization problem.

11

Metaheuristic methods

One day your life will flash before your eyes. Make sure it is worth watching.

Unknown

11.1 Review of key results in metaheuristic methods

In this chapter, we discuss another class of heuristics, which are known as *metaheuristic* methods [36]. An iteration in metaheuristic methods typically aims to improve the current feasible solution, with the initial solution given by the user. Some well-known metaheuristic methods are *iterative improvement, simulated annealing, tabu search,* and *genetic algorithms* [36]. For certain type of problems, metaheuristic methods could be very effective.

The so-called *iterative improvement* (or basic local search) method tries to find a better solution in each iteration by searching in the neighborhood of the current solution, and terminates when a better solution cannot be found. It has been shown that the performance of iterative improvement methods for combinatorial optimization problems may not be satisfactory [19]. This can be explained by the fact that this method tends to stop as soon as it finds a local optimum.

Compared to iterative improvement, *simulated annealing* (SA) [1] has an explicit strategy to escape from local optima. The basic idea of SA is to allow a move (with a probability) even if it may tentatively result in a solution of worse quality than the current solution. There is also a *cooling procedure* in SA, which decreases such randomness (or diversification) as time passes. As the cooling proceeds, SA gradually converges to a simple iterative improvement algorithm, which guarantees convergence. The performance of SA is sensitive to the initial solution and the neighborhood structure (in addition to the cooling procedure).

Figure 11.1 Flow chart of GA.

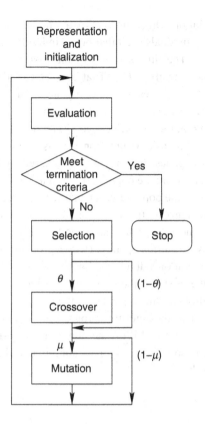

Another well-known technique is *tabu search* (TS), which explicitly uses the history of the search (recorded in a list of solutions declared as "tabu") both to escape from local optima and to implement an exploratory strategy [51]. Similar to SA and iterative improvement, the TS approach is a single-point-search metaheuristic (also called "trajectory methods"). The performance of TS strongly depends on the initial solution and the neighborhood structure, and on the *diversification* and *intensification* schemes that it uses to explore new or previous promising portions of the search region, respectively.

Unlike the above three metaheuristic methods, a *genetic algorithm* (GA) is a *population-based* algorithm that is inspired by the *survival-of-the-fittest* principle, as derived from its natural evolution context. Because of this algorithmic structure, a GA is also classified as an *evolutionary algorithm*. This procedure has the intrinsic strength of dealing with a set of solutions (i.e., a population) at each generation rather than working with a single, current solution. To represent a solution in a GA, we need to first define a data structure (i.e., a *gene*) and then map a solution to a sequence of genes (i.e., a *chromosome*). A GA needs a set of solutions (or individuals) as the first-generation elements. This is the initial step in Fig. 11.1. A GA evaluates each individual by a *fitness* function (in the form of an objective function). If the set of current individuals meets some termination criteria, then the procedure stops and returns the best individual as the final solution. Otherwise, a GA applies a number of genetic operators to the current individuals such that better individuals can be generated for the next generation. The fitness values of the first-generation solutions strongly influence the complexity of GA. That is, if we can identify a set of individuals having good fitness values for the first generation, then we may need a fewer number of iterations to derive a final solution.

In the next iteration, GA selects good individuals as parents based on their fitness values. The basic assumption here is that a good solution often contains some good genes that may also be shared by the optimal solution. To remove bad genes from the population, individuals with low fitness values will be removed out of the population. Then a GA uses a genetic operator known as *crossover* (with a probability θ) to recombine two or more selected individuals to produce new individuals that share good genes from the selected individuals. A GA also uses a *mutation* operator (with a probability μ) to achieve a randomized self-adaptation of individuals. Each new individual is evaluated and may be placed in the next population group if it has a good fitness value. Based on the survival-of-the-fittest principle, the overall quality of the population is likely to improve as the algorithm progresses from one generation to the next.

In this chapter, we will illustrate how a metaheuristic method such as GA can be applied to solve some complex combinatorial optimization problems in wireless networks.

11.2 Case study: Routing for multiple description video over wireless ad hoc networks

As progress in wireless ad hoc networking continues, there is an increasing expectation on enabling content-rich video communications in such networks. However, there are significant technical barriers that hinder the deployment of video applications in wireless ad hoc networks. In fact, what makes traditional single stream coding and layered coding successful in the internet and certain wireless networks is the existence of a relatively stable path during the video session. Consequently, packet loss on important information (e.g., base layer) is kept low, and can be effectively controlled by error control and concealment mechanisms. This is important since, for a layered video, the successful reconstruction of a video relies on the base layer, and the decoding of enhancement layers hinges upon lower enhancement layers as well as upon the base layer. However, this situation hardly holds true in wireless ad hoc networks, where there may not exist any single reliable path, and packet loss may be beyond the recovery capability of most error control mechanisms.

Recently, *multiple description* (MD) coding has become a popular coding technique for media streaming [57]. With MD coding, multiple streams (or descriptions) are generated for a video source and *any* received subset of these streams can be used to reconstruct the original video. Further, the quality of received video commensurate with the number of received descriptions. Such flexibility has positioned MD coding as a perfect candidate for video communications over wireless ad hoc networks [102]. This is because the topology of such networks is intrinsically mesh, within which multiple paths exist between any source and destination pair. Although the links in such networks are fragile, the probability of concurrent loss of all of the descriptions is low. Therefore, MD coding is likely to be very effective.

In this chapter, we study the problem of multi-path routing for double description (DD) video in wireless ad hoc networks. We follow a *cross-layer* approach in problem formulation by considering the application layer performance (i.e., average video distortion) as a function of lower layer parameters (e.g., bandwidth, loss, and path correlation). We show that the objective function is a complex ratio of high-order exponentials that is not decomposable. Consequently, it is futile to develop a tractable analytic solution.

However, we find that a metaheuristic technique such as a GA is eminently suitable in addressing such type of complex cross-layer optimization problems. This is because GAs possess an intrinsic capability of handling a *population* of solutions (rather than working with a single current solution during each iteration). Such capability gives GAs the unique strength of identifying promising regions in the search space (not necessarily convex) and having less of a tendency to be trapped in a local optimum, as compared with other trajectory-based metaheuristics (e.g., *simulated annealing* (SA) and *tabu search* (TS) [19]). Using numerical examples, we show that significant performance gains can be achieved by a GA-based approach over trajectory-based approaches.

The remainder of this chapter is organized as follows. In Section 11.3, we formulate a cross-layer optimization problem for DD video over multiple paths in ad hoc networks. In Section 11.4, we describe our GA-based solution. In Section 11.5, we present numerical examples. Section 11.6 summarizes this chapter.

11.3 Problem description

An ad hoc network can be modeled as a stochastic directed graph $\mathcal{G}\{V, E\}$, where V is the set of vertices and E the set of edges. We assume that nodes are reliable during the video session, but links may fail with certain probabilities. Further, we assume that orthogonal channels are used among the links so that interference among the links can be effectively managed by proper channel assignment. We characterize a link $\{i, j\} \in E$ by:

- b_{ij}: the available bandwidth of link $\{i, j\}$.
- p_{ij}: the probability that link $\{i, j\}$ is "up."
- l_{ij}: average burst length in a packet loss on link $\{i, j\}$.

Table 11.1 Notation

Symbol	Definition
b_{ij}	Bandwidth of link $\{i, j\}$
B_{jnt}	Minimum bandwidth of the shared links
d_0	Distortion when both descriptions are received
d_h	Distortion when only description h is received, $h = 1, 2$
D	Average distortion
E	Set of edges in the network
g_i	An intermediate node in a path
$\mathcal{G}\{V, E\}$	Graph representation of the network
$\{i, j\}$	A link from node i to node j
l_{ij}	Average length of loss burst on link $\{i, j\}$
p_{ij}	Success probability of link $\{i, j\}$
p_{jnt}	Average success probability of joint links
p_{dj}^h	Average success probability of disjoint links on $\mathcal{P}_h$
$\mathcal{P}$	A path from s to t
P_{00}	Probability of receiving both descriptions
P_{01}	Probability of receiving description 1 only
P_{10}	Probability of receiving description 2 only
P_{11}	Probability of losing both descriptions
R_h	Rate of description h in bits/sample
s	Source node
t	Destination node
T_{on}	Average "up" period of the joint links
V	Set of vertices in the network
x_{ij}^h	Routing index variables, defined in (11.10)
α_{ij}	"up" to "down" transition probability of link $\{i, j\}$
β_{ij}	"down" to "up" transition probability of link $\{i, j\}$
θ	GA crossover rate
μ	GA mutation rate

From these link-level statistics, we can explore path-level bandwidth and failure probabilities, which are the key factors to determine video distortion (see (11.2)). Other link characteristics, such as delay, jitter, congestion, and signal strength may also be incorporated into this framework as well (e.g., see [103]). Table 11.1 lists the notation used in this chapter.

11.3.1 Rate-distortion regions for DD coding

Throughout this chapter, we use DD coding for MD video, which is widely used in practice [8; 12; 23; 53; 102]. In general, using more descriptions and paths will increase the robustness to packet loss and path failure. But more descriptions may increase the video bit rate for the same video quality. A study in [135] demonstrates that the most significant performance gain is achieved when the number of descriptions increases from one to two, but with only marginal improvements achieved for further increase in the number of descriptions.

For video coding and communications, a distortion rate model describes the relationship between the achieved distortion and the bit rate (i.e., the quality and the length of the representation). For two descriptions (each generated for a sequence of video frames), denote d_h as the achieved distortion when only description h is received, $h = 1, 2$, and d_0 the distortion when both descriptions are received. Denote R_h as the rate in bits/pixel of description $h, h = 1, 2$. The rate-distortion region for a memoryless independent and identically distributed (i.i.d.) Gaussian source with the square error distortion measure was first introduced in [117]. For computational efficiency, Alasti *et al.* in [6] introduce the following distortion-rate function, which we use in this chapter:

$$\begin{cases} d_0 = \frac{2^{-2(R_1+R_2)}}{2^{-2R_1}+2^{-2R_2}-2^{-2(R_1+R_2)}} \cdot \sigma^2 \\ d_1 = 2^{-2R_1} \cdot \sigma^2 \\ d_2 = 2^{-2R_2} \cdot \sigma^2, \end{cases} \tag{11.1}$$

where σ^2 is the variance of the source. Denote P_{00} as the probability of receiving both descriptions, P_{01} as the probability of receiving description 1 only, P_{10} as the probability of receiving description 2 only, and P_{11} as the probability of losing both descriptions. Then, the average distortion of the received video can be approximated as:

$$D = P_{00} \cdot d_0 + P_{01} \cdot d_1 + P_{10} \cdot d_2 + P_{11} \cdot \sigma^2. \tag{11.2}$$

Our simulation results show that although the distortion-rate function in (11.1) is an approximation for DD video, significant improvement in received video quality could be achieved over alternative approaches by incorporating (11.2) into the optimal routing problem formulation (see Section 11.5). Further, our formulation does not depend on any specific distortion-rate function. A more accurate distortion-rate function for MD video could be incorporated into our formulation should it become available in the future.

11.3.2 Description rates and success probabilities

As a first step to formulate the problem of optimal multi-path routing, we need to know how to compute the average distortion D as a function of link statistics for a *given* pair of paths. That is, we need to compute the end-to-end bandwidth (or rate) for each stream and joint probabilities of receiving the descriptions (see (11.1) and (11.2)).

For a source–destination pair $\{s, t\}$, consider two given paths $[\mathcal{P}_1, \mathcal{P}_2]$ in $\mathcal{G}\{V, E\}$. Since we do not mandate "disjointedness" in routing, $\mathcal{P}_1$ and $\mathcal{P}_2$ may share nodes and links. Similar to the approach in [8] and [12], we classify the links along the two paths into three sets: set one consisting of links shared by both paths, denoted as $\mathcal{J}(\mathcal{P}_1, \mathcal{P}_2)$, and the other two sets consisting of disjoint links on the two paths, denoted as $\bar{\mathcal{J}}(\mathcal{P}_h)$, $h = 1, 2$, respectively. Then, the minimum bandwidth of $\mathcal{J}(\mathcal{P}_1, \mathcal{P}_2)$, B_{jnt}, is

$$B_{jnt} = \begin{cases} \min_{\{i,j\} \in \mathcal{J}(\mathcal{P}_1, \mathcal{P}_2)}\{b_{ij}\}, & \text{if } \mathcal{J}(\mathcal{P}_1, \mathcal{P}_2) \neq \emptyset, \\ \infty & \text{otherwise.} \end{cases} \tag{11.3}$$

The rates of the two video streams, R_1 and R_2, can be computed as

$$\begin{cases} R_h = \rho \cdot B(\mathcal{P}_h), & \text{if } \sum_{m=1}^{2} B(\mathcal{P}_m) \leq B_{jnt}, h = 1, 2, \\ R_1 + R_2 \leq \rho \cdot B_{jnt} & \text{otherwise,} \end{cases} \tag{11.4}$$

where $B(\mathcal{P}_h) = \min_{\{i,j\} \in \mathcal{P}_h}\{b_{ij}\}$, $h = 1, 2$, and ρ is a constant determined by the video format and frame rate. For a video with coding rate f frames/s and a resolution of $W \times V$ pixels/frame, we have $\rho = 1/(\kappa \cdot W \cdot V \cdot f)$, where κ is a constant determined by the chroma subsampling scheme. For the quarter common intermediate format (QCIF) [176×144 Y pixels/frame, 88×72 Cb/Cr pixels/frame], we have $\kappa = 1.5$ and $\rho = 1/(1.5 \cdot 176 \cdot 144 \cdot f)$. The first line in (11.4) is for the case when the joint links are not the bottleneck of the paths. The second line of (11.4) is for the case where one of the joint links is the bottleneck of both paths. In the latter case, we assign the bandwidth to the paths by splitting the bandwidth of the shared bottleneck link in proportion to the mean success probabilities of the two paths, while an alternative approach is to split the bandwidth evenly for balanced descriptions.

We now focus on how to compute the end-to-end success probabilities. For disjoint portion of the paths, it suffices to model packet loss as a Bernoulli event. This is because the loss of the two descriptions is assumed to be independent in the disjoint portions. Thus, the success probability on the disjoint portions of the two paths is

$$p_{dj}^h = \begin{cases} \prod_{\{i,j\} \in \bar{\mathcal{J}}(\mathcal{P}_h)} p_{ij}, & \text{if } \bar{\mathcal{J}}(\mathcal{P}_h) \neq \emptyset, h = 1, 2, \\ 1 & \text{otherwise, } h = 1, 2. \end{cases} \tag{11.5}$$

On the joint portion of the paths, loss of the two streams is correlated. To model such correlation, we assume that each shared link $\{i, j\}$ follows an on–off process modulated by a discrete-time Markov chain, as shown in Fig. 11.2(a). With this model, there is no packet loss when the link is in the "up" state; all packets are dropped when the link is in the "down" state. Transition probabilities, $\{\alpha_{ij}, \beta_{ij}\}$, can be computed from the link statistics as $\beta_{ij} = 1/l_{ij}$ and $\alpha_{ij} = (1 - p_{ij})/(p_{ij}l_{ij})$.

For K shared links, the aggregate failure process of these links is a Markov process with 2^K states. In order to simplify the computation, we follow the well-known Fritchman model [43] to model the aggregate process as an on–off process. Since a packet is successfully delivered on the joint portion if and only if all joint links are in the "up" state, we can lump up all the states with at least one link failure into a single "down" state, while using the remaining state

Figure 11.2 Link and path models.

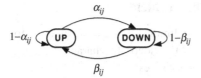

(a) The Gilbert two-state link model.

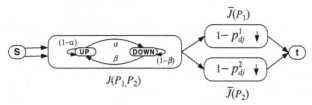

(b) A simplified path model for double-description video.

where all the links are in good condition as the "up" state. Denote T_{on} as the average length of the "up" period. We have

$$T_{on} = \frac{1}{1 - \prod_{\{i,\,j\} \in \mathcal{J}(\mathcal{P}_1,\,\mathcal{P}_2)}(1 - \alpha_{ij})}. \tag{11.6}$$

If the joint link set is not empty, the probability of a successful delivery on the joint links can be written as

$$p_{jnt} = \begin{cases} \prod_{\{i,\,j\} \in \mathcal{J}(\mathcal{P}_1,\,\mathcal{P}_2)} p_{ij}, & \text{if } \mathcal{J}(\mathcal{P}_1, \mathcal{P}_2) \neq \emptyset, \\ 1 & \text{otherwise.} \end{cases} \tag{11.7}$$

Finally, the transition probabilities of the aggregate on–off process are

$$\alpha = \frac{1}{T_{on}} \quad \text{and} \quad \beta = \frac{p_{jnt}}{T_{on}(1 - p_{jnt})}. \tag{11.8}$$

Note that $\alpha = 0$ and $\beta = 0$ if $\mathcal{J}(\mathcal{P}_1, \mathcal{P}_2) = \emptyset$. The consolidated path model is illustrated in Fig. 11.2(b), where $\mathcal{J}(\mathcal{P}_1, \mathcal{P}_2)$ is modeled as a two-state Markov process with parameters $\{\alpha, \beta\}$, and $\bar{\mathcal{J}}(\mathcal{P}_h)$ is modeled as a Bernoulli process with parameter $(1 - p_{dj}^h)$, $h = 1, 2$.

With the consolidated path model, the joint probabilities of receiving the descriptions are

$$\begin{cases} P_{00} = p_{jnt} \cdot (1 - \alpha) \cdot p_{dj}^1 \cdot p_{dj}^2 \\ P_{01} = p_{jnt} \cdot p_{dj}^1 \cdot \left[1 - (1 - \alpha) \cdot p_{dj}^2 \right] \\ P_{10} = p_{jnt} \cdot \left[1 - (1 - \alpha) p_{dj}^1 \right] \cdot p_{dj}^2 \\ P_{11} = 1 - p_{jnt} \cdot \left[p_{dj}^1 + p_{dj}^2 - (1 - \alpha) \cdot p_{dj}^1 \cdot p_{dj}^2 \right]. \end{cases} \tag{11.9}$$

11.3.3 The optimal multi-path routing problem

With the above preliminaries, we now set out to formulate the multi-path routing problem for MD video. To characterize any s–t path $\mathcal{P}_h$, we define the following binary variables:

$$x_{ij}^h = \begin{cases} 1, & \text{if } \{i, j\} \in \mathcal{P}_h, \\ 0 & \text{otherwise.} \end{cases} \tag{11.10}$$

With these variables, an arbitrary path $\mathcal{P}_h$ can be represented by a vector $\mathbf{x}^h$ of $|E|$ elements, each of which corresponds to a link and has a binary value. We formulate the problem as follows:

OPT-MM

Minimize $D = P_{00} \cdot d_0 + P_{01} \cdot d_1 + P_{10} \cdot d_2 + P_{11} \cdot \sigma^2$ $\tag{11.11}$

$$\text{subject to} \sum_{j:\{i,j\}\in E} x_{ij}^h - \sum_{j:\{j,i\}\in E} x_{ji}^h = \begin{cases} 1, & \text{if } i = s, \quad i \in V, \, h = 1, 2, \\ -1, & \text{if } i = t, \quad i \in V, \, h = 1, 2, \\ 0 & \text{otherwise, } i \in V, \, h = 1, 2 \end{cases}$$
$$\tag{11.12}$$

$$\sum_{j:\{i,j\}\in E} x_{ij}^h \begin{cases} \leq 1, & \text{if } i \neq t, \, i \in V, \, h = 1, 2 \\ = 0, & \text{if } i = t, \, i \in V, \, h = 1, 2 \end{cases} \tag{11.13}$$

$$x_{ij}^1 \cdot R_1 + x_{ij}^2 \cdot R_2 \leq \rho \cdot b_{ij}, \quad \{i, j\} \in E \tag{11.14}$$

$$x_{ij}^h \in \{0, 1\}, \quad \{i, j\} \in E, \, h = 1, 2. \tag{11.15}$$

In problem OPT-MM, $\{x_{ij}^h\}$ are binary optimization variables. Constraints (11.12) and (11.13) ensure that the paths are loop-free, while constraint (11.14) ensures the links are stable. For a given pair of paths, the average video distortion D is determined by the end-to-end statistics and the correlation of the paths, as given in (11.1), (11.4), and (11.9). Different statistics of a given pair of paths lead to a different video distortion. Specifically, the larger the end-to-end bandwidth is, the higher the video rate is, and the smaller the distortion is. With a lower end-to-end loss rate, fewer video frames will be corrupted. This is modeled in (11.11), where σ^2 is usually much larger than d_0, d_1, and d_2, and d_h is usually larger than d_0, $h = 1, 2$. Finally, the impact of path correlation is actually considered in the derivation of the joint probabilities of receiving the description. In Problem OPT-MM, all these three elements are integrated in the objective function (11.11), and are jointly optimized in routing.

The objective function (11.11) is a highly complex ratio of high-order exponentials of the x-variables. The objective evaluation of a pair of paths involves identifying the joint and disjoint portions, which is only possible when both paths are completely determined (or can be conditioned on the exceedingly complex products of the binary factors x_{ij}^1 and $(1 - x_{ij}^1)$ with x_{ij}^2 and $(1 - x_{ij}^2)$). In [145], Sherali *et al.* considered a problem that seeks a pair of disjoint paths in a network such that the total travel time over the paths is minimized, where the travel time on a link might be either a constant or a non-decreasing (or unstructured) function of the time spent on the previous links traversed. Even for a simple special case where all the links except one have a constant travel time (and hence linear objective terms), the problem was shown to be NP-hard. Our problem has much more complex relationships pertaining to the contribution of each individual link to the objective function, which depends in general on the other links that are included in a fashion that has no particular structural property such as convexity. Hence, it is likely to be NP-hard as well. However, we do not have a proof in this chapter.

11.4 A metaheuristic approach

In this section, we present a solution procedure that produces a pair of feasible and near-optimal paths.

We find that GAs [9] are eminently suitable for addressing this type of complex combinatorial problems, most of which are multi-modal and nonconvex. GAs are *population-based* metaheuristic inspired by the *survival-of-the-fittest* principle. It has the intrinsic strength of dealing with a set of solutions (i.e., a population) at each step rather than working with a single, current solution. At each iteration, a number of genetic operators are applied to the individuals of the current population in order to generate individuals for the next generation. In particular, GA uses genetic operators known as *crossover* to recombine two or more individuals to produce new individuals, and *mutation* to achieve a randomized self-adaptation of individuals. The driving force in GA is the *selection* of individuals based on their fitness (in the form of an objective function) for the next generation. The survival-of-the-fittest principle ensures that the overall quality of the population improves as the algorithm progresses from one generation to the next.

We follow the flow chart in Fig. 11.1 to design a GA-based solution to the MD multi-path routing problem. In what follows, we use an example ad hoc network shown in Fig. 11.3(a) to illustrate the components in a GA-based solution. The termination condition in Fig. 11.1 could be based on the total number of iterations (generations), maximum computing time, a threshold of desired video distortion, or a threshold based on the lower bound developed in [104].

11.4.1 Solution representation and initialization

In a GA, a feasible solution is encoded in the genetic format. For a routing problem, a natural encoding scheme would be to define a node as a gene. Then, an end-to-end path, consisting of an ordered sequence of nodes (connected by the corresponding wireless links), can be represented as a chromosome [2].

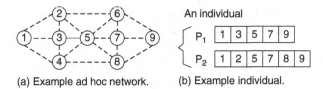

(a) Example ad hoc network.   (b) Example individual.

For problem OPT-MM, each feasible solution consists of a pair of paths (i.e., a pair of chromosomes), denoted as $[\mathcal{P}_1, \mathcal{P}_2]$. An individual in this case is a pair of vectors containing the nodes on paths $\mathcal{P}_1$ and $\mathcal{P}_2$ (see, e.g., Fig. 11.3(b)).

Before entering the main loop in Fig. 11.1, we need to generate an initial population, i.e., a set of solutions. A simple approach would be to generate this set of solutions by randomly appending feasible elements (i.e., nodes with connectivity) to a partial solution. Under this approach, each construction process starts with source node s. Then, the process randomly chooses a link incident to the current end-node of the partial path and appends the link with its corresponding head-node to augment the path, until destination node t is reached. It is important to ensure that the intermediate partial path is loop-free during the process. After generating a certain set of paths for s–t independently, a population of individuals can be constructed by pairing paths from this set. Our numerical results show that a properly designed GA is not very sensitive to the quality of the individuals in the initial population.

11.4.2 Evaluation

The fitness function $f(\bar{x})$ of an individual, $\bar{x} = [\mathcal{P}_1, \mathcal{P}_2]$, is closely tied to the objective function (i.e., distortion D). Since the objective is to minimize the average distortion function D, we use a fitness function which is defined as the inverse of the distortion value, i.e., $f(\bar{x}) = 1/D(\bar{x})$. This simple fitness definition appears to work very well, although other better fitness definitions may be available.

11.4.3 Selection

During this operation, GA selects individuals that have a better chance or potential to produce "good" offspring in terms of their fitness values. By virtue of the selection operation, "good" genes among the population are more likely to be passed to the future generations. We use the so-called *Tournament* selection [9] scheme, which randomly chooses m individuals from the population each time, and then selects the best of these m individuals in terms of their fitness values. By repeating either procedure multiple times, a new population can be selected.

Figure 11.4 An example of the crossover operation.

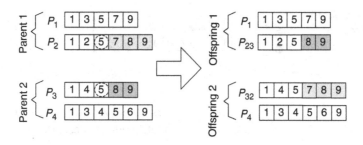

11.4.4 Crossover

Crossover mimics the genetic mechanism of reproduction in the natural world, in which genes from parents are recombined and passed to offspring. The decision of whether or not to perform a crossover operation is determined by the *crossover rate* θ.

Fig. 11.4 illustrates one possible crossover implementation. Suppose that we have two-parent individuals $x_1 = [\mathcal{P}_1, \mathcal{P}_2]$ and $x_2 = [\mathcal{P}_3, \mathcal{P}_4]$. We could randomly pick one path in x_1 and one in x_2, say $\mathcal{P}_2$ and $\mathcal{P}_3$. If one or more common nodes exist in these two chosen paths, we could select the first such common node that exists in $\mathcal{P}_2$, say g_r, where $g_r \notin \{s, t\}$, and we can then concatenate nodes $\{s, \ldots, g_r\}$ from $\mathcal{P}_2$ with nodes $\{g_{r+1}, \ldots, t\}$ in $\mathcal{P}_3$ (where g_{r+1} denotes the next downstream node of g_r in $\mathcal{P}_3$) to produce a new path $\mathcal{P}_{23}$. Likewise, using the first such node $g_{r'}$ in $\mathcal{P}_3$ that repeats in $\mathcal{P}_2$ (which may be different from g_r), we can concatenate the nodes $\{s, \ldots, g_{r'}\}$ from $\mathcal{P}_3$ with the nodes $\{g_{r'+1}, \ldots, t\}$ in $\mathcal{P}_2$ to produce a new path $\mathcal{P}_{32}$. It is important that we check the new paths to be sure that they are loop-free. The two offspring generated in this manner are $[\mathcal{P}_1, \mathcal{P}_{23}]$ and $[\mathcal{P}_{32}, \mathcal{P}_4]$. On the other hand, if $\mathcal{P}_2$ and $\mathcal{P}_3$ are disjoint, we could swap $\mathcal{P}_2$ with $\mathcal{P}_3$ to produce two new offspring $[\mathcal{P}_1, \mathcal{P}_3]$ and $[\mathcal{P}_2, \mathcal{P}_4]$.

11.4.5 Mutation

The objective of the mutation operation is to *diversify* the genes of the current population, which helps prevent the solution from being trapped in a local optimum. Just as some malicious mutations could happen in the natural world, mutation in GA may produce individuals that have worse fitness values. In such cases, some "filtering" operation is needed (e.g., the selection operation) to reject such "bad" genes and to drive GA toward optimality.

Mutation is performed on an individual with probability μ (called the *mutation rate*). For better performance, we propose a schedule to vary the mutation rate within $[\mu_{min}, \mu_{max}]$ over iterations (rather than using a fixed μ).

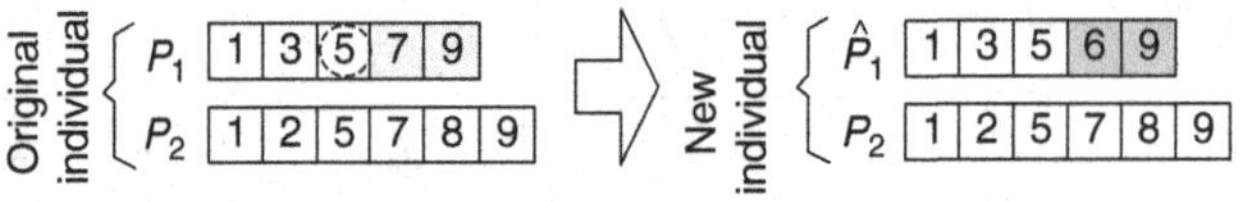

Figure 11.5 An example of the mutation operation.

The mutation rate is first initialized to μ_{max}; then as generation number k increases, the mutation rate gradually decreases to μ_{min}, i.e.,

$$\mu_k = \mu_{max} - \frac{k \cdot (\mu_{max} - \mu_{min})}{T_{max}}, \tag{11.16}$$

where T_{max} is the maximum number of generations. Such schedule of μ is similar to the cooling schedule used in SA, and yields better convergence performance for problem OPT-MM.

Fig. 11.5 illustrates a simple example of the mutation operation. In this example, we could implement mutation as follows. First, we choose a path $\mathcal{P}_h$, $h = 1$ or 2, with equal probabilities. Then, we can randomly pick an integer value k in the interval $[2, |\mathcal{P}_h| - 1]$, where $|\mathcal{P}_h|$ is the cardinality of $\mathcal{P}_h$, and let the partial path $\{s, \ldots, g_k\}$ be $\mathcal{P}_h^u$, where g_k is the kth node along $\mathcal{P}_h$. Finally, we can use any constructive approach to build a partial path from g_k to t, denoted as $\mathcal{P}_h^d$, which does not repeat any node in $\mathcal{P}_h^u$ other than g_k. If no such alternative segment exists between g_k and t, we keep the path intact; otherwise, a new path can now be created by concatenating the two partial paths as $\mathcal{P}_h^u \cup \mathcal{P}_h^d$.

11.5 Numerical examples

In this section, we present some numerical examples for the GA-based solution. In each example, we generate a wireless ad hoc network topology by placing a number of nodes at random locations in a rectangular region, where connectivity is determined by the distance coverage of each node's transmitter. The source–destination nodes s and t are uniformly chosen from the nodes. For each link, the failure probability is uniformly chosen from [0.01, 0.3]; the available bandwidth is uniformly chosen from [100, 400] Kb/s, with 50 Kb/s steps; the mean burst length is uniformly chosen from [2;6]. A DD video codec is implemented and used in the simulations.

We set the GA's parameters as follows: the population size is 15; $\theta = 0.7$; μ is varied from 0.3 to 0.1 using the schedule described in Section 11.4; σ^2 is set to 1, since it does not affect path selection decisions. The GA is terminated after a predefined number of generations or after a prespecified computation time. The best individual found by the GA upon its termination is prescribed as the solution to problem OPT-MM.

Table 11.2 Comparison of the average distortions obtained by the GA-based routing and exhaustive search

Network size	Topo. 1 10-node	Topo. 2 10-node	Topo. 3 15-node	Topo. 4 15-node
Global Opt.	0.3308	0.2004	0.3863	0.2969
GA (average)	0.3330	0.2004	0.3937	0.2972
GA (std. dev.)	7.6e-6	0	2.8e-5	2.9e-6
Lower bound	0.2810	0.1832	0.3527	0.2444

11.5.1 Near-optimality

One important performance concern is the quality of the GA solutions. As discussed, due to the complex nature of problem OPT-MM, a closed-form optimal solution is not obtainable. However, for small networks, an optimal solution may be numerically obtained via an exhaustive search and can be used to compare with the proposed GA-based solutions.

Table 11.2 shows the optimal distortion values found by GA (each is the average of 30 runs) and by exhaustive search for two ten-node and two 15-node networks. We find that the solutions found by GA are very close to the global optimum in all cases. In addition, the standard deviation of the 30 GA results for the same network is very small. The average computational time for GA is 0.29 s for the ten-node network (about 60 generations) and 0.39 s for the 15-node network (about 70 generations) on a Pentium 4 2.4 GHz computer (512 MB memory) with MATLAB 6.5. For exhaustive search, the average computational time is 58.7 s for the ten-node case and 1877 s for the 15-node case. We also compute the lower bound [104] for each of the networks. The results are given in the last row of Table 11.2.

11.5.2 Comparison with trajectory methods

For comparison purposes, we implemented simulated annealing (SA) and tabu search (TS) for the same problem. We used the *geometric cooling schedule* for the SA implementation with a decay coefficient $\omega = 0.99$ [1]. For the TS implementation, we choose a *tabu list* of five for small networks and ten for large networks [51].

SA was initially motivated by an analogy between the way a piece of metal cools and freezes into a minimum energy crystalline structure (annealing process), and the search for a minimum in a more general system [1]. When SA explores the solution space, it accepts a nonimproving solution with a probability, which decreases with iterations. We use a probabilistic acceptance function

$$Pr\{\bar{x} \leftarrow \hat{x}\} = \begin{cases} 1, & \text{if } D(\hat{x}) < D(\bar{x}) \\ \exp\{\frac{-|D(\hat{x})-D(\bar{x})|}{c_k}\} & \text{otherwise,} \end{cases} \qquad (11.17)$$

where c_k is a control parameter analogous to temperature in a physical system, $\bar{x}$ is the current solution, and $\hat{x}$ is a perturbation of $\bar{x}$. The fashion in which c_k is manipulated is called the *cooling schedule*. The following cooling schedule is used in our experiments [1]:

1. $c_0 = 1$: i.e., nearly all transitions will be accepted at the beginning of the search process;
2. $c_{k+1} = \omega \cdot c_k$: i.e., the control parameter is decremented each time when a nonimproving solution is accepted, and remains at each value for a sufficient time for the system to "return to an equilibrium." ω is the decay coefficient. We set $\omega = 0.99$ for all experiments in this chapter.

Compared with SA, TS explicitly uses the history of the search, both to escape from local minima and to implement an exploratory strategy. Specifically, TS uses a *tabu list* to prevent returning to recently visited solutions, therefore avoiding endless cycling and possibly forcing the search process to accept nonimproving solutions [51]. In our experiments, we use a tabu list of five for small networks (e.g., ten-node networks) and ten for large networks (e.g., 50-node networks). The tabu list is implemented using a first-in-first-out queue. An explored solution is always inserted at the tail of the queue. When the queue is full, the head of the queue is removed.

In Fig. 11.6, we plot the evolution of distortion values obtained by GA, SA, and TS for a ten-node network and a 50-node network, respectively. All the three metaheuristics are terminated after running for 1 s. Upon termination, GA has evolved 210 generations in Fig. 11.6(a) and 75 generations in Fig. 11.6(b); SA ran for 1500 iterations in Fig. 11.6(a) and 700 iterations in Fig. 11.6(b); TS ran for 1050 iterations in Fig. 11.6(a) and 550 generations in Fig. 11.6(b). GA has fewer number of iterations than SA and TS. For both networks, the best distortion values found by GA are evidently much better than those by SA or TS. In Fig. 11.6(a), GA quickly converges to the global optimal, while both SA and TS are trapped at local optima (i.e., no further decrease in distortion value after hundreds of iterations). The same trend can be observed in the 50-node network case shown in Fig. 11.6(b), although the global optimum cannot be found here. We also plot the lower bounds in the figures.

An interesting observation from Fig. 11.6 is that for GA, the biggest improvement in distortion is achieved in the initial iterations, while the improvement gets smaller as GA evolves more generations. The initial population is generated using the random construction method discussed in Section 11.4.1, with no consideration of video performance. The initial solutions

Figure 11.6 Comparison of
distortion evolution of three
metaheuristic methods.

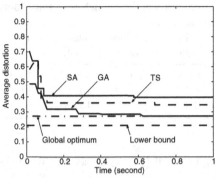

(a) Distortion evolution for a 10-node network.

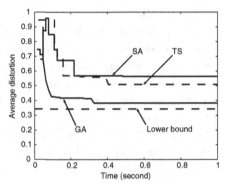

(b) Distortion evolution for a 50-node network.

usually have high distortion values. The distortion value quickly drops over iterations, indicating that the GA performance is not very sensitive to the quality of the initial population. Also note that the SA and TS curves increase at some time instances (e.g, the TS curve at 0.06 s in Fig. 11.6(a) and the SA curve at 0.08 s in Fig. 11.6(b)), which implies that a nonimproving solution is accepted in order to escape from local minima.

In addition to providing better solutions, another strength of GA over trajectory methods is that multiple "good" solutions can be found after a single run. Such extra good paths can be used as alternative (or backup) paths if needed.

11.5.3 Comparison with traditional multi-path routings

In this section, we compare the GA-based solution to traditional multi-path routings, which we call network-centric routing, due to its lack of cross-layer consideration. We implement two popular network-centric multi-path routing algorithms, namely k-shortest path (SP) routing (with $k = 2$ or 2-SP) [38] and disjoint path routing, Disjoint Pathset Selection Protocol (DPSP) [120]. Our 2-SP implementation uses hop count as the routing metric so that the two shortest paths are used. In our DPSP implementation, we set the link costs

Table 11.3 Comparison of GA and network-centric routing

Routing algorithm	$\mathcal{P}_1$ Succ. ratio	$\mathcal{P}_2$ Succ. ratio	Desc. 1 rate	Desc. 2 rate	Average PSNR
GA	0.994	0.952	350 Kb/s	350 Kb/s	29.71 dB
2-SP	0.798	0.782	100 Kb/s	200 Kb/s	23.42 dB
DPSP	0.965	0.793	100 Kb/s	100 Kb/s	25.65 dB

to $\log(1/p_{ij})$, for all $\{i, j\} \in E$, so that two disjoint paths with the highest end-to-end success probabilities are used. We compare the performance of our GA-based multi-path routing with these two algorithms over a 50-node ad hoc network using a video clip.

There are many ways to generate MD video (see [57] for an excellent survey). We choose a time domain partitioning coding scheme, where two descriptions are generated by separating the even-and odd-numbered frames and coding them separately. This simple time domain partitioning method is widely used in many video streaming studies [8; 12; 23; 102]. Compared to a traditional single description coder, this coder has a comparable computational complexity. Its coding efficiency is slightly lower than a single description coder, due to the fact that a longer motion prediction distance is used. However, this reduced coding efficiency is well justified by the resulting enhanced error resilience. We use an H.263+ like codec. Since our approach is quite general, we conjecture that the same trend in performance would be observed for other video codecs, such as H.264 or MPEG-2 or MPEG-4. The QCIF sequence "Foreman" (400 frames) is encoded at 15 fps for each description. A 10% macroblock level intra-refreshment is used. Each Group of Blocks (GOB) is carried in a different packet. The received descriptions are decoded and PSNR values of the reconstructed frames computed. When a GOB is corrupted, the decoder applies a simple error concealment scheme by copying from the corresponding slice in the most recent, correctly received frame.

The quality of the paths found by the three algorithms are presented in Table 11.3. The 2-SP algorithm has the worst performance in terms of path success probabilities. The DPSP algorithm has an improved success probability performance since it uses link success probabilities in routing. However, it may sacrifice path bandwidth while pursuing low-loss paths. As a result, it produces the lowest end-to-end bandwidths. We observe that our GA-based routing yields paths with much higher end-to-end success probabilities and end-to-end bandwidths, resulting in greatly improved video quality.

The PSNR curves of the received video frames are plotted in Fig. 11.7. We observe that the PSNR curve obtained by GA is well above those obtained by the aforementioned network-centric routing approaches. Using GA, the improvements in average PSNR value over 2-SP and DPSP are 6.29 dB and

Figure 11.7 PSNR curves of received video sequences.

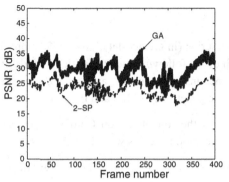

(a) GA-based algorithm versus 2-SP.

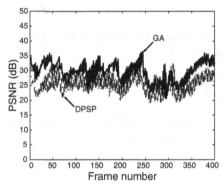

(b) GA-based algorithm versus DPSP.

4.06 dB, respectively. We also experiment with an improved 2-SP algorithm, which also uses link success probabilities as link metric (as in DPSP). In this case, our GA-based routing achieves a 1.27 dB improvement over this enhanced 2-SP version, which is still significant in terms of visual video quality.

11.6 Chapter summary

Metaheuristic methods are an important class of heuristic methods and have been applied to solve some very complex problems in wireless networks. In this chapter, we gave a review of some well-known metaheuristic methods (e.g., basic local search, SA, TS, and GA). In the case study, we focused on developing a GA-based method to solve a multi-path routing problem for MD video. We found that a GA-based solution is eminently suitable to address this particular problem, which involves complex objective functions and exponential solution space. By exploiting the survival-of-the-fittest principle, a GA-based solution is able to evolve to a population of better solutions after each iteration and eventually offers a near-optimal solution.

11.7 Problems

11.1   Describe (in greater detail than in Section 11.1) the basic concepts and approach of the following metaheuristic methods: iterative improvement, simulated annealing, tabu search, and genetic algorithm. What is the difference between the genetic algorithm and these other metaheuristic methods?

11.2   Study the so-called "Ant Colony Optimization" metaheuristics in the literature, and compare this with the GA approach.

11.3   Why is the set of initial solutions important to the performance of GA?

11.4   Why is the population size important in GA? What is the consequence if the population size is too large or too small?

11.5   What is the purpose of "crossover" in a GA? Explain the impact of the crossover rate θ on GA.

11.6   What is the purpose of "mutation" in a GA? Explain the impact of the mutation rate μ on GA.

11.7   We use a dynamic μ_k (see Eq. (11.16)) for mutation. What is the advantage of using such a dynamic μ_k?

11.8   Why is GA eminently suitable for solving a problem such as problem OPT-MM?

11.9   In a GA-based algorithm described in this chapter, how can we ensure that the solutions after crossover and mutation are feasible?

11.10   What are the different randomization steps within GA? Is randomness important to the genetic algorithm? Why?

Other Topics

12 Asymptotic capacity analysis

Luck is dividend of sweat. The more you sweat, the luckier you get.

Unknown

12.1 Review of asymptotic analysis

In previous chapters, we studied how to optimize a performance metric (e.g., a function of throughput) for a given network instance, where the network is of *finite* size. In this chapter, we study a different type of problem, where the goal is to find how much information source nodes can send to their destination nodes as either the node density goes to infinity or, with the same node density, the network area goes to infinity. Such a study is called an asymptotic capacity analysis, also known as capacity scaling law analysis. For such an investigation, the impact of network topology is usually of little interest and the focus is on *random* network topology. This is the approach that we adopt in this chapter. We consider a random network where each node is randomly deployed and each node has a randomly chosen destination node. We study an asymptotic capacity problem under different interference models and derive related results using the entities $\Omega(\cdot)$, $O(\cdot)$, and $\Theta(\cdot)$, which are defined as follows [30]:

- $f(n) = \Omega(g(n))$ if $f(n) \geq C \cdot g(n)$ for all $n > n_0$, where C and n_0 are positive constants;
- $f(n) = O(g(n))$ if $f(n) \leq C \cdot g(n)$ for all $n > n_0$, where C and n_0 are positive constants;
- $f(n) = \Theta(g(n))$ if $C_1 \cdot g(n) \leq f(n) \leq C_2 \cdot g(n)$ for all $n > n_0$, where C_1, C_2, and n_0 are positive constants.

Table 12.1 A summary of references on asymptotic capacity under unicast/multicast/broadcast.

	Protocol model	Physical model
Unicast	[15; 33; 60; 86]	[60]
Multicast	[92; 138]	[63]
Broadcast	[61]	[62; 91]

Since the seminal work of Gupta and Kumar [60] on capacity scaling laws for a wireless network, there has been growing interest in this important area. Related work on random ad hoc networks can be further divided into the following two categories: unicast capacity (e.g., [15; 33; 60; 86]) and multicast/broadcast capacity (e.g., [61–63; 91; 92; 138]). In [60], Gupta and Kumar showed that for a random network, its capacity is $\Theta(\frac{B}{\sqrt{n \ln n}})$ under the protocol model, and is $[\Omega(\frac{B}{\sqrt{n \ln n}}), O(\frac{B}{\sqrt{n}})]$ under the physical model with synchronized power control, where B is the achievable bit rate for a successful transmission. In [15; 33; 86], the unicast capacity of multi-channel ad hoc networks was investigated under the protocol model. Bhandari and Vaidya [15] studied the capacity of multi-channel single-radio (MC-SR) networks where there is a set of channels in the network and each node can only switch to a subset of these channels. Kyasanur and Vaidya [86] studied the unicast capacity of multi-channel multi-radio (MC-MR) networks where the number of bands used at a node is limited by the number of radios at the node. Dai *et al.* [33] extended the work of [86] to MC-MR networks with consideration of directional antennas. Keshavarz-Haddad *et al.* [61] studied the broadcast capacity for a homogeneous dense network under the protocol model. The broadcast capacity under the physical model was studied in [62; 91]. Li *et al.* [92] examined the multicast capacity of wireless ad hoc networks. Shakkottai *et al.* [138] investigated the multicast capacity of large-scale wireless ad hoc networks under the protocol model. The multicast capacity under the physical model was analyzed in [63]. We summarize these efforts in Table 12.1.

Finally, there exists some research related to capacity scaling laws under specialized wireless communication technologies, such as the generalized physical model [42; 181] (via adaptive modulation and coding scheme), ultra-wide band (UWB) [177], multiple-input multiple-output (MIMO) [118], network coding [97], multiple-packet reception (MPR) [165], and cognitive radio networks [75; 162; 176].

12.2 Capacity scaling laws of wireless ad hoc networks

In our case study, we describe how to analyze asymptotic capacities based on the original work in [42; 60]. We strive for clarity in our presentation in order to enhance the accessibility of this intricate material to graduate students. To simplify the discussion, we normalize distance and consider a wireless ad hoc network comprising n nodes randomly deployed in a 1×1 area.[1] Each node acts as a source node and transmits data to a randomly chosen destination node via multi-hop. The per-node throughput $\lambda(n)$ is defined as the data rate that can be transported from each source to its destination. We show how to analyze the maximum asymptotic per-node throughput (or per-node capacity) as node density increases.

Since asymptotic capacity analysis depends on the underlying interference model, we consider the following three interference models in this chapter:

- **Protocol model**: Under this model, a transmission is successful if (i) the receiver is within the transmission range of the transmitter, and (ii) the receiver is outside the interference range of other transmitters. The achievable rate of a successful transmission is a constant B. This model is also called the disk model and assumes the existence of interference is binary (yes or no).
- **Physical model**: Under this model, a transmission is successful if its *signal-to-interference-and-noise ratio* (SINR) is over certain threshold β. The achievable rate of a successful transmission is assumed to be a constant B. This model corresponds to a coding scheme that can support transmission with a data rate B if the SINR threshold is satisfied.
- **Generalized physical model**: Under this model, the achievable rate B of a transmission is determined by the Shannon capacity formula, i.e., $B = W \log_2(1 + \text{SINR})$. Unlike the physical model, there is no SINR threshold β for a successful transmission. As long as the transmission rate is no more than $W \log_2(1 + \text{SINR})$, such a transmission rate can be supported.

We can see that the protocol model is the simplest model, while the generalized physical model is the most challenging one. In the rest of this chapter, we study capacity scaling laws under these three interference models. For each model, we will develop a feasible solution, which will be used as a capacity lower bound. We will also analyze a capacity upper bound. The results for these three models are summarized as follows:

- **Case 1:** Under the protocol model, the capacity of a wireless ad hoc network with n nodes is $\Theta(\frac{B}{\sqrt{n \ln n}})$ almost surely as $n \to \infty$. By "*almost surely,*" we mean that the probability of the event that the capacity is $\Theta(\frac{B}{\sqrt{n \ln n}})$ approaches one as $n \to \infty$.

---

[1] In [60], to avoid the boundary effect, it is assumed that nodes are deployed on the surface of a three-dimensional sphere. As we shall see in this chapter, such an assumption is not needed.

- **Case 2:** Under the physical model, the capacity of a wireless ad hoc network with n nodes is $\lambda(n) \in [\Omega(\frac{B}{\sqrt{n \ln n}}), O(\frac{B}{n^{1/\alpha}})]$ almost surely as $n \to \infty$, where α is the path loss index.
- **Case 3:** Under the generalized physical model, the capacity of a wireless ad hoc network with n nodes is $\Omega(\frac{B}{\sqrt{n}})$ almost surely as $n \to \infty$.

In Sections 12.3, 12.4, and 12.5, we give details on how to develop these results. Section 12.6 summarizes this chapter. Table 12.2 lists the notation used in this chapter.

Table 12.2 Notation.

General notation	
$a(n)$	The size of a small square cell
B	The maximum data rate from a transmitter to a receiver
d_{ij}	The distance between nodes i and j
$\mathcal{E}(t)$	The set of active links at time t
$l(n)$	The side-length of a cell
n	The number of nodes in the network
$p(n)$	The common transmission power at all nodes under synchronized power control or the protocol model
$Pr\{A\}$	The probability of an event A
SINR_{ij}	The SINR on link (i, j)
α	Path-loss index
η	Ambient noise power
$\lambda(n)$	Per-node throughput of a wireless ad hoc network
ψ	A feasible solution

Protocol model-specific notation	
$\bar{D}$	The mean distance between a source node and its destination node
$r(n)$	The common transmission range for all nodes
Δ	A parameter to set the interference range

Physical model-specific notation	
$d(q, b)$	The length of the qth hop traversed by the bth unit of data
g_{ij}	The channel gain on link (i, j)
$h(b)$	The number of hops traversed by the bth unit of data from its source to the destination
H	$= \sum_{b=1}^{\lambda n T} h(b)$, i.e, the total number of hops traversed by all data in a time duration T
$p_{ij}(t)$	The transmission power used by node i to transmit to node j at time t under an independent power control
β	An SINR threshold for successful transmissions

Generalized physical model-specific notation	
B_m	A grid of $m(n)$ horizontal and $m(n)$ vertical paths
$B_{w \times h}$	A box with width $w \cdot l(n)\sqrt{2}$ and height $h \cdot l(n)\sqrt{2}$

Table 12.2 (cont.)

$B_{w \times h}^{\leftrightarrow}$	The event that there is a crossing path between $B_{w \times h}$'s left and right sides
$c_{ij}(d)$	The data rate that a node i in a cell can transmit to a node j in another cell at most d (diagonal) cells away
$m(n)$	The number of cells crossed by a horizontal or vertical path
P	The common transmission power at all nodes in the feasible solution
P_m^i	The maximal number of edge-disjoint crossing paths of rectangle R_m^i between its left and right sides
q	The probability that there is at least one node in a cell
R_m^i	The ith rectangle of size $1 \times \frac{\kappa \ln m(n)}{m(n)}$
$S_{w \times h}$	The dual box of $B_{w \times h}$
$S_{w \times h}^{\updownarrow}$	The event that there is a crossing path between $S_{w \times h}$'s top and bottom sides
W	Channel bandwidth

12.3 Case 1: Asymptotic capacity under the protocol model

In this section, we analyze the capacity scaling law under the protocol model. Under this model, a node i can successfully transmit data with a rate B to a node j if and only if the following constraints hold:

• The receiving node j is within the transmission range of node i, i.e.,

$$d_{ij} \leq r(n),$$

where d_{ij} is the distance between nodes i and j and $r(n)$ is the transmission range. The above constraint implicitly sets a constraint on next-hop routing, i.e., it defines a set of possible candidate nodes (within the transmission range) as the next-hop node.

• For any other link (k, l) that is active at the same time, it is necessary that $d_{kj} \geq (1 + \Delta)r(n)$, where $(1 + \Delta)r(n)$ represents an interference range. This is to keep the concurrently transmitting node k sufficiently far away from node j so that k's interference on j is negligible. More formally, we have

$$d_{kj} \geq (1 + \Delta)r(n) \text{ for each link } (k, l) \in \mathcal{E}(t) \text{ and } (k, l) \neq (i, j), \quad (12.1)$$

where $\mathcal{E}(t)$ is the set of links in the network that are active at time t. Constraint (12.1) implicitly sets a constraint on scheduling, i.e., it defines conflict relationships among all the links in the network.

Note that under the protocol model, we assume that the same (synchronized) transmission power $p(n)$ is used for all nodes. Thus, the transmission range $r(n)$ is the same for all nodes.

We consider a common throughput $\lambda(n)$ for each node (source) in the network to its randomly selected destination. The goal is to find the maximum $\lambda(n)$ that can be transported by the network. In Section 12.3.1, we develop a capacity upper bound. The lower bound analysis is given in Section 12.3.2. The results of this section can be summarized as follows (Theorems 12.1 and 12.2):

Under the protocol model, the capacity of a wireless ad hoc network with n nodes is $\lambda(n) = \Theta(\frac{B}{\sqrt{n \ln n}})$ almost surely as $n \to \infty$.

12.3.1 A capacity upper bound

For a wireless ad hoc network, the required resource to transmit data from a source node to its destination node depends on the number of hops between these two nodes, which is yet to be determined. As a result, it is not easy to analyze the capacity upper bound directly. Instead of analyzing capacity directly, we can analyze the sum of transmission rates over all nodes in the network, which we call an *aggregate rate* (AR). We find that AR should be no less than a minimum required value, which is a function of $\lambda(n)$. We also find that AR is upper bounded by a constant. Combining these two results, we can compute an upper bound on throughput $\lambda(n)$. The details are given in the proof of Theorem 12.1.

Theorem 12.1

Under the protocol model, a capacity upper bound is $O(\frac{B}{\sqrt{n \ln n}})$ almost surely as $n \to \infty$.

Proof. We begin the proof by analyzing the minimum required AR in order to support a throughput $\lambda(n)$ at each node. Let $\bar{D}$ be the mean distance between a source node and its corresponding destination node. Note that under the protocol model, the mean number of hops for each source–destination pair is at least $\frac{\bar{D}}{r(n)}$ and there are n source–destination pairs. Thus, AR is at least $\frac{\bar{D}}{r(n)} n\lambda(n)$.

We now develop an upper bound for AR. Based on the scheduling constraint (12.1), for any two links (i, j) and (k, l) that are active on the same band at the same time, we have $d_{ik} \geq d_{jk} - d_{ij} \geq (1 + \Delta)r(n) - r(n) = \Delta \cdot r(n)$, i.e., the distance between any two concurrent transmitting nodes is at least $\Delta \cdot r(n)$. Thus, if we draw a disk of radius $\frac{\Delta r(n)}{2}$ at each of these transmitting nodes, then these disks must be disjoint. Since the area of each disk is $\frac{\pi \Delta^2 r^2(n)}{4}$ and the center of these disks are within a unit square, the number of these disks is at most $\frac{(1+\Delta r(n))^2}{\pi \Delta^2 r^2(n)/4} = O(r^{-2}(n))$. Thus, the network can support $O(r^{-2}(n))$ transmissions at any time. Therefore, summing the data rates of all the transmissions, and considering the fact that the rate of a transmitting node is B, we have that AR is upper bounded by $O(\frac{B}{r^2(n)})$.

Combining the above two results, we have $\frac{\bar{D}}{r(n)}n\lambda(n) \leq \text{AR} \leq O(\frac{B}{r^2(n)})$. Note that $\bar{D}$ is a constant. Therefore, we have $\lambda(n) = O(\frac{B}{nr(n)})$. It has been shown in [59] that the transmission range $r(n)$ must be greater than $\sqrt{\frac{\ln n}{\pi n}}$ to make the network connected almost surely as $n \to \infty$. Therefore, we have

$$\lambda(n) = O(\frac{B}{nr(n)}) < O\left(\frac{B}{n\sqrt{\frac{\ln n}{\pi n}}}\right) = O\left(\frac{B}{\sqrt{n \ln n}}\right)$$

almost surely as $n \to \infty$. $\qquad\square$

12.3.2 A constructive lower bound

To develop a capacity lower bound, it is sufficient to construct a feasible solution that satisfies all the constraints. A solution contains two parts: a routing scheme and a scheduling scheme. We consider a cell-based approach for the routing scheme. In particular, we divide the unit square area into small cells so that there is at least one node in each cell almost surely as $n \to \infty$ (see Lemma 12.1). For each source–destination pair, the straight line that connects them will pass through a number of cells. We choose a node from each of these cells as a relay for multi-hop routing. To avoid interference, scheduling must be performed along with cell-based routing. That is, under the protocol model, some links cannot be active at the same time. Thus, we need to arrange them into different time slots. To analyze the achieved throughput, we need to identify the number of time slots required for scheduling. We will show that the scheduling problem can be transformed into a vertex-color problem in graph theory. By applying a result from graph coloring, we know how many time slots will be used for scheduling. Consequently, we obtain a complete solution, the throughput of which is stated in the following theorem:

Theorem 12.2
Under the protocol model, we can construct a feasible solution ψ with a throughput of $\lambda(n) = \Omega(\frac{B}{\sqrt{n \ln n}})$ almost surely as $n \to \infty$.

We now give details for the routing and scheduling schemes described in the above discussion.

Routing scheme   We develop a cell-based routing scheme. We divide the unit square into small square cells with width $\sqrt{\frac{\ln n}{n}}$ (see Fig. 12.1). The

Figure 12.1 Multi-hop routing from a source node s to its destination node d.

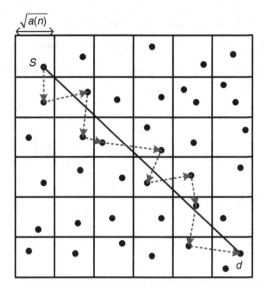

area of a cell is $a(n) = \sqrt{\frac{\ln n}{n}} \times \sqrt{\frac{\ln n}{n}} = \frac{\ln n}{n}$.[2] We set the transmission range $r(n) = \sqrt{5a(n)}$ so that a node in one cell can transmit to a node in any of its four neighboring cells. Next, we draw a line to connect each source–destination pair, which passes through some cells. One node is chosen from each of these cells to relay the traffic from the source node to its destination (see Fig. 12.1 as an example).

Such a routing scheme requires at least one node in each cell. We call a cell without any node an "empty" cell. We have the following result regarding the nonexistence of an empty cell as $n \to \infty$:

Lemma 12.1

If the cell size is $a(n) = \frac{\ln n}{n}$, an empty cell is nonexistent almost surely as $n \to \infty$. Therefore, the routing scheme is feasible almost surely as $n \to \infty$.

Proof. Since there are $\frac{1}{a(n)} = \frac{n}{\ln n}$ cells, each node can be in any cell with an equal probability of $\frac{\ln n}{n}$. For a particular cell i, denote E_i as the event that this cell is an empty cell. Then the probability of E_i is $Pr(E_i) = (1 - \frac{\ln n}{n})^n$. Therefore, the probability that an empty cell exists is $Pr(\bigcup_i E_i) \leq \sum_i Pr(E_i) = \frac{n}{\ln n}(1 - \frac{\ln n}{n})^n$. As $n \to \infty$, this probability is

---

[2] To divide the unit square into an integral number of small cells, we need to set $a(n) = 1/\lfloor n/\ln n \rfloor$. However, to make the proofs in this section easy to understand, we neglect such integrality requirements and simply let $a(n) = 1/(n/\ln n) = \ln n/n$. Note that all the results in this section hold true if we set $a(n) = 1/\lfloor n/\ln n \rfloor$. For example, we ask the readers to prove in Exercise 6.10 that Lemma 12.1 still holds when $a(n) = 1/\lfloor n/\ln n \rfloor$.

Figure 12.2 The shaded area for X_i.

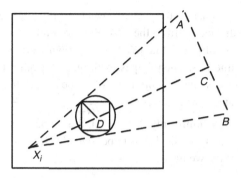

$$\lim_{n \to \infty} \frac{n}{\ln n} \left(1 - \frac{\ln n}{n}\right)^n = \lim_{n \to \infty} \frac{n}{\ln n} \left(1 - \frac{\ln n}{n}\right)^{-\frac{n}{\ln n}(-\ln n)}$$
$$= \lim_{n \to \infty} \frac{n}{\ln n} e^{-\ln n}$$
$$= \lim_{n \to \infty} \frac{1}{\ln n}$$
$$= 0,$$

where the second equality holds because $\lim_{n \to \infty} \frac{n}{\ln n} \to \infty$ and $\lim_{x \to \infty} (1 - \frac{1}{x})^{-x} = e$. $\square$

For this routing scheme, we obtain the following result for the number of S-D lines passing through any cell:

Lemma 12.2

The number of S-D lines passing through any cell is $O(n\sqrt{a(n)})$ almost surely as $n \to \infty$.

To prove this lemma, we need an intermediate result as follows. Denote L_i, $1 \le i \le n$, as the n S-D lines. Denote c_i, $1 \le i \le m = \frac{n}{\ln n}$, as the m small cells. Define p_{ij} as the probability that the S-D line L_i passes through cell Q_i. Then, we have the following result regarding p_{ij}:

Lemma 12.3

For any $1 \le i \le n$ and $1 \le j \le n$, there exists a constant $c_1 > 0$ such that $p_{ij} \le c_1 \sqrt{\frac{\ln n}{n}}$.

Proof. Note that cell Q_j is contained in a disk of radius $d_r = \frac{1}{\sqrt{2}} l(n) = \sqrt{\frac{\ln n}{2n}}$ that is centered at Q_j's center D (see Fig. 12.2). Denote X_i and Y_i as the source and destination nodes of the S-D pair i, respectively. Note that X_i can fall either inside or outside the disk.

We first consider the scenario when X_i falls outside the disk. Suppose that X_i is at distance x from the disk. We extend the two tangent lines origination from X_i equally such that $|X_i A| = |X_i B|$ and $X_i C = \sqrt{2}$, where C is the midpoint of line AB. Then L_i passes through Q_j only if Y_i is in the shaded area. Its area is less than the minimum of 1 (the area of the unit square) and the area of the triangle, which is equal to $\frac{2}{\sqrt{(x+d_r)^2 - d_r^2}} < \frac{2d_r}{x}$.

Since X_i is uniformly distributed, the probability density that it is at a distance x away from the disk is upper bounded by $c_2 \pi (x + d_r)$ for some constant $c_2 > 0$. Thus, we have

$$Pr\{L_i \text{ passes through cell } Q_i | X_i \text{ is outside the disk}\}$$

$$\leq \int_0^{\sqrt{2}} \min\left\{1, \frac{2d_r}{x}\right\} \cdot c_2 \pi (x + d_r) dx$$

$$= \int_0^{2d_r} c_2 \pi (x + d_r) dx + \int_{2d_r}^{\sqrt{2}} \frac{2d_r}{x} \cdot c_2 \pi (x + d_r) dx$$

$$\leq c_3 \sqrt{\frac{\ln n}{n}}, \tag{12.2}$$

where c_3 is a constant.

For the case when X_i falls inside the disk, we have

$$Pr\{L_i \text{ passes through cell } Q_i | X_i \text{ is inside the disk}\} \leq 1. \tag{12.3}$$

Combining (12.2) and (12.3), we have

$$p_{ij} = Pr\{L_i \text{ passes through cell } Q_i | X_i \text{ is outside the disk}\}$$
$$\cdot Pr\{X_i \text{ is outside the disk}\}$$
$$+ Pr\{L_i \text{ passes through cell } Q_i | X_i \text{ is inside the disk}\}$$
$$\cdot Pr\{X_i \text{ is inside the disk}\}$$
$$\leq c_3 \sqrt{\frac{\ln n}{n}} (1 - \pi r_d^2) + 1 \cdot \pi r_d^2$$
$$\leq c_4 \sqrt{\frac{\ln n}{n}}.$$

$\square$

We are now ready to prove Lemma 12.2.

Proof of Lemma 12.2. Denote

$$Y_{ij} = \begin{cases} 1 & \text{if } L_i \text{ passes through cell } Q_j, \\ 0 & \text{otherwise,} \end{cases}$$

for $1 \leq i \leq n$ and $1 \leq j \leq m$. Then, we have $Pr\{Y_{ij} = 1\} = p_{ij}$. Note that for a fixed j, the Y_{ij}-variables are i.i.d.

Denote $Z_j = \sum_{i=1}^{n} Y_{ij}$ as the total number of S-D lines passing through cell Q_j. Consider a fixed j, i.e., a particular cell Q_j. For any $s, a > 0$, based on the Chernoff bound, we have

$$Pr\{Z_j > s\} \le \frac{E[e^{aY_j}]}{e^{as}}. \tag{12.4}$$

For $E[e^{aZ_j}]$, we have

$$E[e^{aZ_j}] = E[e^{a\sum_{i=1}^{n} Y_{ij}}] = \prod_{i=1}^{n} E[e^{aY_{ij}}] = \prod_{i=1}^{n} (e^a p_{ij} + e^0 \cdot (1 - p_{ij}))$$

$$= \prod_{i=1}^{n} (1 + (e^a - 1)p_{ij})$$

$$\le \left[1 + (e^a - 1)c_4 \sqrt{\frac{\ln n}{n}} \right]^n$$

$$\le \left[\exp\left((e^a - 1)c_4 \sqrt{\frac{\ln n}{n}} \right) \right]^n$$

$$= \exp(c_4(e^a - 1)\sqrt{n \ln n}),$$

where the last inequality holds due to the fact that $1 + x \le e^x$. Letting $s = c_5\sqrt{n \ln n}$ in (12.4) (where $c_5 > c_4$), we have

$$Pr\{Z_j > c_5\sqrt{n \ln n}\} \le \exp[\sqrt{n \ln n}(c_4(e^a - 1) - ac_5)].$$

The above inequality holds for any $1 \le j \le m$.

Since $c_5 > c_4$, we can let a be small enough so that

$$Pr\{Z_j > c_5\sqrt{n \ln n}\} \le \exp(-\epsilon\sqrt{n \ln n}),$$

for some constant $\epsilon > 0$. Then, by union bounds, we have

$Pr\{$there exists a cell having more than $c_5\sqrt{n \ln n}$ S-D lines passing through$\}$

$$\le \sum_{j=1}^{m} Pr\{Z_j > c_5\sqrt{n \ln n}\}$$

$$\le m \exp(-\epsilon\sqrt{n \ln n}) = \frac{n}{\ln n} \exp(-\epsilon\sqrt{n \ln n}),$$

which goes to zero as $n \to \infty$.   □

Scheduling scheme We consider a time-slot-based scheduling, i.e., we divide one time frame into multiple time slots to satisfy the protocol model scheduling constraint (12.1).

To analyze the performance of the scheduling scheme, we need to know the number of conflicting links for each link (i, j), which directly affects the number of time slots required for scheduling and throughput. Two links are in conflict if they cannot be active at the same time under the protocol model. We

Figure 12.3 The area that
covers all the transmitting nodes
that may interfere with the
receivers in cell Q.

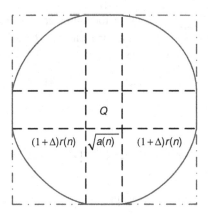

now establish the following lemma on the number of conflicting links for any link in the network:

Lemma 12.4

Under the cell-based routing scheme, the number of conflicting links for any link is upper bounded by $O(n\sqrt{a(n)})$ almost surely as $n \to \infty$.

Proof. Note that for a link (i, j), there are two cases in which a link (k, l) could be its conflicting link. One case is that link (k, l) interferes with link (i, j) and the other case is that link (k, l) is interfered by link (i, j). In this proof, for a given link (i, j), we first analyze the number of conflicting links that interfere with link (i, j). Since the routing scheme is cell-based, the conflict analysis is also based on cells. That is, we first analyze the number of interfering cells and then analyze the number of conflicting links in an interfering cell. We say that a cell Q' is an interfering cell with respect to cell Q if and only if the distance between a transmitting node in cell Q' and a receiving node in cell Q is no more than $(1 + \Delta)r(n)$.

We now show that the number of interfering cells with respect to cell Q is a constant. For a receiving node j in cell Q, the transmitting nodes of the links that interfere with j must be within the area inside the solid line shown in Fig. 12.3, which consists of five rectangles (including the small cell $\sqrt{a(n)} \times \sqrt{a(n)}$ at the center) and four quarter disks with radius $(1 + \Delta)r(n)$. To obtain an upper bound, we define the outermost square area as the interfering area that will not have any transmitting node in it. Hence, the number of interfering cells is no more than $c_6 = (2\lceil \frac{(1+\Delta)r(n)}{\sqrt{a(n)}} \rceil + 1)^2 = (2\lceil \sqrt{5}(1 + \Delta) \rceil + 1)^2$, which is a constant.

Next, we analyze the number of links that interfere with link (i, j) in each interfering cell. Clearly, this number is no more than the number of transmissions in this cell. Based on Lemma 12.2 and the routing scheme discussed earlier, the number of transmissions in a cell is equal to the number of source–destination lines intersecting this cell, which is $O(n\sqrt{a(n)})$ almost surely as $n \to \infty$.

Combining the above results, the number of conflicting links that interfere with link (i, j) is given by $c_6 O(n\sqrt{a(n)}) = O(n\sqrt{a(n)})$ almost surely as $n \to \infty$.

Following a similar analysis, we can obtain the same result on the number of conflicting links that are interfered by link (i, j). Then the number of all conflicting links for a link (i, j) is upper bounded by $O(n\sqrt{a(n)}) + O(n\sqrt{a(n)}) = O(n\sqrt{a(n)})$ almost surely as $n \to \infty$. □

To schedule the conflicting links, we use a conflict graph to model them. Each link in the network corresponds to a vertex in the conflict graph, and any conflict in the network is represented by an edge connecting two corresponding vertices in the conflict graph. If we use a different vertex color to represent each time slot, then the scheduling problem reduces to the well-studied vertex-color problem, which aims to find the minimum number of colors needed for coloring the nodes in a graph such that any two nodes connected by an edge have different colors. Using Lemma 12.4, the degree of each vertex in the conflict graph will be at most $c_7 \cdot n\sqrt{a(n)}$ for some constant c_7 almost surely as $n \to \infty$. Note that by the result for graph coloring [168], it is enough to have $d + 1$ colors for a graph with the maximum degree d. Thus, the required number of colors for the conflict graph is at most $1 + c_7 \cdot n\sqrt{a(n)}$. So we can divide one time frame into at most $1 + c_7 \cdot n\sqrt{a(n)}$ equal length time slots for scheduling. Therefore, the achievable throughput $\lambda(n)$ is given by

$$\lambda(n) \geq \frac{B}{1 + c_7 \cdot n\sqrt{a(n)}} = \frac{B}{1 + c_7\sqrt{n \ln n}} = \Omega\left(\frac{B}{\sqrt{n \ln n}}\right),$$

where the first equality holds because $a(n) = \frac{\ln n}{n}$.

12.4 Case 2: Asymptotic capacity under the physical model

In this section, we analyze the capacity scaling law under the physical model. Under this model, each node is allowed to perform power control. Denote $p_{ij}(t)$ as the power used by node i to transmit to node j.[3] A transmission with a data rate B is successful if and only if the SINR satisfies

$$\text{SINR}_{ij} = \frac{g_{ij} \cdot p_{ij}(t)}{\eta + \sum_{\substack{(k,l)\neq(i,j) \\ (k,l)\in\mathcal{E}(t)}} g_{kj} \cdot p_{kl}(t)} \geq \beta, \tag{12.5}$$

---

[3] In [60], the transmission power at all nodes in a random network is the same and is determined by the number of nodes. A different capacity upper bound $O(\frac{B}{n^{1/2}})$ was obtained for this special case. In this section, we study a general case where independent power control is allowed at each node.

where $g_{ij} = d_{ij}^{-\alpha}$ is the channel gain over link (i, j), $\alpha \geq 2$ is the path-loss index, η is the ambient noise power, and β is the SINR threshold.

In Sections 12.4.1 and 12.4.2, we develop an upper bound and a lower bound, respectively. The main result can be summarized as follows (also see Theorems 12.3 and 12.4):

Under the physical model, the capacity of a wireless ad hoc network with n nodes is $\lambda(n) \in [\Omega(\frac{B}{\sqrt{n \ln n}}), O(\frac{B}{n^{1/\alpha}})]$ almost surely as $n \to \infty$.

12.4.1 Computing an upper bound

Instead of analyzing capacity directly, we analyze the total distance traversed by all data generated over a time duration T (i.e., $\lambda(n)nT$), which we call *aggregate capacity-distance* over T, denoted as ACD_T. Note that ACD_T is similar to the transport capacity in [60]. We find that there exists an inequality relationship between ACD_T and the throughput $\lambda(n)$ and that a constant upper bound can be computed for ACD_T. As a result, we can compute an upper bound for the throughput $\lambda(n)$.

Theorem 12.3

Under the physical model, the capacity of a wireless ad hoc network with n nodes is $\lambda(n) = O(\frac{B}{n^{1/\alpha}})$ almost surely as $n \to \infty$.

Proof. We begin the proof by finding an inequality relationship between ACD_T and the throughput $\lambda(n)$. During T, the network can transport $\lambda(n)nT$ units of data in total. For a particular unit of data b, denote $h(b)$ as the number of hops on the path from its source to its destination and let $d(q, b)$ denote the length of the qth hop. Then we have

$$ACD_T = \sum_{b=1}^{\lambda(n)nT} \sum_{q=1}^{h(b)} d(q, b) \geq \lambda(n)nT\bar{D}, \tag{12.6}$$

where $\bar{D}$ is the average distance between the source and destination nodes.

Next, we develop an upper bound for ACD_T. Since the function $f(x) = x^\alpha$ is convex for any $\alpha \geq 2$, we have

$$\left(\sum_{b=1}^{\lambda(n)nT} \sum_{q=1}^{h(b)} \frac{1}{H} d(q, b) \right)^\alpha \leq \frac{1}{H} \sum_{b=1}^{\lambda(n)nT} \sum_{q=1}^{h(b)} d^\alpha(q, b),$$

where $H = \sum_{b=1}^{\lambda nT} h(b)$ is the total number of hops traversed by all the data over time duration T. This gives us

$$
\begin{aligned}
ACD_T &= \sum_{b=1}^{\lambda(n)nT} \sum_{q=1}^{h(b)} d(q, b) \\
&\le H \left(\frac{1}{H} \sum_{b=1}^{\lambda(n)nT} \sum_{q=1}^{h(b)} d^\alpha(q, b) \right)^{\frac{1}{\alpha}} \\
&= H^{1-\frac{1}{\alpha}} \left(\sum_{b=1}^{\lambda(n)nT} \sum_{q=1}^{h(b)} d^\alpha(q, b) \right)^{\frac{1}{\alpha}}.
\end{aligned}
\tag{12.7}
$$

We now analyze H and $\sum_{b=1}^{\lambda(n)nT} \sum_{q=1}^{h(b)} d^\alpha(q, b)$. We first obtain an upper bound for H. Note that due to half-duplex, a node that is receiving cannot transmit at the same time. Therefore, in any given time slot, at most $\frac{n}{2}$ nodes can transmit simultaneously. For each link (i, j), its capacity is B. Thus, the total bits that can be transmitted by all nodes over T is at most $\frac{nBT}{2}$, i.e.,

$$
H \le \frac{nBT}{2}.
\tag{12.8}
$$

We next find an upper bound for $\sum_{b=1}^{\lambda(n)nT} \sum_{q=1}^{h(b)} d^\alpha(q, b)$. Consider a transmission from node i to node j. Since the SINR constraint (12.5) holds at node j, we have

$$
\frac{p_{ij}(t)/d_{ij}^\alpha}{\eta + \sum_{\substack{(k,l) \ne (i,j) \\ (k,l) \in \mathcal{E}(t)}} p_{kl}(t)/d_{kj}^\alpha} \ge \beta.
$$

Adding the signal power from node i in the denominator, we have

$$
\frac{p_{ij}(t)/d_{ij}^\alpha}{\eta + \sum_{(k,l) \in \mathcal{E}(t)} p_{kl}(t)/d_{kj}^\alpha} \ge \frac{\beta}{\beta + 1}.
$$

Based on this observation, we can develop an upper bound on d_{ij}^α as follows:

$$
\begin{aligned}
d_{ij}^\alpha &\le \frac{\beta + 1}{\beta} \cdot \frac{p_{ij}(t)}{\eta + \sum_{(k,l) \in \mathcal{E}(t)} \frac{p_{kl}(t)}{d_{kj}^\alpha}} \\
&\le \frac{\beta + 1}{\beta} \cdot \frac{p_{ij}(t)}{\eta + \sum_{(k,l) \in \mathcal{E}(t)} \frac{p_{kl}(t)}{(\sqrt{2})^\alpha}} \\
&< \frac{\beta + 1}{\beta} \cdot \frac{p_{ij}(t)}{\sum_{(k,l) \in \mathcal{E}(t)} \frac{p_{kl}(t)}{(\sqrt{2})^\alpha}},
\end{aligned}
$$

where the second inequality holds because $d_{kj} \leq \sqrt{2}$. Summing the above inequality over all simultaneous transmissions, we have

$$\sum_{(i,j)\in\mathcal{E}(t)} d_{ij}^{\alpha} < \frac{\beta+1}{\beta} \cdot \frac{\sum_{(i,j)\in\mathcal{E}(t)} p_{ij}(t)}{\sum_{(k,l)\in\mathcal{E}(t)} \frac{p_{kl}(t)}{(\sqrt{2})^{\alpha}}} = (\sqrt{2})^{\alpha}\frac{\beta+1}{\beta}.$$

Further, summing the above inequality over a time duration T, we have

$$\sum_{t=1}^{T} \sum_{(i,j)\in\mathcal{E}(t)} d_{ij}^{\alpha} < \sum_{t=1}^{T}(\sqrt{2})^{\alpha}\frac{\beta+1}{\beta} = T(\sqrt{2})^{\alpha}\frac{\beta+1}{\beta}. \tag{12.9}$$

In addition, we have

$$\sum_{b=1}^{\lambda(n)nT} \sum_{q=1}^{h(b)} d^{\alpha}(q,b) = \sum_{t=1}^{T} \sum_{(i,j)\in\mathcal{E}(t)} (d_{ij}^{\alpha} \cdot B) = B\sum_{t=1}^{T} \sum_{(i,j)\in\mathcal{E}(t)} d_{ij}^{\alpha}, \tag{12.10}$$

where the first equality holds since each active link (i, j) can carry B bits of data in a unit time. Combining (12.9) and (12.10), we have

$$\sum_{b=1}^{\lambda(n)nT} \sum_{q=1}^{h(b)} d^{\alpha}(q,b) < BT(\sqrt{2})^{\alpha}\frac{\beta+1}{\beta}. \tag{12.11}$$

By (12.7), (12.8), and (12.11), we have

$$\mathrm{ACD}_T < \left(\frac{nBT}{2}\right)^{1-\frac{1}{\alpha}} \left[BT(\sqrt{2})^{\alpha}\frac{\beta+1}{\beta}\right]^{\frac{1}{\alpha}} = \frac{nBT}{\sqrt{2}}\left[\frac{2(\beta+1)}{n\beta}\right]^{\frac{1}{\alpha}}. \tag{12.12}$$

Now we are ready to develop an upper bound for $\lambda(n)$. Based on (12.6) and (12.12), we have

$$\lambda(n)nT\bar{D} \leq \mathrm{ACD}_T < \frac{nBT}{\sqrt{2}}\left[\frac{2(\beta+1)}{n\beta}\right]^{\frac{1}{\alpha}}.$$

Thus,

$$\lambda(n) < \frac{B}{\sqrt{2}\bar{D}}\left[\frac{2(\beta+1)}{n\beta}\right]^{\frac{1}{\alpha}} = O\left(\frac{B}{n^{1/\alpha}}\right)$$

almost surely as $n \to \infty$. $\square$

12.4.2 Computing a lower bound

To obtain a capacity lower bound, we construct a feasible solution. But constructing a feasible solution under the physical model is more difficult than that under the protocol model. This is because under the protocol model, we only need to consider distances in the scheduling constraints (12.1). But under the physical model, we also need to consider power control in the SINR constraints (12.5).

Given that we have developed a feasible solution ψ for the protocol model, we will try to develop a feasible solution for the physical model. We observe that if we set the parameter Δ in the protocol model to be "large enough," then ψ is also a feasible solution for the physical model. This is because a large Δ will impose more constraints within the conflict graph, and thus reduce interference from neighboring nodes in the protocol model solution. As a result, the SINR at a receiver can be made large enough to satisfy the physical model SINR constraint (12.5). In this case, ψ is also a feasible solution for the physical model. The following theorem is based on this insight:

Theorem 12.4

Under the physical model, we can construct a feasible solution ψ with $\lambda(n) = \Omega(\frac{B}{\sqrt{n \ln n}})$ almost surely as $n \to \infty$.

Proof. To prove this result, we show that if we set

$$\Delta \geq \left(\frac{p(n)\beta}{2r^2(n)(p(n) - \beta \eta r(n)^\alpha)} \right)^{\frac{1}{2+\alpha}} - 1, \tag{12.13}$$

where $p(n)$ is the transmission power used in the protocol model solution ψ, then the solution ψ developed for the protocol model (in Section 12.3.2) is also feasible for the physical model.

First, we show that the SINR constraint under the physical model (12.5) is satisfied by ψ. Based on the construction of ψ in Section 12.3.2, we know that once a link (i, j) is active, those nodes that are within a square with a side length of $2(1 + \Delta)r(n) + \sqrt{a(n)}$ (that contains the cell where node j is located) cannot transmit at the same time (see Fig. 12.3). Note that node j may be located at the boundary of the unit square. Thus, the number of links that interfere with link (i, j) is at most

$$
\left\lfloor \frac{2(1 + \Delta)r(n) + 1}{2(1 + \Delta)r(n) + \sqrt{a(n)}} \right\rfloor^2 - 1 \leq \left(\frac{2(1 + \Delta)r(n) + 1}{2(1 + \Delta)r(n) + \sqrt{a(n)}} \right)^2 - 1
$$
$$
= \frac{[1 - \sqrt{a(n)}][4(1 + \Delta)r(n) + \sqrt{a(n)} + 1]}{[2(1 + \Delta)r(n) + \sqrt{a(n)}]^2}
$$
$$
< \frac{2}{[2(1 + \Delta)r(n) + \sqrt{a(n)}]^2}
$$
$$
< \frac{2}{4[(1 + \Delta)r(n)]^2}
$$
$$
= \frac{1}{2[(1 + \Delta)r(n)]^2},
$$

where the second inequality holds because $1 - \sqrt{a(n)} < 1$ and $4(1 + \Delta)r(n) + \sqrt{a(n)} \to 0 < 1$ as $n \to \infty$. Based on the construction of ψ, the interference at the receiving node j from each of these links will be at most $\frac{p(n)}{[(1+\Delta)r(n)]^\alpha}$. Therefore, we have

$$
\begin{aligned}
\text{SINR}_{ij} &> \frac{\frac{p(n)}{r(n)^\alpha}}{\eta + \frac{1}{2[(1+\Delta)r(n)]^2}\frac{p(n)}{[(1+\Delta)r(n)]^\alpha}} \\
&= \frac{r(n)^{-\alpha}}{\eta p(n)^{-1} + \frac{1}{2}[(1+\Delta)r(n)]^{-(2+\alpha)}} \\
&\geq \frac{r(n)^{-\alpha}}{\eta p(n)^{-1} + [\beta^{-1}r(n)^{-\alpha} - \eta p(n)^{-1}]} = \beta,
\end{aligned}
$$

where the second inequality holds by (12.13). Thus, for solution ψ, the SINR constraint (12.5) holds for each link (i, j) that is active at time t. Therefore, the power control and scheduling schemes in ψ are also feasible for the physical model. Furthermore, the achieved capacity on each link under the physical model is equal to that under the protocol model. As a result, the link capacity constraint is still satisfied for each link under the physical model, and the achieved throughput under the physical model will be the same as that under the protocol model, which is $\Omega(\frac{B}{\sqrt{n \ln n}})$ almost surely as $n \to \infty$.   □

12.5 Case 3: Asymptotic capacity lower bound under the generalized physical model

In this section, we analyze the capacity scaling law under the generalized physical model. Under this model, the achievable rate of a transmission is determined by its SINR. That is, the achievable rate from node i to node j is

$$
c_{ij} = W \log_2(1 + \text{SINR}_{ij}).
$$

Under this model, the only requirement is that the transported data rate is no more than this achievable rate c_{ij}.

This model is the most challenging one among the three models. It remains an open problem to compute an asymptotic upper bound for this model. In this section, we present a capacity lower bound by applying the percolation theory based on the materials in [58; 107]. The following theorem is the main result in this section:

> **Theorem 12.5**
> Under the generalized physical model, the capacity of a wireless ad hoc network generated by a Poisson point process with density n in a 1×1 area is $\Omega(\frac{W}{\sqrt{n}})$ almost surely as $n \to \infty$, where W is the channel bandwidth.

Remark 12.1

The above result was established in [42]. Based on this result, the authors of [42] claimed that this result closes the capacity gap for random networks since it matches with the upper bound developed in [60]. Unfortunately, this claim is incorrect for the following reasons:

- The underlying interference models are different. In [60], the upper bound $O(\frac{W}{\sqrt{n}})$ was developed for the physical model, while in [42], the lower bound $\Omega(\frac{W}{\sqrt{n}})$ was developed for the generalized physical model. We cannot assume (and it is usually incorrect to claim) that the same capacity upper bounds can hold under different interference models.
- The random networks under consideration are also different. In [42], it was assumed that each node is the destination of exactly one source, while there is no such assumption in [60]. We also point out that the number of nodes in the random network in [42] may not be n, while the number of nodes in the random network in [60] is n. Given this difference in network settings, we cannot assume that the upper bound result in [60] also holds for the random network in [42].

12.5.1 Main idea

We consider a random network generated by a Poisson point process with density n in a 1×1 area.[4] We assume that each node is the destination of exactly one source. To obtain a capacity lower bound, we assume that all nodes transmit at the same constant power P. We also divide the entire network area into small cells. Our solution on multi-hop routing is based on a highway system in the network. A single node in each cell that is crossed by a highway path is selected to transmit data along this highway. The other nodes transmit to, or receive from, the highway system via a single hop. That is, an end-to-end transmission has four phases: (i) nodes send their data to a node in the highway via one-hop transmissions; (ii) data are carried by a horizontal highway path; (iii) data are carried by a vertical highway path; and (iv) data are delivered from a node in the highway to the destination nodes via one-hop transmissions. Note that if the source node (or destination node) is in the highway, then Phase (i) (or Phase (iv)) is not needed. In each phase, we design a time-slot-based scheduling for transmissions. Since Phases (i) and (iv) have longer transmission distances (and thus a lower peak data rate), there are more time slots to ensure an appropriate data rate in these phases.

---

[4]   In previous sections, the number of nodes in the network is n. In this section, the number of nodes in the network is a random number with an expectation value of n.

Figure 12.4 Construction of the grid.

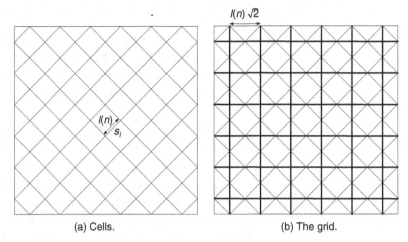

(a) Cells.　　　　　　　(b) The grid.

We now provide a sketch of the results. First, we consider the data rate limit in Phases (ii) and (iii). The constructed highway system has hops with a constant transmission rate. We will also show that the number of nodes that have access to a highway path is at most $O(\sqrt{n})$. Thus, each source node has a data rate no more than $\Omega(W/\sqrt{n})$, where W is the channel bandwidth. Second, we consider the data rate limit in Phases (i) and (iv). We show that the length of these single hops is at most $O(1/\sqrt{n})$, which in turn yields a rate $\Omega(\frac{W}{(\ln n)^3})$. Combining both results, we establish Theorem 12.5.

12.5.2 Construction of the highway

To construct the highway, we select an integer $m(n) = \lfloor\sqrt{\frac{n}{2\ln 6}}\rfloor$ and then partition the area into small cells s_i of side-length $l(n) = \frac{1}{m(n)\sqrt{2}}$, as depicted in Fig. 12.4(a). Denote $X(s_i)$ as the number of nodes inside cell s_i. If $X(s_i) = 0$, then we say that cell s_i is empty. Each cell is nonempty with the same probability $q \equiv Pr\{X(s_i) \geq 1\}$, where $Pr\{A\}$ is the probability of an event A. We have $q = 1 - e^{-(l(n))^2 n} = 1 - e^{-\frac{n}{2(m(n))^2}} > 1 - e^{-\frac{n}{2}\frac{2\ln 6}{n}} = \frac{5}{6}$. Moreover, since $q \to \frac{5}{6}$ as $n \to \infty$, we have $q < 1 - \frac{1}{6e^3}$ as $n \to \infty$. Thus,

$$\frac{5}{6} < q < 1 - \frac{1}{6e^3} \text{ when } n \to \infty. \tag{12.14}$$

For the number of nodes in a cell, we have the following lemma:

Lemma 12.5
There are less than $\ln m(n)$ nodes in each cell almost surely as $n \to \infty$.

To prove this lemma, we need the following result:

Lemma 12.6

For a Poisson random variable X with parameter λ, we have

$$Pr(X \geq x) \leq \frac{e^{-\lambda}(e\lambda)^x}{x^x} \quad \text{for } x > \lambda.$$

Proof. For $s > 0$ and $x > \lambda$, based on Markov's inequality, we have

$$Pr(X \geq x) \leq \frac{E(e^{sX})}{e^{sx}}, \tag{12.15}$$

where $E(e^{sX})$ can be computed as follows:

$$
\begin{aligned}
E(e^{sX}) &= \sum_{k=0}^{\infty} \frac{e^{-k}\lambda^k}{k!} e^{sk} \\
&= e^{\lambda(e^s - 1)} \sum_{k=0}^{\infty} \frac{e^{-\lambda e^s}(\lambda e^s)^k}{k!} \\
&= e^{\lambda(e^s - 1)}. \tag{12.16}
\end{aligned}
$$

Combining (12.15) and (12.16), we obtain

$$Pr(X \geq x) \leq \frac{E(e^{sX})}{e^{sx}} = e^{\lambda(e^s - 1) - sx}.$$

Letting $s = \ln \frac{x}{\lambda} > 0$ in the above inequality, we have

$$Pr(X \geq x) \leq e^{x - \lambda - x \ln(x/\lambda)} = \frac{e^{-\lambda}(e\lambda)^x}{x^x}$$

for $x > \lambda$. $\qquad\square$

We are now ready to prove Lemma 12.5.

Proof of Lemma 12.5. We compute the probability Pr that there is at least one cell that has no fewer than $\ln m(n)$ nodes. By the union bounds and Lemma 12.6, we have

$$
\begin{aligned}
Pr &\leq [m(n)]^2 Pr\{X(s_i) \geq \ln m(n)\} \\
&\leq [m(n)]^2 e^{-(l(n))^2 n} \left[\frac{(l(n))^2 ne}{\ln m(n)} \right]^{\ln m(n)} \\
&= e^{-\frac{n}{2(m(n))^2}} \left[\frac{ne^3}{2(m(n))^2 \ln m(n)} \right]^{\ln m(n)} \\
&= e^{-\Theta(1)} \Theta(\ln n)^{-\ln n},
\end{aligned}
$$

which goes to 0 as $n \to \infty$. $\qquad\square$

Figure 12.5 There are many
paths between left and right
sides in each rectangle R_m^i.

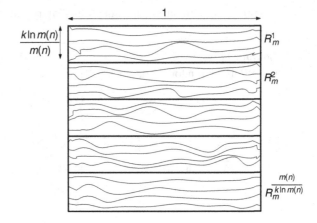

We draw $m(n)$ horizontal lines and $m(n)$ vertical lines across half of the
cells, as shown in Fig. 12.4(b). A path includes some segments of these lines. If
a path does not cross any empty cell, then we call it an *open* path. We will show
that there are at least $\Theta(\sqrt{n})$ open paths crossing the network area between left
and right sides and at least $\Theta(\sqrt{n})$ crossing paths between top and bottom sides
almost surely as $n \to \infty$. We call these paths the *highway system*. Along the
paths of the highway system, we select one node per cell that acts as a relay.

We now present a proof for crossing paths between left and right sides. The
proof for crossing paths between top and bottom sides is similar and is thus
omitted. We divide the network area into rectangles R_m^i of size $1 \times \frac{\kappa \ln m(n)}{m(n)}$
(see Fig. 12.5), where the constant κ is given by $\kappa = \frac{m(n)}{\ln m(n) \left\lceil \frac{-m(n)\ln(6-6q)}{3\ln m(n)} \right\rceil}$. We
note that

$$\kappa > \frac{m(n)}{\ln m(n) \frac{-m(n)\ln(6-6q)}{3\ln m(n)}} = \frac{3}{-\ln(6-6q)} > 1, \qquad (12.17)$$

where the last equality holds because $q < 1 - \frac{1}{6e^3}$ by (12.14). Denote P_m^i as
the maximal number of edge-disjoint crossing paths (between left and right
sides) within the rectangle R_m^i, and let $P_m = \min\left\{ P_m^i : 1 \le i \le \frac{m(n)}{\kappa \ln m(n)} \right\}$. We
have the following result:

Theorem 12.6

For a constant $\delta = \frac{-\ln(6-6q)\kappa - 3}{\ln \frac{q}{1-q}} > 0$, we have that

$$\lim_{n \to \infty} Pr_q\{P_m \le \lfloor \delta \ln m(n) \rfloor\} = 0,$$

where $Pr_q\{A\}$ is the probability of event A under a nonempty probability
q, $m(n) = \lfloor \sqrt{\frac{n}{2\ln 6}} \rfloor$, and $\kappa = \frac{m(n)}{\ln m(n) \left\lceil \frac{-m(n)\ln(6-6q)}{3\ln m(n)} \right\rceil}$. That is, there are at least

$\lfloor \delta \ln m(n) \rfloor + 1$ edge-disjoint crossing paths within each rectangle R_m^i of size $1 \times \frac{\kappa \ln m(n)}{m(n)}$ between left and right sides almost surely as $n \to \infty$.

In Theorem 12.6, we can now better understand why we set $\kappa = \dfrac{m(n)}{\ln m(n) \left\lfloor \frac{-m(n) \ln(6-6q)}{3 \ln m(n)} \right\rfloor}$. Since δ should be a positive number, we need $\kappa > \frac{3}{-\ln(6-6q)}$. Furthermore, the number of rectangles, $\frac{m(n)}{\kappa \ln m(n)}$, should be an integer. The κ that we choose satisfies both requirements.

Note that based on Theorem 12.6, it is clear that there are at least $\frac{m(n)}{\kappa \ln m(n)} \cdot (\lfloor \delta \ln m(n) \rfloor + 1) = \Theta(m(n)) = \Theta(\sqrt{n})$ crossing paths in the network area between left and right sides almost surely as $n \to \infty$.

The proof of Theorem 12.6 needs some additional lemmas. We denote by $B_{w \times h}^{\leftrightarrow}$ the event that there is a crossing path within $B_{w \times h}$ between its left and right sides, where $B_{w \times h}$ is a box with width $w \cdot l(n)\sqrt{2}$ and height $h \cdot l(n)\sqrt{2}$. The following lemma provides an upper bound for the probability of the event $B_{w \times h}^{\leftrightarrow}$:

Lemma 12.7

For a nonempty probability $\hat{q} < \frac{1}{3}$, the probability $Pr_{\hat{q}}\{B_{w \times h}^{\leftrightarrow}\}$ of the event $B_{w \times h}^{\leftrightarrow}$ satisfies

$$Pr_{\hat{q}}\{B_{w \times h}^{\leftrightarrow}\} \le (\lfloor h \rfloor + 1)e^{\ln(3\hat{q})\lceil w \rceil}.$$

Proof. Denote A_i as the event that there exists a crossing path starting from the ith vertex on the left side of box $B_{w \times h}$, where vertices on the left side are ordered from the bottom to the top. Thus, we have $\sum_i Pr_{\hat{q}}\{A_i\} = Pr_{\hat{q}}\{B_{w \times h}^{\leftrightarrow}\}$. Since the number of vertices is at most $\lfloor h \rfloor + 1$, there is at least one index i_0 such that $Pr_{\hat{q}}\{A_{i_0}\} \ge \frac{1}{\lfloor h \rfloor + 1} Pr_{\hat{q}}\{B_{w \times h}^{\leftrightarrow}\}$, i.e.,

$$Pr_{\hat{q}}\{B_{w \times h}^{\leftrightarrow}\} \le (\lfloor h \rfloor + 1)Pr_{\hat{q}}\{A_{i_0}\}.$$

Denote $N_{i_0}(\lceil w \rceil)$ as the number of open paths with $\lceil w \rceil$ hops starting at the i_0-th vertex. Since a crossing path starting from the i_0-th vertex to the right side of $B_{w \times h}$ has at least $\lceil w \rceil$ hops, we have

$$Pr_{\hat{q}}\{A_{i_0}\} \le Pr_{\hat{q}}\{N_{i_0}(\lceil w \rceil) \ge 1\}.$$

Note that paths with $\lceil w \rceil$ hops are open with probability $q^{\lceil w \rceil}$ and the number of paths with $\lceil w \rceil$ hops starting at the i_0-th vertex is at most $3^{\lceil w \rceil}$. We have

$$Pr_{\hat{q}}\{N_{i_0}(\lceil w \rceil) \ge 1\} \le \hat{q}^{\lceil w \rceil} 3^{\lceil w \rceil} = (3\hat{q})^{\lceil w \rceil}$$

Figure 12.6 The grid B_m (with continuous lines) and its dual grid S_m (with dashed lines).

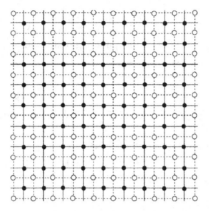

almost surely for a large number of open paths. Combining the above inequalities, we have

$$Pr_{\hat{q}}\{B^{\leftrightarrow}_{w\times h}\} \leq (\lfloor h \rfloor + 1)Pr_{\hat{q}}\{A_{i_0}\}$$
$$\leq (\lfloor h \rfloor + 1)Pr_{\hat{q}}\{N_{i_0}(\lceil w \rceil) \geq 1\}$$
$$\leq (\lfloor h \rfloor + 1)(3\hat{q})^{\lceil w \rceil}$$
$$= (\lfloor h \rfloor + 1)e^{\ln(3\hat{q})\lceil w \rceil}$$

for a nonempty probability $\hat{q} < \frac{1}{3}$. $\qquad\square$

The above lemma asserts that if the nonempty probability is too small, then a box $B_{w\times h}$ cannot be crossed between its left and right sides. We consider the dual question of the existence of a crossing path by exploiting the concept of a *dual grid* from the Percolation theory. A dual grid S_m for the grid B_m is constructed by placing a vertex in each cell of B_m, and connecting two neighboring vertices by an dashed line, as shown in Fig. 12.6. For each edge in S_m that does not cross an open edge of the original grid, we call it an open edge in S_m. We have the following relationship between the original grid and its dual: if the original grid has a nonempty probability q, then the dual grid has a nonempty probability $1 - q$. For the existence of a crossing path, we have the following lemma:

Lemma 12.8
As long as $q > \frac{2}{3}$, we have

$$Pr_q\{B^{\leftrightarrow}_{w\times h}\} \geq 1 - (\lfloor w \rfloor + 1)e^{\ln(3-3q)\lceil h \rceil}.$$

Proof. For the box $B_{w\times h}$ and the corresponding dual box $S_{w\times h}$, the complement of $B^{\leftrightarrow}_{w\times h}$ is event $S^{\updownarrow}_{w\times h}$ that there exists an open crossing path in $S_{w\times h}$ between its top and bottom sides [80].

Note that the dual lattice has a nonempty probability $\hat{q} = 1 - q < \frac{1}{3}$. After rotating the box by 90°, we can apply Lemma 12.7 to the dual lattice and obtain

$$Pr\{S_{w \times h}^{\updownarrow}\} \leq (\lfloor w \rfloor + 1)e^{\ln(3\hat{q})\lceil h \rceil} = (\lfloor w \rfloor + 1)e^{\ln(3-3q)\lceil h \rceil}.$$

This implies the result in this lemma.   □

We need the following definitions from [58]. An event A is an increasing event if for any network where event A occurs, event A still occurs after adding an edge in this network. Next we define the interior of A of depth r, $I_r(A)$. Event $I_r(A)$ is an event such that for any network where $I_r(A)$ occurs, event A occurs after we change up to r arbitrary edges.

Now consider event A of having a crossing path within $B_{w \times h}$ between its left and right sides. It is easy to see that A is an increasing event. For this event, $I_r(A)$ is the event that $(r + 1)$ edge-disjoint crossing paths exist. Theorem 2.45 in [58] asserts the following result:

For an increasing event A and any $0 \leq \hat{q} < q \leq 1$, we have

$$1 - Pr_q\{I_r(A)\} \leq \left(\frac{q}{q - \hat{q}}\right)^r (1 - Pr_{\hat{q}}\{A\}). \tag{12.18}$$

Based on the above results, we are now ready to give a proof of Theorem 12.6.

Proof of Theorem 12.6. Denote $R_m^{i \leftrightarrow}$ as the event that there exists a crossing path within the rectangle R_m^i between its left and right sides and let $\hat{q} = 2q - 1$. Since $q > \frac{5}{6}$, we have $\frac{2}{3} < \hat{q} < q$. By Lemma 12.8, we have

$$Pr_{\hat{q}}\{R_m^{i \leftrightarrow}\} \geq 1 - (m(n) + 1)\exp\left(\ln(3 - 3\hat{q})\left\lceil \frac{\kappa \ln m(n)}{m(n)} / (\sqrt{2}l(n)) \right\rceil\right)$$

$$= 1 - (m(n) + 1)\exp\left(\ln(6 - 6q)\lceil \kappa \ln m(n) \rceil\right). \tag{12.19}$$

Thus, we have

$$Pr_q\{P_m^i \leq \lfloor \delta \ln m(n) \rfloor\} = 1 - Pr_q\{I_{\lfloor \delta \ln m(n) \rfloor}(R_m^{i \leftrightarrow})\}$$

$$\leq \left(\frac{q}{q - \hat{q}}\right)^{\lfloor \delta \ln m(n) \rfloor} (1 - Pr_{\hat{q}}\{R_m^{i \leftrightarrow}\})$$

$$\leq \left(\frac{q}{q - \hat{q}}\right)^{\delta \ln m(n)} (m(n) + 1)e^{\ln(6-6q)\lceil \kappa \ln m(n) \rceil}$$

$$= e^{\delta \ln m(n) \ln \frac{q}{1-q} + \ln(m(n)+1) + \ln(6-6q)\lceil \kappa \ln m(n) \rceil}, \tag{12.20}$$

where the first inequality holds by (12.18), and the second inequality holds by using $\lfloor \delta \ln m(n) \rfloor \leq \delta \ln m(n)$ and (12.19).

We now compute the probability of having at most $\delta \ln m(n)$ edge-disjoint crossing paths in any rectangle R_m^i between its left and right sides. By (12.20), we have[5]

$$Pr_q\{P_m \leq \delta \ln m(n)\} = 1 - Pr_q\{P_m > \delta \ln m(n)\}$$

$$= 1 - (Pr_q\{P_m^i > \delta \ln m(n)\})^{\frac{m(n)}{\kappa \ln m(n)}}$$

$$= 1 - (1 - Pr_q\{P_m^i \leq \delta \ln m(n)\})^{\frac{m(n)}{\kappa \ln m(n)}}$$

$$\leq 1 - (1 - e^{\delta \ln m(n) \ln \frac{q}{1-q} + \ln(m(n)+1) + \ln(6-6q)\lceil \kappa \ln m(n)\rceil})^{\frac{m(n)}{\kappa \ln m(n)}}. \quad (12.21)$$

To show that the above probability tends to zero as $n \to \infty$, we need the following result:

$$1 - e^{\delta \ln m(n) \ln\left(\frac{q}{1-q}\right) + \ln(m(n)+1) + \ln(6-6q)\lceil \kappa \ln m(n)\rceil} > e^{-\frac{\kappa}{m(n)}}. \quad (12.22)$$

To show this, note that $e^{-x} \approx 1 - x$ for a small positive value x. We have

$$\ln(1 - e^{-\frac{\kappa}{m(n)}}) \approx \ln \frac{\kappa}{m(n)} = \ln \kappa - \ln m(n) > -\ln m(n), \quad (12.23)$$

where the last inequality holds since $\kappa > 1$ by (12.17). We also have

$$-\ln(m(n)+1) > -2\ln m(n) \quad (12.24)$$

$$-\ln(6-6q)\lceil \kappa \ln m(n)\rceil > -\ln(6-6q)\kappa \ln m(n). \quad (12.25)$$

Thus, by (12.23), (12.24), and (12.25), we have

$$\frac{\ln(1 - e^{-\frac{\kappa}{m(n)}}) - \ln(m(n)+1) - \ln(6-6q)\lceil \kappa \ln m(n)\rceil}{\ln m(n) \ln \frac{q}{1-q}}$$

$$> \frac{-\ln m(n) - 2\ln m(n) - \ln(6-6q)\kappa \ln m(n)}{\ln m(n) \ln \frac{q}{1-q}}$$

$$= \frac{-\ln(6-6q)\kappa - 3}{\ln \frac{q}{1-q}} = \delta.$$

Therefore, we have

$$\delta \ln m(n) \ln \frac{q}{1-q} + \ln(m(n)+1) + \ln(6-6q)\lceil \kappa \ln m(n)\rceil < \ln(1 - e^{-\frac{\kappa}{m(n)}}),$$

which is equivalent to (12.22).

By (12.21) and (12.22), we have

$$Pr_q\{P_m \leq \delta \ln m(n)\} < 1 - e^{-\frac{\kappa}{m(n)} \frac{m(n)}{\kappa \ln m(n)}} = 1 - e^{-\frac{1}{\ln m(n)}},$$

which goes to zero as $n \to \infty$. $\qquad\square$

---

[5] Incidentally, we fix an error in [42] (Eq. (15)), where it was incorrectly stated that
$$Pr_q\{P_m \leq \delta \ln m(n)\} = (Pr_q\{P_m^i \leq \delta \ln m(n)\})^{\frac{m(n)}{\kappa \ln m(n)}}.$$

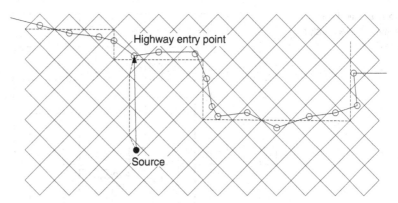

Figure 12.7 The entry point for the draining phase.

12.5.3 Deriving a feasible solution

Routing scheme In Section 12.5.2, we built a highway system for data transmissions in Phases (ii) and (iii), where the hop length (the number of hops) for each transmission is at most $\sqrt{2}l(n) \cdot 2$.

For Phases (i) and (iv), we now design a routing scheme and show that the hop length is at most $\sqrt{2}l(n) \cdot (d + 1)$, where $d = \lfloor \kappa \ln m(n) \rfloor + 1$. We adopt the following routing scheme for Phase (i) (the routing scheme for Phase (iv) is similar). By Theorem 12.6, there are at least $\lfloor \delta \ln m(n) \rfloor + 1$ crossing paths in each rectangle with height $\frac{\kappa \ln m(n)}{m(n)}$ almost surely as $n \to \infty$, where δ and κ are constants, and $m(n) = \lfloor \sqrt{\frac{n}{2\ln 6}} \rfloor$. Thus, we can slice each rectangle into horizontal strips of height $\frac{h}{m(n)}$ (where h is an appropriate constant) so that each strip corresponds to one crossing path. Note that a path may not be fully contained in its corresponding strip. We can then impose that the nodes in a strip communicate directly with the corresponding horizontal path. Although a path may be outside its corresponding strip, it is never farther than $\frac{\kappa \ln m(n)}{m(n)}$ from its corresponding strip. For each source in a strip, we identify an *entry point* on the corresponding horizontal path such that the cell that contains the entry point has the same horizontal position as the cell that contains the source (see Fig. 12.7). Since each source finds its highway within the same rectangle and at the same horizontal position, these two cells are at most $d = \lfloor \frac{\kappa \ln m(n)}{m(n)}/(\sqrt{2}l(n)) \rfloor + 1 = \lfloor \kappa \ln m(n) \rfloor + 1$ (diagonal) cells away.

Thus, in all phases, transmissions between two cells is within at most d (diagonal) cells away, where $d = \lfloor \kappa \ln m(n) \rfloor + 1$ for Phases (i) and (iv) and $d = 1$ for Phases (ii) and (iii).

Scheduling scheme We now design a time-slot-based scheduling. The idea is that when a node transmits, other nodes that are within a certain distance of this node cannot transmit simultaneously, whereas nodes that are outside this distance can transmit. In particular, we use k^2 slots for scheduling, where $k = 2(d + 1)$. A set of cells that are allowed to transmit simultaneously in a time slot is depicted in Fig. 12.8.

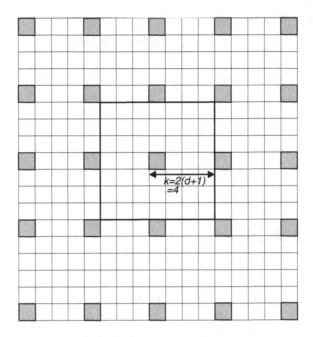

A lower bound for per-node throughput We can derive a lower bound for per-node throughput by analyzing the achievable per-node throughput by the above routing and scheduling schemes.

For the achievable rate at each hop, we have the following result:

Theorem 12.7

Denote $c_{ij}(d)$ as the data rate at which node i in a cell can transmit to node j in another cell that is at most d (diagonal) cells away. Then $c_{ij}(d)$ is at least $\Omega(W)$ almost surely as $n \to \infty$, where $d = 1$ for Phases (ii) and (iii) and $d = \lfloor \kappa \ln m(n) \rfloor + 1$ for Phases (i) and (iv).

Proof. We focus on a particular cell. To analyze $c_{ij}(d)$, we first find an upper bound for the interference at the receiver j. We consider the transmitters in the eight closest cells that are located at a Euclidean distance of at least $l(n)(d + 1)$ from the receiver (see Fig. 12.8), along with the 16 next closest cells that are located at a Euclidean distance of at least $l(n)(3d + 3)$, and so on. By extending the sum of the interferences to the entire two-dimensional area, we have the following bound:

$$I(d) \leq \sum_{i=1}^{\infty} \frac{8i P}{(l(n)(2i - 1)(d + 1))^\alpha} = \frac{8P}{(l(n)(d + 1))^\alpha} \sum_{i=1}^{\infty} \frac{i}{(2i - 1)^\alpha}.$$

$$(12.26)$$

The sum in (12.26) converges because $\alpha > 2$. Thus, if $d = 1$, we have

$$I(d) \leq \Theta((l(n)d)^{-\alpha}). \tag{12.27}$$

Next, we find a lower bound for the signal received from transmitter i. Since the distance between a transmitter and its receiver is at most $\sqrt{2}l(n)(d + 1)$, the signal $S(d)$ at the receiver satisfies

$$S(d) \geq \frac{P}{(\sqrt{2}l(n)(d + 1))^{\alpha}} = \Theta((l(n)d)^{-\alpha}). \tag{12.28}$$

By combining (12.27) and (12.28), a lower bound on $\frac{S(d)}{\eta + I(d)}$ is given by

$$\frac{S(d)}{\eta + I(d)} \geq \frac{\Theta((l(n)d)^{-\alpha})}{\eta + \Theta((l(n)d)^{-\alpha})} = \Theta(1).$$

As a result, the achievable data rate $c_{ij}(d) = W \log_2 \left(1 + \frac{S(d)}{\eta + I(d)}\right)$ has a constant lower bound $\Theta(W)$ almost surely as $n \to \infty$. $\qquad\square$

We can now analyze the achievable per-node throughput, which is bounded by the bottleneck phases among Phases (i)–(iv). We will show that the communication bottleneck resides in the highway Phases (ii) and (iii) with a per-node throughput of $\Omega(\frac{1}{\sqrt{n}})$.

For Phase (i), we have the following lemma:

Lemma 12.9
For transmissions in Phase (i), each source node can achieve at least a rate $\Omega\left(\frac{W}{(\ln n)^3}\right)$ almost surely as $n \to \infty$.

Proof. To compute the rate at which nodes can communicate to the entry points, we apply the second part of Theorem 12.7, which says that each node can communicate to its entry point at least at a rate $\Theta(W)$. For transmissions in Phase (i), where $d = \lfloor \kappa \ln m(n) \rfloor + 1$, there are $(2(d + 1))^2$ time slots required for scheduling. By Lemma 12.5, the number of nodes in each cell is less than $\ln m(n)$ almost surely as $n \to \infty$. Therefore, the actual rate available for each node is at least $\Theta(W)/((2(d + 1))^2 \cdot \ln m(n)) = \Theta\left(\frac{W}{d^2 \ln m(n)}\right)$, i.e., the per-node throughput is $\Omega\left(\frac{W}{(\ln n)^3}\right)$ almost surely as $n \to \infty$. $\qquad\square$

Since each node is the destination of exactly one source, we can prove the following lemma by adopting the same approach as that in Lemma 12.9:

Lemma 12.10
For transmissions in Phase (iv), each destination node can receive data from the highway at least at a rate $\Omega\left(\frac{W}{(\ln n)^3}\right)$ almost surely as $n \to \infty$.

Next, we consider the data rate along the highway. We first have the following lemma:

Lemma 12.11
There are fewer than $\frac{2nh}{m(n)}$ nodes on each strip almost surely as $n \to \infty$.

Proof. Let A_n be the event that there is at least one strip with no fewer than $\frac{2nh}{m(n)}$ nodes. Since the number of nodes in each strip is a Poisson random variable with parameter $\frac{nh}{m(n)}$, by the union bounds and Lemma 12.6, we have

$$Pr\{A_n\} \leq \frac{1}{h} e^{-\frac{nh}{m(n)}} \left(\frac{e \frac{nh}{m(n)}}{\frac{2nh}{m(n)}} \right)^{\frac{2nh}{m(n)}} = \frac{1}{h} e^{-\frac{nh}{m(n)}} \left(\frac{e}{2} \right)^{\frac{2nh}{m(n)}} = \frac{1}{h} \left(\frac{e}{4} \right)^{\frac{nh}{m(n)}},$$

which goes to 0 as $n \to \infty$. $\square$

The following lemma gives the achievable per-node throughput along the highway:

Lemma 12.12
For transmissions in Phases (ii) and (iii), each node along the highway may forward data for many source nodes. The achievable per-node throughput for each source node is at least $\Omega(\frac{W}{\sqrt{n}})$ almost surely as $n \to \infty$.

Proof. We prove the result for the horizontal traffic in Phase (ii) (the proof for Phase (iii) is similar and is thus omitted). For transmissions in Phase (ii), we have $d = 1$ and the number of time slots required for scheduling is a constant $(2(d + 1))^2 = 16$. Further, the number of nodes selected for the highway in each cell is one. Such a node can relay at most all the traffic generated in the corresponding strip of height $\frac{h}{m(n)}$. According to Lemma 12.11, a node on a horizontal highway may relay traffic for all nodes in this strip, which is at most $\frac{2nh}{m(n)}$ nodes. By applying the first part of Theorem 12.7, we conclude that one source node's data along the highway are at least $\Theta(\frac{W}{16 \cdot 1 \cdot \frac{2nh}{m(n)}})$, i.e., the per-node rate is $\Omega(\frac{W}{\frac{2nh}{m(n)}}) = \Omega(\frac{Wm(n)}{n}) = \Omega(\frac{W}{\sqrt{n}})$, almost surely as $n \to \infty$. $\square$

By Lemmas 12.9, 12.12, and 12.10, the per-node throughput is $\Omega(\frac{W}{\sqrt{n}})$ almost surely as $n \to \infty$. This proves Theorem 12.5.

12.6 Chapter summary

In this chapter, we have described an asymptotic capacity analysis for wireless ad hoc networks. Such an analysis addresses an achievable per-node through-put when the number of nodes goes to infinity. We focused on so-called random networks where each node is randomly deployed and each node has a randomly chosen destination node. For such an asymptotic capacity analysis, the results were derived in the form of $\Omega(\cdot)$, $O(\cdot)$, and $\Theta(\cdot)$, where the underlying analysis greatly differed from that in the previous chapters, which focused on optimization problems for *finite-sized* networks. We showed that the asymptotic capacity analysis heavily depends on the underlying interference model. In this chapter, we considered three such models (i.e., the protocol model, the physical model, and the generalized physical model) and showed how to develop asymptotic capacity bounds for each model.

12.7 Problems

12.1 Analyze the $\Omega(\cdot)$, $O(\cdot)$, and $\Theta(\cdot)$ relationships between the following $f(n)$ and $g(n)$ functions:
(a) $f(n) = 3n^2 + 2n$, $g(n) = n^2$;
(b) $f(n) = \ln n$, $g(n) = n$;
(c) $f(n) = 1/n$, $g(n) = 1$.

12.2 Describe the details for the protocol model, the physical model, and the generalized physical model that we used in the asymptotic analysis in this chapter.

12.3 For any two simultaneously active links $i \to j$ and $k \to l$ under the protocol model, what is the maximum ratio between the interference from node k to node j and the signal from node i to node j?

12.4 In the proof of Theorem 12.1, we established that "the network can support $O(r^{-2}(n))$ transmissions at any time." Why is this not n (the number of all source nodes)?

12.5 For a set of links whose receivers are inside a square of side-length $1/\lceil \frac{\sqrt{2}}{\Delta \cdot r(n)} \rceil$, show that these links interfere with each other under the protocol model.

12.6 If the cell size is changed to $a(n) = 1/\lfloor \frac{n}{\ln n} \rfloor$ in Lemma 12.1, prove that there is still no empty cell almost surely as $n \to \infty$. (Hint: You can prove this by following a similar approach to that in the proof of Lemma 12.1.)

12.7 Display a possible route for the case that the S-D line crosses a grid point in Fig. 12.9.

Figure 12.9 The straight line connecting a source node s to its destination node d.

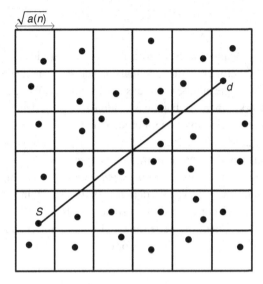

Figure 12.10 A set of links in a network. The solid lines represent links and the dotted lines represent interference.

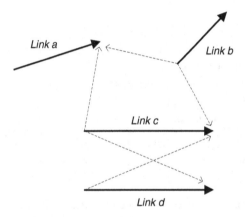

12.8   Under the protocol model, use a conflict graph to represent the interference relationships between the links shown in Fig. 12.10, where the solid lines represent links and the dotted lines represent interference. For example, there exists a dotted line between the transmitter of link c and the receiver of link a, which means that the receiver of link a is within the interference range of the transmitter of link c. What is maximum node degree in the resulting conflict graph?

12.9   What is the relationship between B and β under the physical model?

12.10   Show that a capacity upper bound is $O(\frac{B}{\sqrt{n}})$ when all nodes use the same transmit power under the physical model.

12.11   Suppose that all nodes use the same transmit power under the physical model. Show that for any two successful simultaneous transmissions $i \rightarrow j$

and $k \to l$ under the physical model, we have $d_{kj} \geq (1 + \Delta')d_{ij}$, where $\Delta' = (\beta^{1/\alpha} - 1)$.

12.12   Verify that $f(x) = x^{\alpha}$ (considered in the proof of Theorem 12.3) is a convex function for $x \geq 0$, where $\alpha \geq 2$.

12.13   Prove that if $f(x)$ is a convex function, then $f\left(\sum_{i=1}^{n} \lambda_i x_i\right) \leq \sum_{i=1}^{n} \lambda_i f(x_i)$, where $\sum_{i=1}^{n} \lambda_i = 1$, $\lambda_i \geq 0$ for $i = 1, \ldots, n$.

12.14   Compare the upper bound for a general case of independent power control (in Theorem 12.3) and the upper bound for a special case of synchronized power control (in Footnote 3). Which case has a tighter bound? Why can we obtain a tighter bound for the other case?

12.15   For a random network generated by a Poisson point process with density n in a 1×1 area, show that if we divide the 1×1 area into small regions of area $\frac{\ln n}{n}$, then every small region has at least one node almost surely as $n \to \infty$.

12.16   Why do we develop bounds for q in (12.14)? (Hint: Consider where we have used q.)

12.17   Lemma 12.7 considers the case of $\hat{q} < \frac{1}{3}$. Does the same result hold for the case of $\hat{q} \geq \frac{1}{3}$? If yes, why do we focus on the case of $\hat{q} < \frac{1}{3}$ in this lemma? (Hint: Consider where we have used Lemma 12.7.)

Bibliography

1   E. Aarts and J. Korst, *Simulated Annealing and Boltzman Machines*, New York, NY, John Wiley & Sons, 1989.

2   C. Ahn and R. Ramakrishna, "A genetic algorithm for shortest path routing problem and the sizing of populations," *IEEE Transactions on Evolutionary Computation*, vol. 6, no. 6, pp. 566–579, December 2002.

3   R.K. Ahuja, "Algorithms for the minimax transportation problem," *Naval Research Logistics Quarterly*, vol. 33, pp. 725–739, 1986.

4   I.F. Akyildiz, W. Su, Y. Sankarasubramaniam, and E. Cayirci, "Wireless sensor networks: a survey," *Computer Networks (Elsevier)*, vol. 38, no. 4, pp. 393–422, 2002.

5   I. Akyildiz and X. Wang, "A survey on wireless mesh networks," *IEEE Communications Magazine*, vol. 43, no. 9, September 2005.

6   M. Alasti, K. Sayrafian-Pour, A. Ephremides, and N. Farvardin, "Multiple description coding in networks with congestion problem," *IEEE Transactions on Information Theory*, vol. 47, no. 3, pp. 891–902, March 2001.

7   M. Alicherry, R. Bhatia, and L. Li, "Joint channel assignment and routing for throughput optimization in multi-radio wireless mesh networks," in *Proc. ACM MobiCom*, pp. 58–72, Cologne, Germany, August 28–September 2 2005.

8   J.G. Apostolopoulos, T. Wong, W. Tan, and S. Wee, "On multiple description streaming in content delivery networks," in *Proc. IEEE INFOCOM*, pp. 1736–1745, New York, NY, June 2002.

9   T. Back, D. Fogel, and Z. Michalewicz (eds), *Handbook of Evolutionary Computation*, New York, NY, Oxford University Press, 1997.

10 M.S. Bazaraa, J.J. Jarvis, and H.D. Sherali, *Linear Programming and Network Flows*, 4th edition, New York, NY, John Wiley & Sons Inc., 2010.

11 M.S. Bazaraa, H.D. Sherali, and C.M. Shetty, *Nonlinear Programming: Theory and Algorithms,* 3rd edition, New York, John Wiley & Sons, 2006.

12 A.C. Begen, Y. Altunbasak, and O. Ergun, "Multi-path selection for multiple description encoded video streaming," *EURASIP Signal Processing: Image Communications*, vol. 20, no. 1, pp. 39–60, January 2005.

13 A. Behzad and I. Rubin, "Impact of power control on the performance of ad hoc wireless networks," in *Proc. IEEE INFOCOM*, pp. 102–113, Miami, FL, March 13–17, 2005.

14   D. Bertsekas and R. Gallager, *Data Networks*, Englewood Cliffs, NJ, Prentice Hall, 1992.

15   V. Bhandari and N.H. Vaidya, "Capacity of multi-channel wireless networks with random (c, f) assignment," in *Proc. ACM MobiHoc*, pp. 229–238, Montreal, QC, Canada, September 9–14, 2007.

16   R. Bhatia and M. Kodialam, "On power efficient communication over multi-hop wireless networks: joint routing, scheduling and power control," in *Proc. IEEE INFOCOM*, pp. 1457–1466, Hong Kong, China, March 7–11, 2004.

17   E. Biglieri, R. Calderbank, A. Constantinides, A. Goldsmith, A. Paulraj, and H.V. Poor, *MIMO Wireless Communications*, Cambridge, Cambridge University Press, January 2007.

18   A. Bletsas, A. Khisti, D. Reed, and A. Lippman, "A simple cooperative diversity method based on network path selection," *IEEE Journal on Selected Areas in Communications*, vol. 24, no. 3, pp. 659–672, March 2006.

19   C. Blum and A. Roli, "Metaheuristics in combinatorial optimization: overview and conceptual comparison," *ACM Computing Surveys*, vol. 35, no. 3, pp. 268–308, September 2003.

20   T.X. Brown, H.N. Gabow, and Q. Zhang, "Maximum flow-life curve for a wireless ad hoc network," in *Proc. ACM MobiHoc*, pp. 128–136, Long Beach, CA, October 4–5, 2001.

21   N. Bulusu and S. Jha (eds.), *Wireless Sensor Networks: A Systems Perspective*, Norwood, MA, Artech House, 2005.

22   J. Cai, S. Shen, J.W. Mark, and A.S. Alfa, "Semi-distributed user relaying algorithm for amplify-and-forward wireless relay networks," *IEEE Transactions on Wireless Communication*, vol. 7, no. 4, pp. 1348–1357, April 2008.

23   J. Chakareski, S. Han, and B. Girod, "Layered coding vs. multiple descriptions for video streaming over multiple paths," in *Proc. ACM Multimedia*, pp. 422–431, Berkeley, CA, November 2003.

24   J.H. Chang and L. Tassiulas, "Maximum lifetime routing in wireless sensor networks," *IEEE/ACM Transactions on Networking*, vol. 12, issue 4, pp. 609–619, August 2004.

25   P. Charnsethikul, "The constrained minimax linear assignment problem," *Optimization Methods and Software*, vol. 14, issues 1 & 2, pp. 37–48, 2000.

26   C.C. Chen and D.S. Lee, "A joint design of distributed QoS scheduling and power control for wireless networks," in *Proc. IEEE INFOCOM*, Barcelona, Catalunya, Spain, April 23–29, 2006.

27   L. Chen, S.H. Low, M. Chiang, and J.C. Doyle, "Cross-layer congestion control, routing and scheduling design in ad hoc wireless networks," in *Proc. IEEE INFOCOM*, Barcelona, Catalunya, Spain, April 23–29, 2006.

28   Concorde TSP Solver, http://www.tsp.gatech.edu/concorde/.

29   N. Clemens and C. Rose, "Intelligent power allocation strategies in an unlicensed spectrum," in *Proc. IEEE DySPAN*, pp. 37–42, Baltimore, MD, November 8–11, 2005.

30   T.J. Cormen, C.E. Leiserson, R.L. Rivest, and C. Stein, *Introduction to Algorithms, 2nd edition*, The MIT Press, 2001.

31   IBM ILOG CPLEX Optimizer, http://www-01.ibm.com/software/integration/optimization/cplex-optimizer/.

32   R.L. Cruz and A.V. Santhanam, "Optimal routing, link scheduling and power control in multi-hop wireless networks," in *Proc. IEEE INFOCOM*, pp. 702–711, San Francisco, CA, March 30–April 3, 2003.

33   H. Dai, K. Ng, R.C. Wong, and M. Wu, "On the capacity of multi-channel wireless networks using directional antennas," in *Proc. IEEE INFOCOM*, pp. 1301–1309, Phoenix, AZ, April 13–18, 2008.

34   I. Dietrich and F. Dressler, "On the lifetime of wireless sensor networks," *ACM Transactions on Sensor Networks*, vol. 5, no. 1, pp. 1–39, February 2009.

35   R. Draves, J. Padhye, and B. Zill, "Routing in multi-radio, multi-hop wireless mesh networks," in *Proc. ACM MobiCom*, pp. 114–128, Philadelphia, PA, September 26– October 1, 2004.

36   J. Dreo, A. Petrowski, P. Siarry, and E. Taillard, *Metaheuristics for Hard Optimization: Methods and Case Studies*, New York, Springer-Verlag, 2006.

37   T. Elbatt and A. Ephremides, "Joint scheduling and power control for wireless ad-hoc networks," in *Proc. IEEE INFOCOM*, pp. 976–984, New York, NY, June 23–27, 2002.

38   D. Eppstein, "Finding the k shortest paths," *SIAM Journal on Computing*, vol. 28, no. 2, pp. 652–673, August 1999.

39   R. Etkin, A. Parekh, and D. Tse, "Spectrum sharing for unlicensed bands," in *Proc. IEEE DySPAN*, pp. 251–258, Baltimore, MD, November 8–11, 2005.

40   FCC (Federal Communications Commission), "Facilitating opportunities for flexible, efficient, and reliable spectrum use employing cognitive radio technologies, notice of proposed rule making and order," FCC 03-322.

41   G. J. Foschini and M. J. Gans, "On limits of wireless communications in a fading environment when using multiple antennas," *Wireless Personal Communications*, vol. 6, pp. 311–355, March 1998.

42   M. Franceschetti, O. Dousse, D.N.C. Tse, and P. Thiran, "Closing the gap in the capacity of wireless networks via percolation theory," *IEEE Transaction on Information Theory*, vol. 53, no. 3, pp. 1009–1018, March 2007.

43   D. Fritchman, "A binary channel characterization using partitioned markov chains," *IEEE Transactions on Information Theory*, vol. 13, no. 2, pp. 221–227, April 1967.

44   A.E. Gamal, J. Mammen, B. Prabhakar, and D. Shah, "Throughput-delay trade-off in wireless networks," in *Proc. IEEE INFOCOM*, pp. 464–475, Hong Kong, China, March 7–11, 2004.

45   G. Ganesan and Y.G. Li, "Cooperative spectrum sensing in cognitive radio networks," in *Proc. IEEE DySPAN*, pp. 137–143, Baltimore, MD, November 8–11, 2005.

46   M.R. Garey and D.S. Johnson, *Computers and Intractability: A Guide to the Theory of NP-Completeness*, New York, NY, W.H. Freeman and Company, 1979.

47   A. Gashemi and E. Sousa, "Collaborative spectrum sensing for opportunistic access in fading environments," in *Proc. IEEE DySPAN*, pp. 131–136, Baltimore, MD, November 8–11, 2005.

48   D. Gesbert, M. Shafi, D. Shiu, P. J. Smith, and A. Naguib, "From theory to practice: an overview of MIMO space-time coded wireless systems," *IEEE Journal on Select Areas in Communications*, vol. 21, no. 3, pp. 281–302, April 2003.

49   E. Giler, "Eric Giler demos wireless electricity," URL:http://www.ted.com/talks/eric_ giler_demos_wireless_electricity.html.

50   A. Giridhar and P.R. Kumar, "Maximizing the functional lifetime of sensor networks," in *Proc. ACM/IEEE International Symposium on Information Processing in Sensor Networks*, pp. 5–12, Los Angeles, CA, April 2005.

51   F. Glover and M. Laguna, *Tabu Search*. Boston, MA, Kluwer-Academic, 1997.

52   GNU Linear Programming Kit, http://www.gnu.org/software/glpk/.

53   N. Gogate, D. Chung, S.S. Panwar, and Y. Wang, "Supporting image/video applications in a multihop radio environment using route diversity and multiple description coding," *IEEE Transactions on Circuits and Systems for Video Technology*, vol. 12, no. 9, pp. 777–792, September 2002.

54   A.J. Goldsmith and S.-G. Chua, "Adaptive coded modulation for fading channels," *IEEE Transactions on Communications*, vol. 46, no. 5, pp. 595–602, May 1998.

55   A. Goldsmith, S. A. Jafar, N. Jindal, and S. Vishwanath, "Capacity limits of MIMO channels," *IEEE Journal on Select Areas in Communications*, vol. 21, no. 1, pp. 684–702, June 2003.

56   A. Goldsmith, *Wireless Communications*, Cambridge, Cambridge University Press, 2005.

57   V. Goyal, "Multiple description coding: Compression meets the network," *IEEE Signal Processing Magazine*, vol. 18, pp. 74–93, September 2001.

58   G. Grimmett, *Percolation*, 2nd edition, New York, Springer-Verlag, 1999.

59   P. Gupta and P.R. Kumar, "Critical power for asymptotic connectivity in wireless networks," in *Stochastic Analysis, Control, Optimization and Applications: A Volume in Honor of W.H. Fleming*, W.M. McEneany, G. Yin, and Q. Zhang (eds.), Boston, MA, Birkhauser, pp. 547–566, 1998.

60   P. Gupta and P.R. Kumar, "The capacity of wireless networks," *IEEE Transactions on Information Theory*, vol. 46, no. 2, pp. 388–404, March 2000.

61   A. Keshavarz-Haddad, V. Ribeiro, and R. Riedi, "Broadcast capacity in multihop wireless networks," in *Proc. ACM MobiCom*, pp. 239–250, Los Angeles, CA, September 23–26, 2006.

62   A. Keshavarz-Haddad and R. Riedi, "On the broadcast capacity of multihop wireless networks: interplay of power, density and interference," in *Proc. IEEE SECON*, pp. 314–323, San Diego, CA, June 18–21, 2007.

63   A. Keshavarz-Haddad and R. Riedi, "Multicast capacity of large homogeneous multihop wireless networks," in *Proc. International Symposium on Modeling and Optimization in Mobile, Ad Hoc, and Wireless Networks (WiOpt)*, pp. 116–124, Berlin, Germany, April 2008.

64   S. Haykin, *Adaptive Filter Theory*, Englewood Cliffs, NJ, Prentice-Hall, 1996.

65   S. Haykin, "Cognitive radio: brain-empowered wireless communications," *IEEE Journal on Selected Areas in Communications*, vol. 23, no. 2, pp. 201–220, February 2005.

66   S. He, J. Chen, F. Jiang, D.K.Y. Yau, G. Xing, and Y. Sun, "Energy provisioning in wireless rechargeable sensor networks," in *Proc. IEEE INFOCOM*, pp. 2006–2014, Shanghai, China, April 2011.

67   W. Heinzelman, "Application-specific protocol architectures for wireless networks," Ph.D. thesis, Department of Electrical Engineering and Computer Science, Massachusetts Institute of Technology, June 2000.

68   J.-B. Hiriart-Urruty and C. Lemaréchal, *Fundamentals of Convex Analysis*, Berlin, Springer-Verlag, 2001.

69   Y.T. Hou, Y. Shi, and H.D. Sherali, "On node lifetime problem for energy-constrained wireless sensor networks," *ACM/Springer Mobile Networks and Applications*, vol. 10, no. 6, pp. 865–878, December 2005.

70   Y.T. Hou and Y. Shi, "Variable bit rate flow routing in wireless sensor networks," *IEEE Transactions on Wireless Communications*, vol. 56, no. 6, pp. 2140–2148, June 2007.

71   Y.T. Hou, Y. Shi, and H.D. Sherali, "Rate allocation and network lifetime problems for wireless sensor networks," *IEEE/ACM Transactions on Networking*, vol. 16, no. 2, pp. 321–334, April 2008.

72   Y.-C. Hu and D. Johnson, "Design and demonstration of live audio and video over multihop wireless ad hoc networks," in *Proc. IEEE MILCOM*, pp. 7–10, Anaheim, CA, October 2002.

73   J. Huang, R.A. Berry, and M.L. Honig, "Spectrum sharing with distributed interference compensation," in *Proc. IEEE DySPAN*, pp. 88–93, Baltimore, MD, November 8–11, 2005.

74   O. Ileri, D. Samardzija, T. Sizer, and N.B. Mandayam, "Demand responsive pricing and competitive spectrum allocation via a spectrum server," in *Proc. IEEE DySPAN*, pp. 194–202, Baltimore, MD, November 8–11, 2005.

75   S.-W. Jeon, N. Devroye, M. Vu, S.-Y. Chung, and V. Tarokh, "Cognitive networks achieve throughput scaling of a homogeneous network," in *Proc. ICST International Symposium on Modeling and Optimization in Mobile, Ad Hoc, and Wireless Networks (WiOpt)*, Seoul, Korea, June 26–27, 2009.

76   X. Jiang, J. Polastre, and D. Culler, "Perpetual environmentally powered sensor networks," in *Proc. ACM/IEEE International Symposium on Information Processing in Sensor Networks*, pp. 463–468, Los Angeles, CA, April 2005.

77   A. Kansal, J. Hsu, S. Zahedi, and M.B. Srivastava, "Power management in energy harvesting sensor networks," *ACM Transactions Embedded Computing Systems*, vol. 6, no. 4, article 32, September 2007.

78   K. Kar, M. Kodialam, T.V. Lakshman, and L. Tassiulas, "Routing for network capacity maximization in energy-constrained ad-hoc networks," in *Proc. IEEE INFOCOM*, pp. 673–681, March 30–April 3, 2003, San Francisco, CA.

79   F. Kelly, A. Maulloo, and D. Tan, "Rate control for communication networks: shadow prices, proportional fairness and stability," in *Journal of the Operational Research Society*, vol. 49, no. 3, pp. 237–252, March 1998.

80   H. Kesten, *Percolation Theory for Mathematicians*, Boston, MA, Birkhauser, 1982.

81   J. Kleinberg and E. Tardos, *Algorithm Design*, Boston, MA, Addison-Wesley Longman Publishing, 2005.

82   M. Kodialam and T. Nandagopal, "Characterizing the capacity region in multi-radio multi-channel wireless mesh networks," in *Proc. ACM MobiCom*, pp. 73–87, Cologne, Germany, August 28–September 2, 2005.

83   M. Kodialam and T. Nandagopal, "Characterizing achievable rates in multi-hop wireless mesh networks with orthogonal channels," *IEEE/ACM Transactions on Networking*, vol. 13, no. 4, pp. 868–880, August 2005.

84   G. Kramer, I. Maric, and R.D. Yates, "Cooperative communications," *Foundations and Trends in Networking*, Now Publishers, June 2007.

85   A. Kurs, A. Karalis, R. Moffatt, J.D. Joannopoulos, P. Fisher, and M. Soljacic, "Wireless power transfer via strongly coupled magnetic resonances," *Science*, vol. 317, no. 5834, pp. 83–86, 2007.

86   P. Kyasanur and N.H. Vaidya, "Capacity of multi-channel wireless networks: impact of number of channels and interfaces," in *Proc. ACM MobiCom*, pp. 43–57, Cologne, Germany, August 28–September 2, 2005.

87   J.N. Laneman, D.N.C. Tse, and G.W. Wornell, "Cooperative diversity in wireless networks: efficient protocols and outage behavior," *IEEE Transactions on Information Theory*, vol. 50, no. 12, pp. 3062–3080, December 2004.

88   L. Lasdon, *Optimization Theory for Large Scale Systems*, New York, NY, Macmillan., 1970.

89   E. Lawler, *Combinatorial Optimization: Networks and Matroids*, Mineola, NY, Dover Publications Inc., 2001.

90   J.-W. Lee, M. Chiang, and R.A. Calderbank, "Price-based distributed algorithm for optimal rate-reliability tradeoff in network utility maximization," *IEEE Journal on Selected Areas in Communications*, vol. 24, issue 5, pp. 962–976, May 2006.

91   X.-Y. Li, J.-Z. Zhao, Y.-W. Wu, S.-J. Tang, X.-H. Xu, and X.F. Mao, "Broadcast capacity for wireless ad hoc networks," in *Proc. IEEE MASS*, pp. 114–123, Atlanta, GA, September 29–October 2, 2008.

92   X.-Y. Li, "Multicast capacity of wireless ad hoc networks," *IEEE/ACM Transactions Networking*, vol. 17, no. 3, pp. 950–961, June 2009.

93   Z. Li, Y. Peng, W. Zhang, and D. Qiao, "J-RoC: A joint routing and charging scheme to prolong sensor network lifetime," in *Proc. IEEE ICNP*, pp. 373–382, Vancouver, Canada, October 17–20, 2011.

94   D. Linden and T.B. Reddy (eds.), *Handbook of Batteries*, 3rd edition, McGraw-Hill, 2002.

95   LINDO Systems–Optimization Software, http://www.lindo.com/.

96   X. Liu and W. Wang, "On the characteristics of spectrum-agile communication networks," in *Proc. IEEE DySPAN*, pp. 214–223, Baltimore, MD, November 8–11, 2005.

97   J. Liu, D. Goeckel, and D. Towsley, "Bounds on the gain of network coding and broadcasting in wireless networks," in *Proc. IEEE INFOCOM*, pp. 724–732, Anchorage, AK, May 6–12, 2007.

98   S.H. Low, "A duality model of TCP and queue management algorithms," *IEEE/ACM Transactions on Networking*, vol. 11, issue 4, pp. 525–536, August 2003.

99   H. Luss and D.R. Smith, "Resource allocation among competing activities: a lexicographic minimax approach," *Operations Research Letters*, vol. 5, no. 5, pp. 227–231, November 1986.

100   J. R. Magnus and H. Neudecker, *Matrix Differential Calculus with Applications in Statistics and Economics*, New York, NY, Wiley, 1999.

101   N. Malpani and J. Chen, "A note on practical construction of maximum bandwidth paths," *Information Processing Letters*, vol. 83, pp. 175–180, August 2002.

102   S. Mao, S. Lin, S.S. Panwar, Y. Wang, and E. Celebi, "Video transport over ad hoc networks: multistream coding with multipath transport," *IEEE Journal on Select Areas in Communications*, vol. 21, no. 10, pp. 1721–1737, December 2003.

103   S. Mao, S. Kompella, Y.T. Hou, H.D. Sherali, and S.F. Midkiff, "Routing for concurrent video sessions in ad hoc networks," *IEEE Transactions on Vehicular Technology*, vol. 55, issue 1, pp. 317–327, January 2006.

104   S. Mao, Y.T. Hou, X. Cheng, H.D. Sherali, S.F. Midkiff, and Y.-Q. Zhang, "On routing for multiple description video over wireless ad hoc networks," *IEEE Transactions on Multimedia,* vol. 8, issue 5, pp. 1063–1074, October 2006.

105   I. Maric and R.D. Yates, "Cooperative multihop broadcast for wireless networks," *IEEE Journal on Selected Areas in Communications*, vol. 22, issue 6, pp. 1080–1088, August 2004.

106   M. McHenry and D. McCloskey, "Spectrum occupancy measurements: Republican National Convention, New York, NY, August 30–September 3, 2004," available at http://www.sharedspectrum.com/wp-content/uploads/4_NSF_NYC_Report.pdf.

107 R. Meester and R. Roy, *Continuum Percolation*, Cambridge, Cambridge University Press, 1996.

108 N. Megiddo, "Linear-time almorithm for linear programming in r^3 and related problems," *SIAM Journal on Computing*, vol. 12, pp. 759–776, 1983.

109 E.C. van der Meulen and P. Vanroose. "The capacity of a relay channel, both with and without delay," *IEEE Transactions on Information Theory*, vol. 53, no. 10, pp. 3774–3776, October 2007.

110 Z. Michalewicz and D.B. Fogel, *How to Solve It: Modern Heuristics*, 2nd edition, New York, Springer-Verlag, 2004.

111 S.M. Mishra, A. Sahai, and R.W. Brodersen, "Cooperative sensing among cognitive radios," in *Proc. IEEE International Conference on Communications*, pp. 1658–1663, Istanbul, Turkey, June 11–15, 2006.

112 J. Mitola III, Cognitive radio: an integrated agent architecture for software defined radio, Ph.D. thesis, KTH Royal Institute of Technology, 2000.

113 G.L. Nemhauser and L.A. Wolsey, *Integer and Combinatorial Optimization*, New York, NY, John Wiley & Sons, 1999.

114 A. Nemirovski, *Interior Point Polynomial Time Methods in Convex Programming*, 2004, available at http://www2.isye.gatech.edu/ñemirovs/Lect_IPM.pdf.

115 N. Nie and C. Comaniciu, "Adaptive channel allocation spectrum etiquette for cognitive radio networks," in *Proc. IEEE DySPAN*, pp. 269–278, Baltimore, MD, November 8–11, 2005.

116 R. Ogier, F. Templin, and M. Lewis, "Topology dissemination based on reverse-path forwarding (TBRPF)," February 2004, IETF RFC 3684.

117 L. Ozarow, "On a source coding problem with two channels and three receivers," *Bell System Technical Journal*, vol. 59, no. 10, pp. 84–91, December 1980.

118 A. Özgür, O. Lévêque, and D.N.C. Tse, "Hierarchical cooperation achieves optimal capacity scaling in ad hoc networks," *IEEE Transaction on Information Theory*, vol. 53, no. 10, pp. 3549–3572, October 2007.

119 M. Padberg and G. Rinaldi, "A branch-and-cut algorithm for the resolution of large-scale symmetric traveling salesman problems," *SIAM Review*, vol. 33, no. 1, pp. 60–100, 1991.

120 P. Papadimitratos, Z. Haas, and E. Sirer, "Path set selection in mobile ad hoc networks," in *Proc. ACM MobiHoc*, pp. 1–11, Lausanne, Switzerland, June 2002.

121 G. Park, T. Rosing, M.D. Todd, C.R. Farrar, and W. Hodgkiss, "Energy harvesting for structural health monitoring sensor networks," *Journal of Infrastructure Systems*, vol. 14, no. 1, pp. 64–79, March 2008.

122 A.J. Paulraj, D.A. Gore, R.U. Nabar, and H. Bolcskei, "An overview of MIMO communications — a key to gigabit wireless," *Proceedings of the IEEE*, vol. 92, no. 2, pp. 198–218, February 2004.

123 C. Peng, H. Zheng, and B.Y. Zhao, "Utilization and fairness in spectrum assignment for opportunistic spectrum access," *ACM/Springer Mobile Networks and Applications*, vol. 11, issue 4, pp. 555–576, August 2006.

124 Y. Peng, Z. Li, W. Zhang, and D. Qiao, "Prolonging sensor network lifetime through wireless charging," in *Proc. IEEE RTSS*, pp. 129–139, San Diego, CA, November 30–December 3, 2010.

125 PowerCast, http://www.powercastco.com.

126 A.P. Punnen and K.P.K. Nair, "Improved complexity bound for the maximum cardinality bottleneck bipartite matching problem," *Elsevier Discrete Applied Mathematics*, vol. 44, pp. 91–93, 1994.

127 R. Ramanathan and R. Rosales-Hain, "Topology control of multihop wireless networks using transmit power adjustment," in *Proc. IEEE INFOCOM*, pp. 404–413, 2000, Tel Aviv, Israel, March 26–30.

128 A. Raniwala and T. Chiueh, "Architecture and algorithms for an IEEE 802.11-based multi-channel wireless mesh network," in *Proc. IEEE INFOCOM*, pp. 2223–2234, Miami, FL, March 13–17, 2005.

129 T.S. Rappaport, *Wireless Communications: Principles and Practice*, Upper Saddle River, NJ, Prentice Hall, 1996.

130 J.H. Reed, *Software Radio: A Modern Approach to Radio Engineering*, Upper Saddle River, NJ, Prentice Hall, May 2002.

131 V. Rodoplu and T.H. Meng, "Minimum energy mobile wireless networks," *IEEE Journal on Selected Areas in Communications*, vol. 17, no. 8, pp. 1333–1344, August 1999.

132 J. B. Rosen, "The gradient projection method for nonlinear programming, part I: linear constraints," *SIAM Journal on Applied Mathematics*, vol. 8, pp. 181–217, 1960.

133 A. Sankar and Z. Liu, "Maximum lifetime routing in wireless ad-hoc networks," in *Proc. IEEE INFOCOM*, pp. 1089–1097, Hong Kong, China, March 7–11, 2004.

134 Wireless Innovation Forum, http://www.sdrforum.org.

135 E. Setton, Y. Liang, and B. Girod, "Adaptive multiple description video streaming over multiple channels with active probing," in *Proc. IEEE ICME*, Baltimore, MD, July 2003.

136 A. Sendonaris, E. Erkip, and B. Aazhang, "User cooperation diversity – part I: system description," *IEEE Transactions on Communications*, vol. 51, no. 11, pp. 1927–1938, November 2003.

137 A. Sendonaris, E. Erkip, and B. Aazhang, "User cooperation diversity – part II: implementation aspects and performance analysis," *IEEE Transactions on Communications*, vol. 51, no. 11, pp. 1939–1948, November 2003.

138 S. Shakkottai, X. Liu, and R. Srikant, "The multicast capacity of large multihop wireless networks," in *Proc. ACM MobiHoc*, pp. 247–255, Montreal, Quebec, Canada, September 9–14, 2007.

139 S. Sharma, Y. Shi, Y.T. Hou, and S. Kompella, "An optimal algorithm for relay node assignment in cooperative ad hoc networks," *IEEE/ACM Transactions on Networking*, vol. 19, issue 3, pp. 879–892, June 2011.

140 H.D. Sherali and W.P. Adams, "A hierarchy of relaxations between the continuous and convex hull representations for zero-one programming problems," *SIAM Journal on Discrete Mathematics*, vol. 3, no. 3, pp. 411–430, 1990.

141 H.D. Sherali and C.H. Tuncbilek, "A global optimization algorithm for polynomial programming problems using a Reformulation-Linearization Technique," *Journal of Global Optimization*, vol. 2, pp. 101–112, 1992.

142 H.D. Sherali and W.P. Adams, "A hierarchy of relaxations and convex hull characterizations for mixed-integer zero-one programming problems," *Discrete Applied Mathematics*, vol. 52, pp. 83–106, 1994.

143 H.D. Sherali and C.H. Tuncbilek, "Comparison of two Reformulation-Linearization Technique based linear programming relaxations for polynomial programming problems," *Journal of Global Optimization*, vol. 10, pp. 381–390, 1997.

144   H.D. Sherali, "Global optimization of nonconvex polynomial programming problems having rational exponents," *Journal of Global Optimization*, vol. 12, no. 3, pp. 267–283, 1998.

145   H. Sherali, K. Ozbay, and S. Subramanian, "The time-dependent shortest pair of disjoint paths problem: complexity, models, and algorithms," *Networks*, vol. 31, no. 4, pp. 259–272, December 1998.

146   H.D. Sherali and W.P. Adams, *A Reformulation-Linearization Technique for Solving Discrete and Continuous Nonconvex Problems*, Dordrecht/Boston/London, Kluwer Academic Publishers, 1999.

147   H.D. Sherali, "On mixed-integer zero-one representations for separable lower-semicontinuous piecewise-linear functions," *Operations Research Letters*, vol. 28, no. 4, pp. 155–160, 2001.

148   H.D. Sherali and L. Liberti, "Reformulation-Linearization Technique for global optimization," in *Encylopedia of Optimization*, 2nd edition, P. Pardalos and C. Floudas (eds.) Springer, Berlin, pp. 3263–3268, 2008.

149   H.D. Sherali, "Reformulation-Linearization Technique for MIPs," in *Wiley Encyclopedia of Operations Research and Management Science*, Invited Advanced Topic paper, February 2011.

150   Y. Shi and Y.T. Hou, "Optimal power control for multi-hop software defined radio networks," in *Proc. IEEE INFOCOM*, pp. 1694–1702, Anchorage, AL, May 6–12, 2007.

151   Y. Shi and Y.T. Hou, "Theoretical results on base station movement problem for sensor network," in *Proc. IEEE INFOCOM*, pp. 376–384, Phoenix, AZ, April 14–17, 2008.

152   Y. Shi, Y.T. Hou, J. Liu, and S. Kompella, "Bridging the gap between protocol and physical models for wireless networks," *IEEE Transactions on Mobile Computing*, 2012.

153   K. Sohrabi, J. Gao, V. Ailawadhi, and G. Pottie, "Protocols for self-organizing of a wireless sensor network," *IEEE Personal Communications Magazine*, vol. 7, pp. 16–27, October 2000.

154   M.E. Steenstrup, "Opportunistic use of radio-frequency spectrum: a network perspective," in *Proc. IEEE DySPAN*, pp. 638–641, Baltimore, MD, November 8–11, 2005.

155   I. E. Telatar, "Capacity of multi-antenna Gaussian channels," *European Transactions on Telecommunications*, vol. 10, no. 6, pp. 585–596, November 1999.

156   N. Tesla, "Apparatus for transmitting electrical energy," US patent number 1,119,732, issued in December 1914.

157   B. Tong, Z. Li, G. Wang, and W. Zhang, "How wireless power charging technology affects sensor network deployment and routing," in *Proc. IEEE ICDCS*, pp. 438–447, Genoa, Italy, June 2010.

158   D. Ugarte and A.B. McDonald, "On the capacity of dynamic spectrum access enabled networks," in *Proc. IEEE DySPAN*, pp. 630–633, Baltimore, MD, November 8–11, 2005.

159   T.A. Vanderelli, J.G. Shearer, and J.R. Shearer, "Method and apparatus for a wireless power supply," US patent number 7,027,311, issued in April 2006.

160   E. Visostky, S. Kuffner, and R. Peterson, "On collaborative detection of TV transmissions in support of dynamic spectrum sharing," in *Proc. IEEE DySPAN*, pp. 338–345, Baltimore, MD, November 8–11, 2005.

161   A. Volgenant and C.W. Duin, "Improved polynomial algorithms for robust bottleneck problems with interval data," *Elsevier Computers and Operations Research*, vol. 37, issue 5, pp. 909–915, May, 2010.

162   M. Vu, N. Devroye, M. Sharif, and V. Tarokh, "Scaling laws of cognitive networks," in *Proc. ICST International Conference on Cognitive Radio Oriented Wireless Networks and Communications (CrownCom)*, pp. 2–8, Orlando, FL, July 31–August 3, 2007.

163   W. Wang, V. Srinivasan, and K.C. Chua, "Using mobile relays to prolong the lifetime of wireless sensor networks," in *Proc. ACM MobiCom*, Cologne, pp. 270–283, Germany, August 2005.

164   W. Wang, V. Srinivasan, and K.C. Chua, "Extending the lifetime of wireless sensor networks through mobile relays," *IEEE/ACM Transactions on Networking*, vol. 16, no. 5, pp. 1108–1120, October 2008.

165   Z. Wang, H.R. Sajadpour, and J.J. Garcia-Luna-Aceves, "The capacity and energy efficiency of wireless ad hoc networks with multi-packet reception," in *Proc. ACM Mobi-Hoc*, pp. 179–188, Hong Kong, China, May 26–30, 2008.

166   R. Wattenhofer, L. Li, P. Bahl, and Y.-M. Wang, "Distributed topology control for power efficient operation in multihop wireless ad hoc networks," in *Proc. IEEE INFOCOM*, pp. 1388–1397, Anchorage, AK, April 22–26, 2001.

167   E. Welzl, "Smallest enclosing disks," *Lecture Notes in Computer Science (LNCS)*, vol. 555, pp. 359–370, 1991.

168   D.B. West, *Introduction to Graph Theory*, Upper Saddle River, NJ, Prentice Hall, 2001.

169   J. Winters, "On the capacity of radio communication systems with diversity in a Rayleigh fading environment," *IEEE Journal on Selected Areas in Communications*, vol. 5, no. 5, pp. 871–878, June 1987.

170   J. H. Winters, "Smart antenna techniques and their application to wireless ad hoc networks," *IEEE Wireless Communications Magazine*, vol. 13, no. 4, pp. 77–83, August 2006.

171   WiTricity Corp., http://www.witricity.com.

172   Wireless Power Consortium, http://www.wirelesspowerconsortium.com/.

173   A. Wyglinski, M. Nekovee, and Y.T. Hou (eds.), *Cognitive Radio Communications and Networks: Principles and Practice*, Elsevier, December 2009.

174   C. Xin, B. Xie, and C.-C. Shen, "A novel layered graph model for topology formation and routing in dynamic spectrum access networks," in *Proc. IEEE DySPAN*, pp. 308–317, Baltimore, MD, November 8–11, 2005.

175   S. Ye and R. S. Blum, "Optimized signaling for MIMO interference systems with feedback," *IEEE Transactions on Signal Processing*, vol. 51, no. 11, pp. 2839–2848, November 2003.

176   C. Yin, L. Gao, and S. Cui, "Scaling laws of overlaid wireless networks: a cognitive radio network vs. a primary network," *IEEE/ACM Transactions on Networking*, vol. 18, no. 4, pp. 1317–1329, August 2010.

177   H. Zhang and J.C. Hou, "Capacity of wireless ad-hoc networks under ultra wide band with power constraint," in *Proc. IEEE INFOCOM*, pp. 455–465, Miami, FL, March 13–17, 2005.

178   F. Zhang, X. Liu, S.A. Hackworth, R.J. Sclabassi, and M. Sun, "In vitro and in vivo studies on wireless powering of medical sensors and implantable devices," in *Proc. IEEE/NIH Life Science Systems and Applications Workshop (LiSSA)*, pp. 84–87, Bethesda, MD, April 2009.

179   J. Zhao, H. Zheng, and G. Yang, "Distributed coordination in dynamic spectrum allocation networks," in *Proc. IEEE DySPAN*, pp. 259–268, Baltimore, MD, November 8–11, 2005.

180   Y. Zhao, R.S. Adve, and T.J. Lim, "Improving amplify-and-forward relay networks: optimal power allocation versus selection," in *Proc. IEEE International Symposium on Information Theory*, pp. 1234–1238, Seattle, WA, July 9–14, 2006.

181   R. Zheng, "Information dissemination in power-constrained wireless network," in *Proc. IEEE INFOCOM*, Barcelona, Catalunya, Spain, April 23–29, 2006.

Index

Printed in the United States
by Baker & Taylor Publisher Services